ACOUSTICAL HOLOGRAPHY

Volume 6

ACOUSTICAL HOLOGRAPHY

A Continuation Order Plan is available for this series. A continuation order will bring delivery of each new volume immediately upon publication. Volumes are billed only upon actual shipment. For further information please contact the publisher.

ACOUSTICAL HOLOGRAPHY

Volume 6

Edited by

Newell Booth

Naval Undersea Center
San Diego, California

PLENUM PRESS • NEW YORK AND LONDON

The Library of Congress cataloged the first volume of this title as follows:

International Symposium on Acoustical Holography.
Acoustical holography; proceedings. v. 1–
New York, Plenum Press, 1967–

    v. illus. (part col.), ports. 24 cm.

Editors: 1967–   A. F. Metherell and L. Larmore (1967 with H. M. A. el-Sum)
Symposiums for 1967–   held at the Douglas Advanced Research Laboratories, Huntington Beach, Calif.

    1. Acoustic holography—Congresses—Collected works.   I. Metherell, Alexander A., ed. II. Larmore, Lewis, ed. III. el-Sum, Hussein Mohammed Amin, ed. IV. Douglas Advanced Research Laboratories. v. Title.

QC244.5.I 5   69–12533

Library of Congress   [r70n3] rev

Library of Congress Catalog Card Number 69-12533
ISBN 0-306-37726-8

Proceedings of the Sixth International Symposium on Acoustical
Holography and Imaging held in San Diego, California,
February 4-7, 1975

© 1975 Plenum Press, New York
A Division of Plenum Publishing Corporation
227 West 17th Street, New York, N.Y. 10011

United Kingdom edition published by Plenum Press, London
A Division of Plenum Publishing Company, Ltd.
Davis House (4th Floor), 8 Scrubs Lane, Harlesden, London, NW10 6SE, England

Printed in the United States of America

PREFACE

This volume contains the Proceedings of the Sixth International Symposium on Acoustical Holography and Imaging, held in San Diego, California, February 4-7, 1975. The title of this symposium differs from that of the first four by the addition of the word "Imaging", reflecting an increase in emphasis on nonholographic methods of acoustical visualization. For convenience, no change has been made in the title of this published series.

The 38 papers presented here define the state-of-the-art in the rapidly developing field of acoustical holography and imaging. Many of them describe applications in such fields as medical diagnostics, microscopy, nondestructive testing, underwater viewing, and seismology.

The Editor recognizes the diligent efforts of the authors in advancing the technology of Acoustical Imaging and thanks them for preparing and submitting descriptions of their work. The papers were selected with the able assistance of the Program Committee that consisted of P.S. Green, Stanford Research Institute; J.F. Havlice, Stanford Microwave Laboratory; B.P. Hildebrand, Battelle Northwest; D.R. Holbrooke, Children's Hospital of San Francisco; P.N. Keating, Bendix Research Laboratories; A. Korpel, Zenith Radio Corporation; B.J. McKinley, Lawrence Livermore Laboratory; A.F. Metherell, University of Miami; J. Powers, Naval Postgraduate School; and F.L. Thurstone, Duke University. The Editor appreciates the help of the session chairmen: D.R. Holbrooke, Children's Hospital of San Francisco; Mahfuz Ahmed, Zenith Radio Corporation; R.K. Mueller, University of Minnesota; G. Wade, University of California at Santa Barbara; B.P. Hildebrand, Battelle Northwest; and G.S. Kino, Stanford University. A particular note of thanks is due to Ben Saltzer, Naval Undersea Center, for serving as local arrangements chairman for the symposium, and to Rand Kuhl, Technical Engineering Services, for assistance in handling the manuscripts.

The Symposium was sponsored by the Office of Naval Research, in cooperation with the IEEE and the Acoustical Society of America. The Naval Undersea Center, San Diego, California hosted the event.

The SEVENTH INTERNATIONAL SYMPOSIUM ON ACOUSTICAL HOLOGRAPHY AND IMAGING is planned for Chicago, Illinois in mid-1976 under the Chairmanship of L. W. Kessler, Sonoscan Inc., 752 Foster Avenue, Bensenville, IL 60106.

CONTENTS

BIOMEDICAL IMAGING WITH THE SRI ULTRASONIC CAMERA*

J. R. Suarez, K. W. Marich, J. F. Holzemer,
J. Taenzer, P. S. Green
Stanford Research Institute
333 Ravenswood Avenue
Menlo Park, CA 94025

ABSTRACT

Researchers at SRI have developed a real-time ultrasonic imaging system for the visualization of anatomical structures in an orthographic projection rather than the more common B-scan presentation. This system produces high resolution images of a 15-cm x 15-cm field by employing pulsed/range-gated operation and a 192-element linear array of piezoelectric detectors. The system's high sensitivity (10^{-11} W/cm^2) permits imaging through the abdomen of adult humans with incident energies well below any known toxic level. Both transmission and reflection imaging are provided. Images are displayed on a television monitor and may be recorded on video tape.

Results of _in vitro_ and _in vivo_ experimentation are presented. _In vitro_ studies have been made of both normal and abnormal excised organs such as the spleen, uterus, liver, and kidney. _In vivo_ studies have resulted in preclinical images of various anatomical structures and appendages of the human adult. These studies indicate the diagnostic potential of this instrument.

---

*This work was supported in part by Public Health Service Grant GM18780.

1

INTRODUCTION

In 1973, a real-time ultrasonic imaging system, developed at Stanford Research Institute, was reported (1). This paper will serve as a progress report, describing further development of the ultrasonic camera and the results of some pre-clinical experimentation.

The camera system incorporates many design features that are desirable in a clinical imaging instrument. Real-time, orthographic grey-scale images of a 15-cm x 15-cm field of view are displayed at 15 frames per second on a standard television monitor. Dynamic viewing of the real-time image allows the examiner to position the patient for maximal diagnostic information. Indeed, the ability of the human brain to reconstruct three-dimensional information from moving two-dimensional images--in addition to localized palpation--greatly aids in the interpretation and identification of anatomical structures.

The resolution attained with this instrument is approximately 1.3 mm, with a depth of focus of 1 cm. Because the image is an orthographic projection similar to a standard X-ray view, the zone of focus is perpendicular to the direction of propagation of the ultrasonic waves and may be positioned at any desired depth in the patient. The camera system can be used in either the transmission or reflection mode. For both modes the sound is pulsed, rather than continuous, which reduces the average power and minimizes patient exposure. The high detection sensitivity ($\sim 10^{-11} W/cm^2$) allows through-transmission imaging of the adult abdomen using average intensities that are typically less than 300 $\mu W/cm^2$ with peak intensities of less than 18 mW/cm^2. Degradation of transmission images by unwanted multiple reflections is diminished by range gating the receiver. While in the reflection mode, range gating allows selection of only those reflections from the zone of focus. When imaging in the reflection mode, a transducer located to the side of the lens package "illuminates" the object. (See Figure 1.)

APPARATUS

Briefly, the components of the SRI imaging system (Figure 1) are the transmitter, a transmitting transducer, ultrasonic lenses and prisms, a 192-element receiving array, and a display. An ultrasound chirp (6 µsec duration and 800 KHz bandwidth

centered at 2 MHz) is propagated through the water and is
scattered by the object. The perturbed sound field is collected and focused by a pair of polystyrene lenses onto the
piezoelectric receiving array. Therefore, each pulse results
in one line of the image. To generate the entire ultrasonic
image, two polystyrene prisms located between the two identical lens elements are counter-rotated to sweep the ultrasonic field across the receiving array. The resultant image
consists of approximately 400 interlaced lines, each containing 192 picture elements. Separate amplifiers are provided for each of the array elements. These amplifiers were
designed for both linear and logarithmic operation. Other
circuitry is provided for range gating, detection, storage,
and commutation.

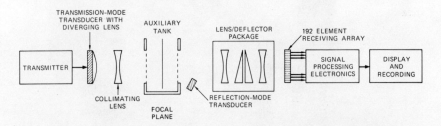

Figure 1. Simplified block diagram of the SRI real-time
imaging system.

        The laboratory version of the ultrasonic imaging system
employs a large water tank in which the patient is partially
immersed during the examination. The in vitro studies, involving imaging of excised organs and phantoms, were performed in a small auxiliary water tank equipped with Mylar
windows and mounted in the larger tank in the position
indicated in Figure 1. In the laboratory system, the ultrasonic illuminator, the lens package, and the receiver array
are in fixed positions within the water tank. The examiner
moves either the patient or, as in the in vitro studies, the
organ, to bring the area of interest into the fixed zone of
focus.

        A new clinical version of the laboratory imaging system
has been designed, constructed, and installed in the Veterans
Administration Hospital, Palo Alto, California, where Dr.
Leslie M. Zatz is conducting the clinical evaluation. The

essential differences between the laboratory and clinical
systems are that the second instrument does not require
patient immersion for examination, and that the means of selecting the plane of examination (zone of focus) within the
patient is easily controlled by the physician. In this instrument, the patient is held relatively immobile between two
flexible plastic, water-filled membranes (Figure 2). The
physician selects the general area to be visualized by ori-
enting the patient between the coupling bags. The plane of
focus is selected by moving the lens-package/receiver-array
assembly back and forth with respect to the patient by means
of a panel-mounted electronic control (Figure 3).

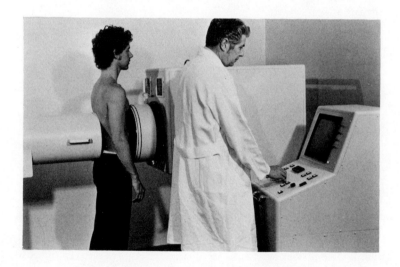

Figure 2. The SRI ultrasonic clinical imaging system. The
patient is positioned between the water-filled coupling bags
while the examiner can change the plane of focus visualized
by moving a panel-mounted "joy stick" coupled to the lens-
package/receiver-array assembly.

Several months of preliminary clinical trials have produced encouraging results, and further technical improvements
will increase the camera's clinical usefulness. Results of
both in vivo and in vitro real-time ultrasonic investigations
have been recently reported in the literature (2-4).

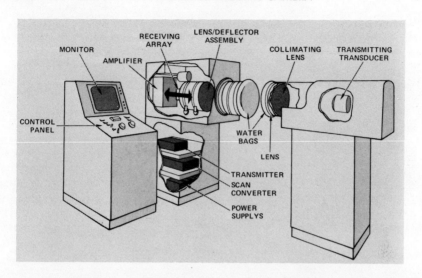

Figure 3. Cutaway artist drawing showing the components of the SRI ultrasonic clinical camera. (From L. M. Zatz (4) 1975)

RESULTS

Preliminary *in vitro* and *in vivo* investigations conducted at SRI using the laboratory version of the imaging system have resulted in useful information that has led to a better assessment of the potential clinical applications of this technique. The biological *in vitro* imaging studies were performed using freshly excised animal and human organs. These studies are important because they allow a direct correlation between the acoustic image and known anatomical structures, thus leading to a better understanding and interpetation of the *in vivo* real-time ultrasonic images.

Specific organs and pathological conditions were chosen for the *in vitro* studies to define, in a model environment, the imaging capabilities of this real-time technique. The intricate details of ducts and vessels in the parenchyma of the excised liver and kidney can be seen in Figure 4. A comparison of transmission and reflection mode imaging is shown in Figures 5 and 6. An application that may have diagnostic potential is the noninvasive detection of renal stones and soft tissue tumors. Transmission images of these pathological

conditions in an excised sheep kidney and an excised human
uterus are shown in Figures 7 and 8. An ultrasonic image of
an excised sheep heart is presented in Figure 9. These
figures are a small but representative sample of the many
images we have produced with many different organs and objects.

In vivo visualization of soft tissue structures of the
appendages, including details of the joints of the fingers,
wrists, elbows, knees, and shoulders were easily accomplished
in both infants and adults. Typical images of the hand and
foot are shown in Figures 10 and 11. A bifurcation of a
superficial vein in the lower leg muscle can be clearly seen
in Figure 12. Vessels have been seen in transmission images
of many parts of the body; this modality may have potential
for the diagnosis of vascular disorders, such as localizing
a thrombus or atherosclerotic plaque.

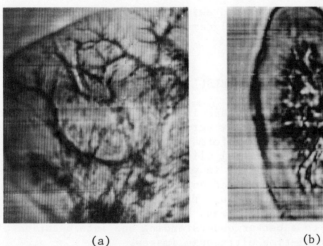 

(a)   (b)

Figure 4. Ultrasonic transmission images of a freshly excised
normal human liver (a) and kidney (b). Air was entrapped in
the larger ducts and vessels of the liver (a) as it was sub-
merged into the water tank, thus forcing the air into the
smaller ductal networks and accentuating the fine structure.
In the kidney (b) the calycial structures are clearly dis-
cernable without pretreatment or injection of contrast agents.
(Reprinted from J. Clin. Ultrasound, Marich et al. (2), March,
1975).

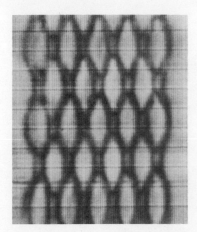 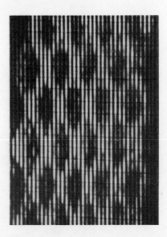

Figure 5. Ultrasonic transmission (left) and reflection (right) images of a metal grid with holes approximately 5 x 10 mm in size.

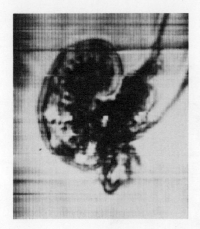 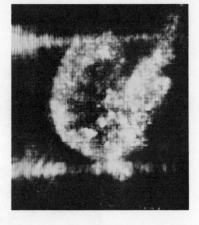

Figure 6. Ultrasonic transmission image (left) and a reflection image (right) of a lamb kidney in which 4 ml of a 33% barium solution was injected into the ureter. Horizontal lines are the metal wires holding the specimen. Catheters can be seen going into the ureter and renal artery. Note the clarity of the calycial structures in both images.

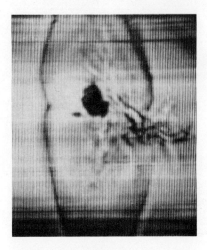

Figure 7. Ultrasonic transmission image of a sheep kidney
containing a large renal stone.

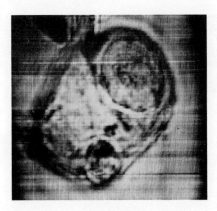

Figure 8. Ultrasonic transmission image of an excised human
uterus containing two large fibroid tumors. Note the sharp
edge outline of the capsule surrounding the fibroid.

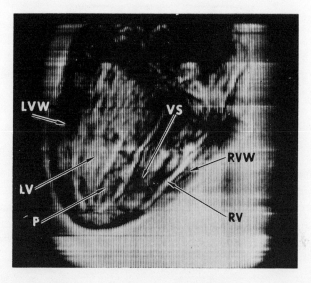

Figure 9. An ultrasonic through-transmission image of an excised bovine heart showing: left ventricular wall (LVW), left ventricle (LV) with inserted metal probe (P), intra-ventricular septum (VS), right ventricle (RV), and right ventricular wall (RVW). The structures seen in the left upper ventricle are highly suggestive of the chordae tendineae of the mitral valve.

Figure 10. Ultrasonic transmission image of the hand showing the metacarpal-phalangeal and interphalangeal joints. The linear structures visualized between the bones are probably related to muscles, tendons, and neurovascular bundles.

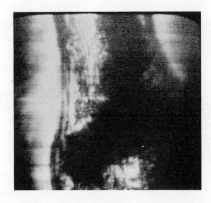

Figure 11. Ultrasonic transmission image of the foot showing
the calcaneus with the attached Achilles tendon. Articulation
of bony joints is easily observed with this real-time imaging
technique.

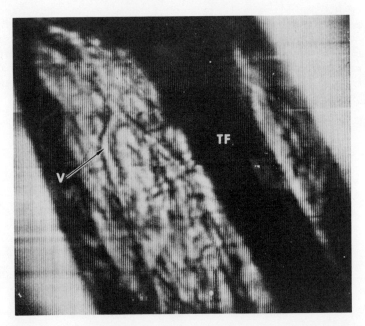

Figure 12. Ultrasonic through-transmission image of the lower
leg showing the tibia and fibula superimposed (TF) and a bifurcation of a superficial vein (V) in the gastrocnemius
muscle.

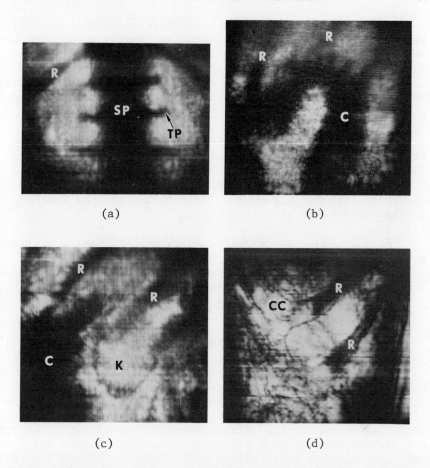

(a) (b)

(c) (d)

Figure 13. Ultrasonic transmission images showing an antero-
posterior projection of the spine (a), colon (b), and kidney
(c), and a posteroanterior projection showing the costal cartilage (d). Image (a) shows the 12th rib (R) and the lumbar
spine (SP) with the transverse processes (TP) of the vertebrae
projecting laterally. Image (b) clearly shows the haustral
pattern of the splenic flexure of the gas-filled colon (C).
Out-of-focus ribs (R) which appear as dark bands, are seen in
the upper section of the picture. Image (c) shows the outline
of the left kidney (K) with two out-of-focus ribs (R) overlying
the image. The dark area in the lower left of the picture is
the gas-filled colon (C). Image (d) shows the junction of the
calcified ribs (R) and the costal cartilage (CC).

Intraabdominal imaging studies have shown that some internal structures could be readily and clearly visualized in most subjects primarily by their features. Figure 13 shows several anatomical details that are not readily visualized by the use of X-rays. Such soft tissue imaging by ultrasound may become a very important feature of the orthographic ultrasonic camera.

Preliminary qualitative studies of the thoracic cavity revealed a transmission acoustic "window" within the chest that allows the visualization of moving cardiovascular structures. Images of the neck demonstrate the esophagus, trachea, and large vascular structures, and studies are now underway to assess their clinical applicability.

Finally, in vivo transmission through the skull has been achieved in a very cursory experiment. However, it is expected that the interpretation of intracranial images, although potentially feasible, will present one of the most difficult problems in transmission imaging.

SUMMARY

In summary, very significant advances in ultrasound imaging have been made in the last three years, both in equipment development and in acquiring a basic understanding of ultrasonic transmission imaging as a modality. Further development of the real-time C-scan in the reflection mode, now possible with the SRI instrument, is being given increased attention. The real-time orthographic presentation made possible by this focused imaging system, together with the accompanying grey-scale capability, suggests that the advantages of this approach can play a major role in clinical medicine.

ACKNOWLEDGMENTS

The authors would like to thank Drs. Leslie M. Zatz and Albert Macovski for their active collaboration and many helpful discussions during the instrument development and experimentation.

REFERENCES

1. Green, P. S., Schaefer, L. F., Jones, E. D., and Suarez, J. R.: A New High-Performance Ultrasonic Camera. In Acoustical Holography. Vol. 5, P. S. Green (Ed.), pp. 493-503, Plenum Press, New York, 1974.

2. Marich, K. W., Zatz, L. M., Green, P. S., Suarez, J. R., and Macovski, A.: Real-Time Imaging With a New Ultrasonic Camera: Part I, In Vitro Experimental Studies on Transmission Imaging of Biological Structures. J. Clin. Ultrasound, March, 1975.

3. Zatz, L. M., Marich, K. W., Green, P. S., Lipton, M. J., Suarez, J. R., and Macovski, A.: Real-Time Imaging With a New Ultrasonic Camera: Part II, Preliminary Studies in Normal Adults. J. Clin. Ultrasound, March, 1975.

4. Zatz, L. M.: Initial Clinical Evaluation of a New Ultrasonic Camera. Presented at the Radiological Society of North America Meeting, Chicago, Illinois, December 1-6, 1974 (submitted to Radiology for publication).

THROUGH-TRANSMISSION ACOUSTICAL HOLOGRAPHY FOR MEDICAL IMAGING—A STATUS REPORT

K.R. Erikson[1], B.J. O'Loughlin[2], J.J. Flynn[2],
E.J. Pisa[1], J.E. Wreede[1], R.E. Greer[1], B. Stauffer[1],
and A.F. Metherell[1]

1. Actron, A Division of McDonnell Douglas Corp.
Monrovia, California 91016

2. California College of Medicine, University
of California at Irvine, Irvine, California 92664

The findings reported here represent the current status
of an evaluation of through-transmission holographic acoustical
imaging in a clinical setting. The major objective of this
effort is the investigation of the diagnostic usefulness of
holographic images of soft tissue structures and organs in
the human body, particularly the abdomen. This work involves
a series of experiments including the imaging of organs in
vitro, animals, patients, pregnant women, and a series of
test objects. The 1 MHz holographic system used in this
evaluation is described, together with images of several
test objects. Selected clinical results, including in vivo
images of several near term fetuses, are presented and discussed. Finally, we comment on through-transmission acoustical holography as a modality for soft tissue diagnosis.

INTRODUCTION

The findings reported here represent selected results
of an attempt to evaluate the diagnostic value of holographic
acoustical imaging in a clinical setting. The usefulness of
holographic images of soft tissue structures in the human
body, particularly in the abdomen, is currently under investigation as a part of a program begun in December 1973. This
clinical demonstration makes use of an acoustical holography
system developed by the Actron Division of the McDonnell
Douglas Corporation for abdominal imaging.

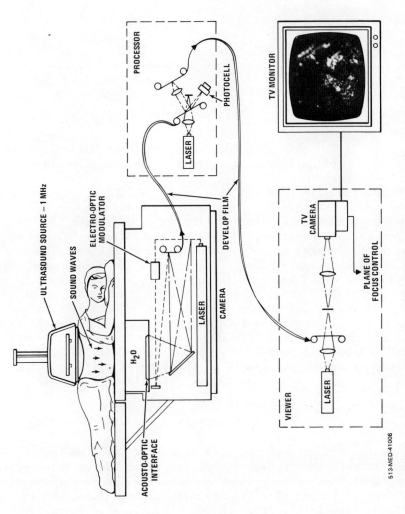

513-MED-41006

Figure 1. Acoustical Holography System Block Diagram

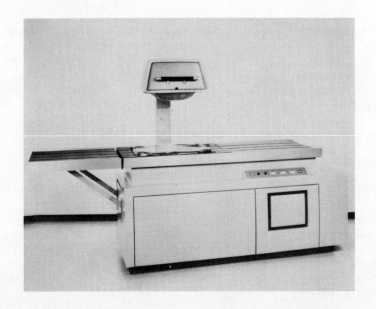

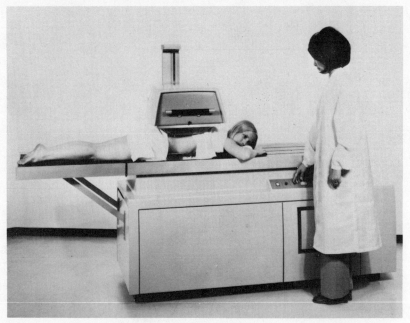

Figure 2.   Acoustical Holography Camera System

While a detailed technical description of this acoustical holography system has appeared elsewhere (e.g., Refs 1 and 2), several desirable features are worth noting here. The acoustical holography system (Figures 1, 2 and 3) record the entire patient volume within a 30 cm diameter cylindrical field-of-view on each frame of a holographic movie with a resolution of 3 mm. Each frame requires only a 90 microsecond pulse of 1 MHz ultrasound, insuring a time-averaged maximum acoustical power level of less than 3.6 milliwatts/cm^2 at the standard 4 frame per second rate. With the holographic viewer (Figure 3), a radiologist can either run the movie focused in a selected plane or stop to focus on different depths of the subject volume in any single movie frame. In the viewer any 5 mm thick image plane within the 30 cm diameter cylindrical patient volume can be brought into focus in any one frame. This is a unique property of this holography system. Specific engineering details of the holography system are discussed in Appendix I and safety aspects in Appendix II.

Figure 3. Viewer

The optical detection technique used to record the acoustical hologram in this system is described in detail in Refs. 2 and 4 and is not discussed here. While this detection technique and especially its implementation in the current system (see Appendix I) offer certain advantages as listed above, they also impose certain restrictions. The system is limited to through-transmission imaging because of the inherently low system sensitivity. As discussed in Refs 2 and 4, the optical detection technique reduces the signal to noise ratio of the optical hologram by a factor of at least five thus making the system quite susceptable to noise. Also, this detection technique requires the user to compromise between optimum operating sensitivity and optimum dynamic range (gray scale) in the final image. The reader should bear in mind that all of these restrictions are unique to the particular system discussed in this paper and are not inherent to acoustical holography per se.

IMAGES OF TEST OBJECTS

Through-transmission images of complicated objects such as biological tissue are unfamiliar and inherently difficult to interpret. We have imaged several simple, well understood test objects in order to check holography system performance. Test objects were devised for measurement of resolution, plane of focus, depth of focus and critical angle effects.

Resolution Test Target

The resolution test target is a 0.375 mm thick laminate of stainless steel plates through which enlarged images of the standard Air Force 1956 resolution target have been photoetched (Figure 4a). (The overall target dimension is 33 cm square.) In the acoustic image (Figure 4b) the smallest bar pattern normally resolved is equivalent to 0.22 line pairs/mm (bars and spaces each 2.3 mm). The high contrast resolution target is not representative of the biological tissue intended to be imaged, but does serve as an effective monitor of system performance. It is imaged as part of the daily warmup procedure, as well as before and after each patient.

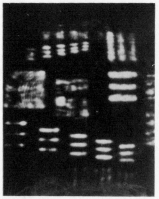

(a) OPTICAL IMAGE
(OUTSIDE DIMENSION, 33 CM)

(b) ACOUSTICAL HOLOGRAPHIC IMAGE
(APPARENT RESOLUTION, 2.3 MM)

Figure 4.   Acoustic Resolution Target

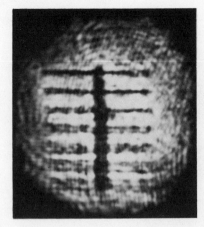

(b) FOCUSED ON UPPER BAR

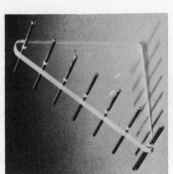

(c) OPTICAL IMAGE
(1 INCH BETWEEN BARS)

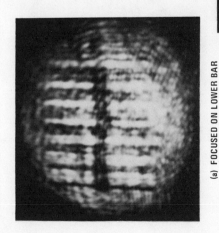

(a) FOCUSED ON LOWER BAR

Figure 5. Plane of Focus Test Object

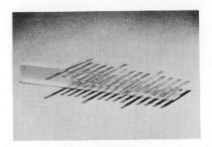

(a) OPTICAL IMAGE

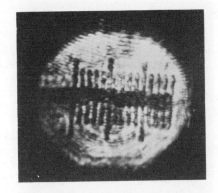

(b) FOCUSED ON UPPER BARS

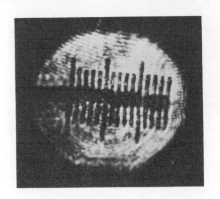

(c) FOCUSED ON MIDDLE BARS

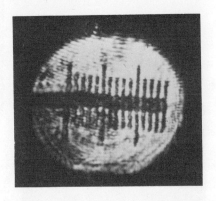

(d) FOCUSED ON LOWER BARS

Figure 6.   Depth Of Focus Test Object

Plane of Focus Test Object

The plane of focus test object shown in Figure 5c is a Lucite plate with 4.5 mm diameter steel rods inserted at 25.4 mm depth intervals. Its purpose was to calibrate the focus knob on the viewer system which was shown in Figure 3. Representative images focused at the two extreme bars are shown in Figures 5a and b. As described below, it was also used to determine the distorting effect of biological tissue.

Depth of Focus Test Object

The depth of focus test object (Figure 6a) is similar to the previous object but with a 1 mm depth spacing of the 1 mm thick aluminum plates (center to center spacing of 12 mm). The depth of focus at three planes in the image volume is shown in Figures 6b, c and d. Approximately five bars are in focus at any time over the 20 mm range of the test object. This 5 mm depth of focus is constant throughout the depth of the 30 cm image volume.

Fluid Filled Balloons

A series of cylindrical balloons (45 mm diameter) were filled with isopropyl alcohol/water or ethylene glycol/water mixtures to provide the range of sound velocities exhibited by soft biological tissues. An acoustical image of three of these balloons is shown in Figure 7. The purpose of this test was to determine if critical angle reflection in this range of sound velocities is sufficient to cause edge enhancement and subsequent visualization of boundaries as discussed by Holbrooke, et al (Ref 3).

Figure 8 presents the experimentally observed edge thickness of these balloons as a function of the sound velocity of the internal fluid when focused on the center plane of the balloon. The surrounding medium in all cases was water. The balloons are visible when the velocity difference is greater than about 20 meters/sec. Below that, there is an indication that the balloon is present only when it is moved; however, the edges are indistinct and not always present. The balloon disappeared completely when filled with water, implying that the rubber itself is thin enough

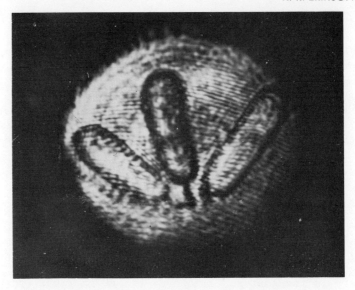

Figure 7.   Fluid Filled Test Objects

to have no appreciable effect on the sound wave. As the
velocity difference decreases the apparent edge thickness
also decreases until a constant value of about 1.5 mm is
reached. The only effect of further reduction of velocity
difference is a decrease of edge contrast. This is the
resolution limit for a 1 MHz system.

A theoretical curve for apparent edge thickness due
to critical angle reflection is also shown in Figure 8.
Because our experimental data indicate edge thicknesses
greater than those expected from critical angle reflection
alone, another explanation for edge enhancement was sought.
A second curve is shown for twice refracted sound which
misses the detector, but this also is not the complete
answer. Refraction at the upper surface of the acousto-
optic interface together with the subjective measurement
of the edge thickness (because it is not a hard edge) may
account for the discrepancy.

Another type of edge enhancement is shown in Figure 9.

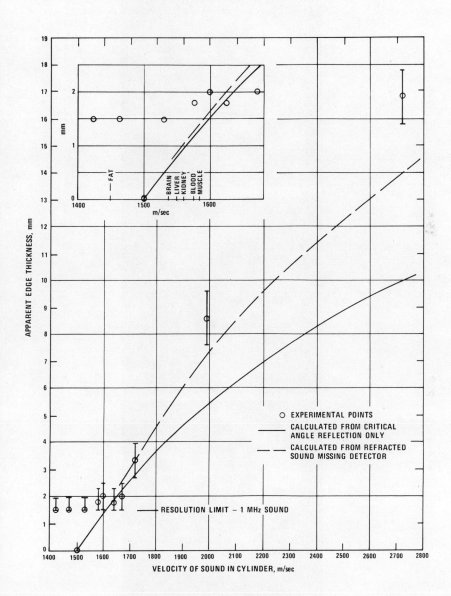

Figure 8.   Edge Thickness Data

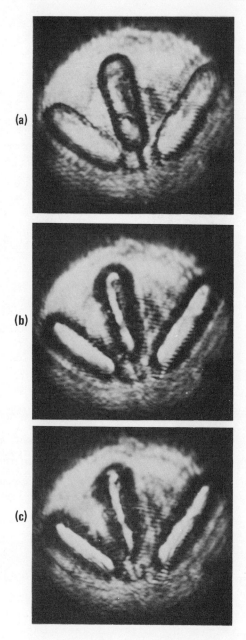

Figure 9. Effects Of Refracting Media

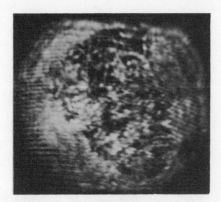

(a) IN FOCUS

(c) B-SCAN IMAGE

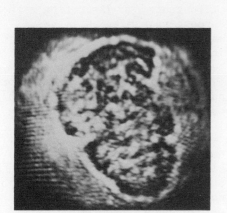

(b) OUT-OF-FOCUS

Figure 10.   Excised Fat Layer

These are the three balloons of Figure 8 focused in three
different planes. The minimum edge thickness is presumably
the median plane of the balloons where they are in best
focus (Figure 9a). Slightly out of focus balloons tend to
have thicker edges (Figure 9b). If we focus far enough
behind the balloon, a virtual line image of the acoustic
source appears (Figure 9c). This is easily explained by
the fact that the balloon acts as a negative cylindrical
lens. If the balloon is filled with a medium of lower
acoustic velocity than the surrounding water, the virtual
source is not seen.

Fatty Tissue

An excised piece of abdominal fat (a panniculus
removed in cosmetic surgery) was imaged with the holography
system. In Figure 10a the fat is nearly in focus and the
edges are quire indistinct since the velocity of sound in
fat is close to that of water. Focusing in a different
plane (Figure 10b) caused enhancement of all edges in the
fat, which is composed of globules in a membranous matrix.
Figure 10c is a B-scan of the fat for comparison.

This specimen was used along with the plane of focus
test object (Figure 5c) to illustrate the distorting effect
of fat tissue on acoustic through-transmission images.
Figure 11 shows images focused on the third bar which
was about 50 mm from the fat. In Figure 11a, the fat is
on the source side of the test object. Figure 11b shows
the notably greater effect when the fat is on the detector
side. The fat layer was approximately 50 mm thick, which
is more than ordinarily encountered in ultrasonic imaging.
The complexity possible in the fat layers of the body is
shown in Figure 12. These are B-scans of an obese woman.
The fat structure can be complicated, consisting both of
peritoneal and properitoneal fat of changing dimensions
and position. Imaging through fatty tissue with a through-
transmission system is by no means trivial and may be the
single most important subject dependent problem.

CLINICAL RESULTS

The clinical investigation involves a series of
experiments including imaging of animal organs, normal and

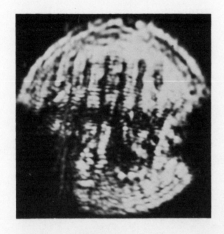

(a) FAT ON SOURCE SIDE OF TEST OBJECT

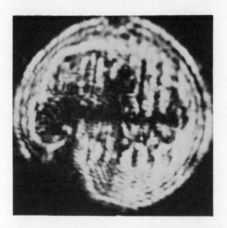

(b) FAT ON DETECTOR SIDE OF TEST OBJECT

Figure 11. Distortion Effect Of Fat Layer

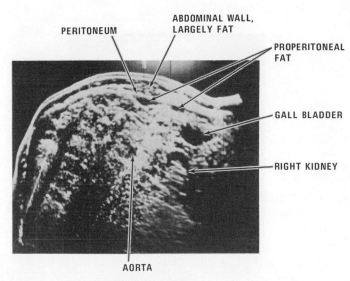

(a) TRANSVERSE B-SCAN

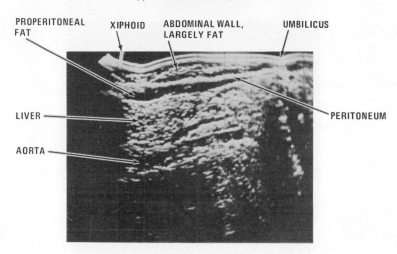

(b) LONGITUDINAL B-SCAN

Figure 12.   Appearance Of Fat Layers In B-Scans

abnormal organs in vitro, living animals, volunteers, patients, and healthy pregnant women. Because of space
limitations, only a few of the in vivo images available
are included here. We have chosen in vivo images of three
near term pregnancies in order to demonstrate the nature
of images obtained from thick biological media. Also,
the holography system was designed primarily for use in
obstetrics and gynecology.

When examining the results presented here, the reader
should continually bear in mind that image interpretation
on the actual system is greatly facilitated by one's
ability to focus while stopped on a frame as well as
by one's ability to view dynamic information. Both of
these advantages are lost when viewing the static images
which appear here, and consequently image interpretation
is far more difficult.

The images suffer from three sources of artifacts:
tissue dependent phenomena, ever present out-of-focus
information, and coherent effects resulting from the use
of coherent acoustics and optics. These artifacts make
discovering the object like finding a chameleon camouflaged
in heavy foliage. That is why displaying the image in
motion picture format is so important. Just as a jackdaw
cannot recognize a beetle until it moves, so the motion of
a structure such as a fetal extremity aids considerably in
its indentification.

Case 1. Near Term Pregnancy

This case, a 41-year old, 212 pound woman in the 36th
week of gestation, demonstrates that a fetus in a large
woman can be imaged reliabily (Figure 13a and b). The
fetus was in the vertex position and considerable detail
of the spine and pelvis can be observed. The outline of
the abdominal wall and buttocks can also be seen clearly.
Longitudinal and transverse B-scans (Figure 13c and d)
were taken concurrently, and are included for comparison.

Case 2. Near Term Pregnancy

This case, a 26-year old diabetic woman in the 35th

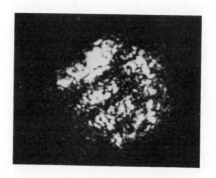

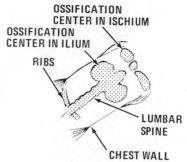

(a) HOLOGRAPHIC IMAGE   (b) SCHEMATIC DIAGRAM

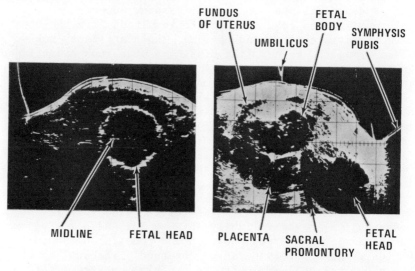

(c) TRANSVERSE B-SCAN   (d) LONGITUDINAL B-SCAN

Figure 13.   Near Term Fetus In Utero (Vertex Position)

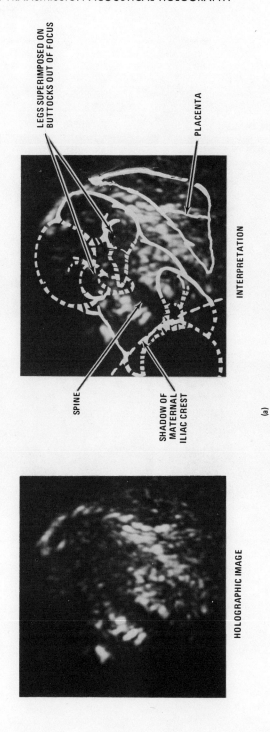

Figure 14. Motion Sequence Of Fetus In Utero

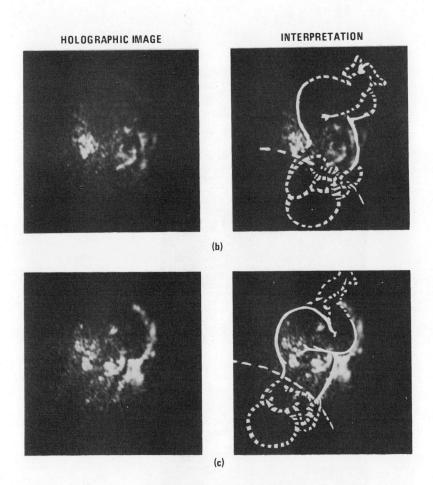

Figure 14. Continued

week of gestation, demonstrates the typical appearance of
a near-term fetus in vivo in the vertex position (Figure
14a). The sensitivity of the system is noted to be
sufficient. Fetal parts which can be identified include
the abdominal wall, the dorsolumbar spine, lower ribs,
pelvis, and lower extremities. Fetal motion was observed
in the movie reconstruction and is shown here in Figures
14a, b and c. These are three consecutive frames from
the movie reconstruction, together with some interpretation
to aid the reader. Figure 14a is prior to the motion. In
Figure 14b, the fetus has turned and started to arch its
back. Finally, in Figure 14c, the fetus has finished its
turn and proceeded to kick.

Case 3.   Near Term Pregnancy

This case, a 35-year old woman in the 35th week of
gestation, demonstrates the potential of the system for
imaging near-term breech fetuses in vivo. Figure 15 shows
the orientation of the woman and fetus with respect to
the holography system, together with the resulting holographic image. The system sensitivity was sufficient not
only to image the fetus, but even to penetrate the calcified
cranium to display the fetal brain. The brain was well
outlined, showing structures which we think may be gyri
and sulci of the frontal, parietal, occipital and temporal
lobes (Figure 17a). Figure 17b is taken from the same
frame of the holographic movie focused in a different plane.
We think the centrally located dark structure is the
corpus callosum. The skull appears thickened because the
calcification in the bony tables produces a decreased
critical angle resulting in total reflection from the
curved edges of the skull. The soft tissues of the face
appear well outlined and in focus. Fetal motion was also
observed in the movie.

The brain structures of Figures 17a and 17b show
dramatic resemblance to those structures of a 19-week
old fetal cadaver previously imaged in vitro (c.f.Ref 4),
and shown here in Figure 16a. Figure 16b shows the dissected
fetal cadaver for comparison. These structures, to the
best of our knowledge, have never before been demonstrated
in vivo by any type of imaging modality including x-ray
and other types of ultrasonic imaging.

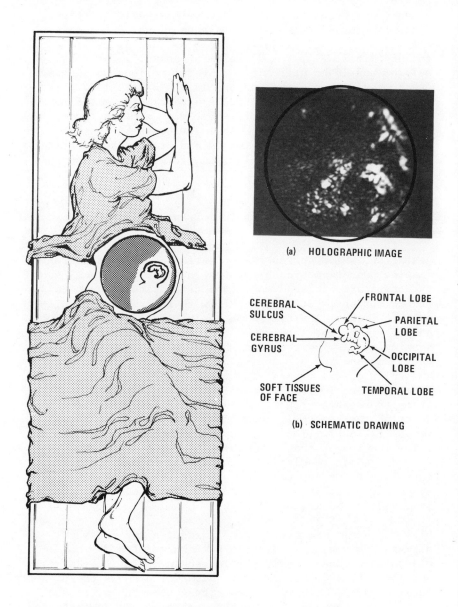

(a) HOLOGRAPHIC IMAGE

(b) SCHEMATIC DRAWING

Figure 15. Near Term Breech Fetus

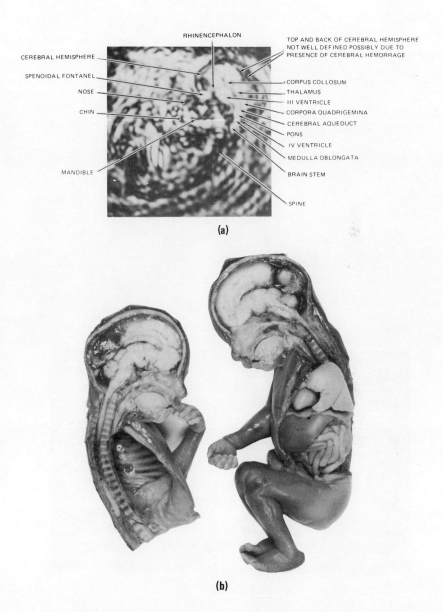

(a)

(b)

Figure 16. Fetal Cadaver Imaged In Vitro

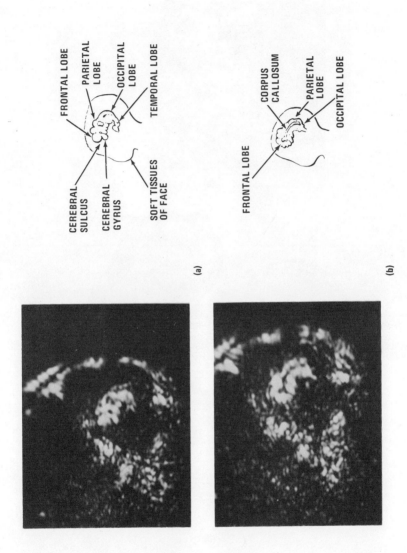

Figure 17.   Near Term Breech Fetus In Utero

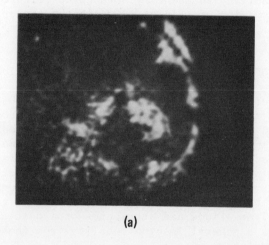

(a)

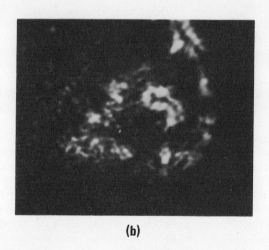

(b)

Figure 18. Defocused Version Of Figure 17

We have often found that by defocusing the image
slightly we can smooth out the background noise and
improve our ability to identify structure. Figure 18 is
a defocused version of Figure 17 which the reader may
find helpful.

THE PLACE OF ACOUSTICAL HOLOGRAPHY IN MEDICAL IMAGING

After being intimately involved with an acoustical
holography system in the clinic for over a year now and
having imaged a wide variety of subjects, both medical
and engineering in nature, we have arrived at some personal
opinions regarding the place of acoustical holography in
medical imaging. Although these opinions are certainly
biased by our experience with the particular holography
system described here, we will make every effort to
distinguish limitations and attributes of acoustical
holography as a general imaging modality from those
unique to the Actron system.

The single most unique characteristic of an acoustical
hologram is that the entire object volume is recorded in
the hologram, thus allowing the viewer to focus throughout
the volume using only a single hologram. The incredibly
high information bandwidth of the hologram also allows
considerable flexibility in a posteriori image processing.
However, because a coherent wave is required to generate
a hologram, the resulting images are plagued by edge
ringing and other diffraction artifacts. Because holographic images are different in nature than conventional
B-scans, they require considerable experience to interpret.

There are a number of additional factors quite independent of holography per se which greatly influence the
utility of acoustical holography as a medical imaging
modality. For example, the choice of through-transmission
versus reflection imaging significantly alters the character of the images, whether or not the system is holographic. Through-transmission imaging of tissue is
dependent upon differential absorption and scattering at
edges to produce images. Reflection imaging is primarily
dependent on impedance changes at interfaces. Through-
transmission images are inherently small variations
superimposed in a large background. Reflection images are
plagued by specular reflection problems (most objects

appear smooth at the acoustic frequencies of interest),
through-transmission by shadowing, refraction, and critical
angle effects.

Holography does not help the viewer overcome the
complexities (i.e., inhomogeneities, multiple scattering,
etc.) of the human body as a propagation medium. While
acoustical holography is advantageous for hard scatterers
in a homogeneous medium, it is much less well suited for a
complicated, inhomogeneous medium such as the body.

CONCLUSION

Acoustical imaging of biological tissue has many of
the aspects of optical microscopy of metallurgical sections,
where the wavelength of light becomes an important resolution parameter. The conventional metallographic microscope
has three light sources. One, directly "overhead" for
reflection imaging, is introduced through a beam splitter;
one directly behind is used for through-transmission; and
the third is free to move above the stage in any direction.
The microscopist then chooses any combination of intensities
from these sources. The correct combination shows brilliant
colors through differing absorptions in through-transmission.
The overhead light shows surfaces and planes in reflection.
Most importantly, the moveable light source is arranged
to produce highlights from boundaries or to make the light-
ing more uniform (as the portrait photographer does). The
practiced microscopist can produce excellent images through
trial and error.

Consider the physician trying to produce images with
current acoustical systems which only allow through-
transmission or reflection and to this time do not allow
the flexibility to exercise judgement and trial and error.

Implicitly we have defined a number of system parameters
in this analogy which can now be made explicit.
1) Reflection and transmission imaging are both needed
in the same system. This necessarily implies a piezoelectric detector or one of equivalent sensitivity
for reflection imaging.
2) A real time system is needed to allow the insonification and/or subject to be adjusted for an optimum

(or even any) image, through trial and error.

3) Multi-frequency or "white" sound is important to take advantage of differential sound absorption with frequency.

4) The whiteness of the sound we wish to employ implies both temporal and spatial incoherence; i.e., we need the acoustical analog of the incandescent light bulb. This tends to make acoustical holography less attractive in its present implementations.

5) The "white" sound will also help reduce, or will eliminate, coherent artifacts present in most acoustical imaging systems.

6) Use of white sound has some disadvantages, since the depth of focus of such a low f/number, high resolution system becomes very small and the ability to maintain a biological specimen in focus becomes correspondingly more difficult.

7) Finally it is clear that a superior detector such as the eye prefers a directly viewed microscopic image because it can use a great deal more dynamic range than a CRT can provide. Such an optimum display may be hard to find.

In summary, this acoustical holography system has been a technological challenge overcome. The evaluation at Orange County Medical Center has improved our understanding of the problem of acoustical imaging of the human body. We hope that in reporting this work as fully and honestly as we can, we have contributed to others in this exciting, rewarding field.

ACKNOWLEDGEMENTS

This work was partially supported by NIGMS Grant No. 20834-01.

Special thanks are due R. Anwyl of Eastman Kodak; H. Walh of Replica Optics; R. Carlsen of Channel Industries; H. Zierhut and G. Vetter of Z.V.S.; and R. Norton, R. Watts, T. Stoddard, R. Hill, and J. Samaras for their contributions to the holography system. The authors would also like to thank L. Cunningham and L. Axelrod of O.C.M.C. for their help with the clinical aspects of this work.

REFERENCES

1. A.F.Metherell, et. al., "A Medical Imaging Acoustical
   Holography System Using Linearized Subfringe Holo-
   graphic Interferometry," pp 453-470 in Acoustical
   Holography, Vol V,P.Green, editor, Plenum Press, 1974.

2. A.F.Metherell, "Linearized Subfringe Interferometric
   Holography," pp 41-58 in Acoustical Holography, Vol V,
   P.Green, editor, Plenum Press, 1974.

3. D.R.Holbrooke, et. al., "Medical Uses of Acoustical
   Holography," pp 415-451 in Acoustical Holography,
   Vol V, P.Green, editor, Plenum Press, 1974.

4. K.R.Erikson, "Acoustical Holography Using Temporally
   Modulated Optical Holography," pp 59-81 in Acoustical
   Holography, Vol V, P.Green, editor, Plenum Press, 1974.

5. W.O'Brien, M.Shore, R.Fred and W.Leach, "On Assessment
   of Risk to Ultrasound," Proc. IEEE Symp. Sonics
   Ultrasonics, pp 486-490, 1972.

6. M.Ziskin, "Survey of Patient Exposure to Diagnostic
   Ultrasound," Proceedings of a Workshop held at Battelle
   Seattle Research Center, Seattle, Washington, November
   8-11, 1971; D.H.E.W. Pub. (FDA) 73-8008 BRH/DBE 73-1;
   pp 203-205, September 1972.

7. F.Dunn and F.J.Fry, "Ultrasonic Threshold Dosages for
   the Mammalian Central Nervous System," IEEE Trans.
   Biomed. Eng., BME-18, 253-256, 1971.

8. K.R.Erikson, F.J.Fry and J.P.Jones, "Ultrasound in
   Medicine-A Review," IEEE Trans. Sonics and Ultrasonics,
   SU-21, No.3, pp 144-170, July 1974.

9. C.R.Hill, "Acoustic Intensity Measurements on Ultrasonic
   Diagnostic Devices," Ultrasonographic Medica, Vol 2,
   J.Bock and K.Ossoining, editors, pp 21-27, 1971,
   Vienna Academy of Medicine.

APPENDIX I: DETAILED DESCRIPTION OF THE ACOUSTICAL

HOLOGRAPHY SYSTEM

CAMERA

Figure 19 is an optical schematic of the camera portion
of the acoustical holography system. The system is
basically very simple; however, a number of added devices
are used for adjustment and alignment of the optical system
and monitoring various parameters.

Acoustic Transducer

The acoustic transducer consists of four six-inch
square piezoelectric ceramic plates mounted close together
to form a 30 by 30 cm aperture (approximately 200 wavelengths at 1 MHz.) The subject is in the extreme near
field of the transducer, where the beam is a well collimated,
uniform plane wave with only minor fluctuations in phase
across the central area of the beam. At the edge of the
beam where classical Fresnel edge diffraction occurs, there
are large fluctuations in phase and amplitude.

The transducer material is a modified barium titanate
made by Channel Industries, Santa Barbara, California. An
integral glaze ceramic on the front face allows the transducer to be operated directly into water without the
requirement for an additional insulating or corrosion
resistant medium. The transducer is highly resonant (about
10 kHz bandwidth) at 1.1 MHz where each plate has an electrical impedance of 16 ohms which is ideally suited to
transistor power amplifiers. A solid state amplifier
capable of producing approximately 300 volts peak into
16 ohms at a low duty cycle was designed for this system.
Each plate of the transducer has its own power amplifier.

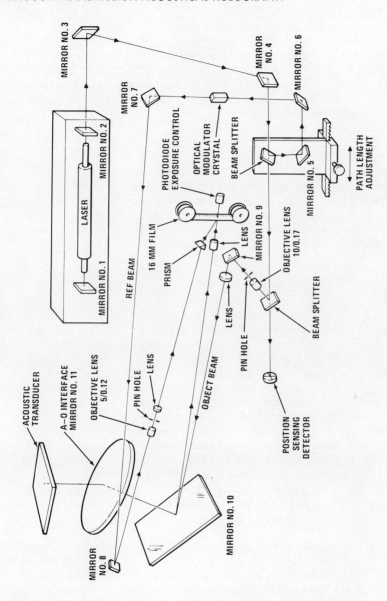

Figure 19. Camera System Optical Schematic

Patient Coupling Interface

The waterbags were vacuum formed from 10 mil polyethylene. After some experimentation an optimum bag shape
was evolved. The coupling bags produced no noticeable
artifacts or other problems except when imaging an object
with a small radius of curvature such as an arm. In the
case of a large relatively flat area such as the pregnant
abdomen, excellent contact was achieved routinely. The
waterbag was made horizontal in order to allow a large
degree of flexibility for easy conformance to the body.
Patient acceptance of the waterbags was very good. Indeed,
many patients in the third trimester actually preferred to
lie on the waterbed rather than the conventional hospital
gurney (mobile table) often used for regular B-scan examination.

Acousto-Optic Interface

The acousto-optic interface (labeled AO Interface in
Figure 19) is the heart of the acoustical holography system
and occupied the largest part of the engineering development.
It is here that sound waves create a minute deflection of
the solid-air interface which is detected by the optical
holography system. Originally a highly stretched thin
Mylar membrane was used. In the early engineering model
this presented no problem because the interface and optical
system were on top of the water chamber. In the clinical
prototype, however, conservation of space and other design
criteria required that the optical system be placed underneath the water tank. This required pressurization of the
optical side of the Mylar interface to balance the water
load on the top surface and at the same time adjust for any
changing patient loads.

As the engineering development proceeded, it became
clear that the optical quality which could be achieved with
such a Mylar membrane was not sufficient. A solid interface made of syntactic foam, formulated by 3M Corporation,
Minneapolis, Minnesota from an Epon 828 epoxy resin and
glass microballoons averaging less than 50 microns in diameter was used. The acoustic impedance of the composite
was designed to be equal to that of water (density = .61
gm/cc, velocity of sound = 2200 meters/sec.) This

provided a solid surface which at normal incidence produced a very small reflection, i.e., it was acoustically matched to the water. This rigid structure also allowed the water tank to be placed above the optical system. The interface is planar on the acoustic side and concave on the optical side. It is tilted with respect to the acoustic transducer to produce an off-axis acoustical hologram.

The concave optical surface on the bottom side of the interface proved to be the most difficult single part of the system to fabricate. An optical replication technique was used and developed to a high degree by Replica Optics, Inc., Whittier, California. In the replication process an optical quality master is made. The blank acoustic material syntactic form was lapped to an approximate concave spherical shape. Several layers of epoxy were applied in sequence and allowed to cure against the optical master, thus providing a reflecting surface of optical quality.

The optical quality of the acousto-optic interface remains the greatest single source of noise in the camera system itself as opposed to subject dependent noise or acoustical speckle.

Optical Holographic System

A pulsed xenon laser, operated at 5353 Å, with a 6 microsecond Gaussian pulse and a peak power of approximately 35 watts (0.2 millijoules per pulse) was used. At the four pulse per second frame rate no liquid cooling was required. The laser tube and power supply were built by Hughes Electron Dynamics Division, Torrance, California. Long term optical output stability required the construction of an Invar frame and special mirror mounts. Originally Kodak 649 GH film was used, which has very high resolution, but very low sensitivity, placing severe constraints on the laser. In the clinical prototype an experimental, higher sensitivity film was used (discussed below), thus relaxing the power requirements on the laser and improving reliability.

The optical system is fairly conventional. Mirrors 3-9 are required for folding the system for compactness. The path length adjustment between mirrors 4 and 5, and the beam-splitter was needed to adjust the relative path

lengths of the object and reference beam since the coherence
length of the laser is only a few centimeters. This did
not prove to be a ciritical adjustment.

The reference beam passes through an optical phase
modulator (discussed below), a spatial filter, a collimator
lens and then passes through a prism which deflects the
light to an angle of 30 degrees at the film plane. This
rather large reference beam angle was needed so that
another lens could be placed in the object beam immediately
in front of the film.

The object beam is also passed through a spatial filter,
then diverged and reflected from mirror number 10 to the
bottom side of the acousto-optic interface. The spherical
A-0 interface surface is used in a concentric configuration
with the film plane on the diverging side of focus. Using the
spherical surface in such a concentric manner produces
minimal aberrations; however, because the replicated surface
deviates from a sphere the object beam exhibits intensity
variations which greatly exceed the dynamic range of the film.
The lens immediately in front of the film plane in the object
beam was required to smooth out these variations as well as
improve the shape of the refocused spot. A photodiode was
included behind the film plane as an exposure meter.

Optical Phase Modulator

Originally a lithium noibate crystal manufactured
by Crystal Technology, Mountain View, California, was used
as a phase modulator. The lithium noibate was subject to
optical damage from the pulsed laser even though the peak
power is considerably below the damage threshold previously
reported. Undoubtedly, the large number of pulses produced
a cumulative effect heretofore unreported. In the clinical
prototype, a lithium tantalate crystal was used (also made
by Crystal Technology). The crystal is roughly 3x3x15 mm
and has a half-wave voltage of 200 v. No degradation of
the optical quality of this phase modulator has been
detected in more than one year of use of the system.

Hologram Size And Film

The original concept of the system required the use
of movie film for both the optical and the acoustical holo-

gram. 16 mm was the largest format size feasible from a
laser power, material and equipment cost standpoint. Super
8 was the cost preference for both holograms but resolution,
speckle and aberrations dictated a larger size for the
optical hologram. The optical configuration requires a
resolution in the optical hologram of approximately 1200
lp/mm. In order to provide a high quality reconstruction
of the acoustical hologram, the film must also have a
high MTF at 1200 lp/mm. For normal holographic work an
MTF of 0.5 would be adequate. However, for subfringe
holographic interferometry (at our normal optical phase
modulator settings) the brightness of the image is reduced
5 to 10 times from a normal hologram, whereas other optical
noise is constant, (Ref 4). This requires a film with a
MTF higher than 0.5 at 1200 lp/mm, which excludes most
holographic films. Both Kodak 649 GH and FE4210 were
found to be adequate. Although the acoustical hologram
only requires ∿600 lp/mm and much less sensitivity, the
same film was used for both holograms for convenience.

FE4210 is a direct positive, experimental Kodak film,
sensitized for the 5353 Å xenon line and has adequate MTF
at 1200 lp/mm to obtain good quality subfringe interferometric optical holograms.

System Operation

The sequence of events in the camera system is as
follows:
1) The 16 mm film is advanced. When the film is fully
advanced, a sync signal is sent from the film transport
to the electronics. The freely running 1 MHz oscillator is gated on for 90 microseconds.
2) The power amplifiers drive the acoustic transducers
producing a 90 microsecond burst of 1 MHz ultrasound.
3) The sound travels through the patient and strikes
the acousto-optic interface thereupon coming to equilibrium.
4) The laser is triggered exposing the film.
5) At the start of the sync pulse, the 1 MHz sinusoidal
voltage is applied to the optical phase modulator which
remains on until well after the laser has fired. A
very high Q circuit is used for the optical modulator

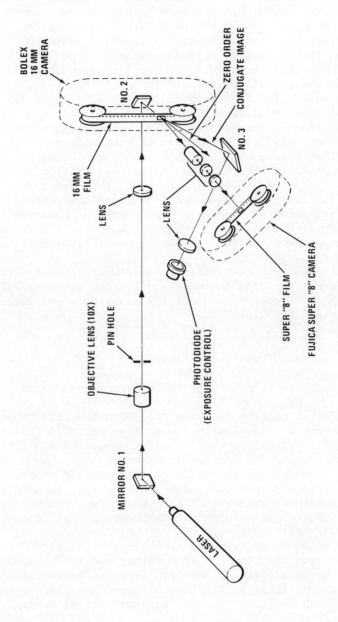

Figure 20.   Optical Schematic Of Copy System

which allows it to continue working properly during the large discharge of the laser.
6) The film is advanced and the next frame sequence begins.

COPY CAMERA

Figure 20 is an optical schematic of the copy system in which the optical hologram is reconstructed to form an image of the mirrored surface of the acousto-optic interface.

The reconstructed interface has bright and dark fringes across it as shown in Figure 21. These fringes are a result of the detection system which is discussed in greater detail in Refs 2 and 4. This is a true acoustical hologram in that both amplitude and phase are encoded in the gray scale of the interface image.

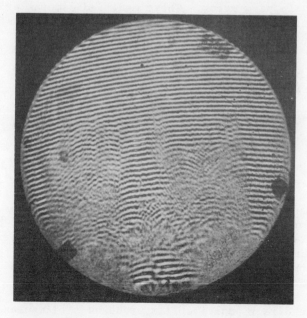

Figure 21. Typical Acoustical Hologram

Optical System

Originally a 10 milliwatt CW helium neon laser was used. However, even this relatively small amount of wavelength change (viz. 6328 versus 5353 Å) was enough to produce significant optical aberrations due to the tilted interface configuration. In the clinical prototype a pulsed argon laser (5145 Å) made by Britt Electronics was used. This change in wavelength produced acceptably small optical aberrations. The reconstruction was done using the 16 mm optical hologram allowing the zero order or undiffracted beam to pass out of the system. The true image was refocused with a high resolution lens onto the Super 8 film. The conjugate image was reflected by mirror number 3 onto a photodiode for exposure control.

Film Transport

The 16 mm film transport was a modified Bolex 16 mm movie camera which had a stepping motor drive. The Super 8 film transport was a modified Fujica Super 8 camera with a similar stepping motor. These modified cameras proved to be a continual problem due to jitter and other instabilities since they were designed to be run continuously.

SYSTEM VIEWER

Figure 22 is an optical schematic of the viewer system. In this system the Super 8 acoustical hologram is reconstructed to form an acoustical image.

The laser in this system is a 1 milliwatt CW helium neon laser manufactured by Hughes Aircraft Company, Electron Dynamics Division, Torrance, California. The film transport is a modified Bell and Howell Super 8 projector intended for amateur use. It is a self-threading projector. Following the spatial filter and hologram is a transform lens which allows spatial filtering of the holographic image to remove the zero order and one conjugate image.

The direct positive film produces a film plane aperture, significantly reducing coherent noise in the viewer system.

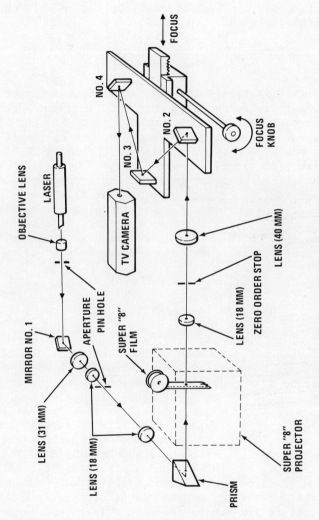

Figure 22. Optical Schematic Of Viewer System

Following the spatial filter a second lens focuses the
acoustical image on a TV vidicon. Mirrors 2, 3 and 4 are
placed on a rack and pinion focusing device to allow the
optical path length between the hologram and the TV camera
to be changed, thus selecting a plane of focus in the
reconstructed image. The TV picture is then displayed on
a conventional television monitor.

FILM PROCESSOR

A modified Eastman Kodak Prostar processor is used.
The same chemistry is used for both the 16 mm and Super 8 mm
films. The processor is run at a high speed (10 ft. per
minute) allowing rapid dry-to-dry processing of the film.
A major inconvenience in the system is the 15 to 20 minute
image access time caused primarily by film handling. This
was not thought to be a detriment to the early evaluation
of this technique, but is clearly not suitable for a
successful clinical product. Experiments have been made
at Actron to reduce this image access time to less than
ten seconds through the use of thermoplastic recording
systems. Space limitations do not allow further discussion
of this work.

APPENDIX II: SAFETY ASPECTS

Any medical procedure involves risks if improperly
used or if adequate standards are not employed. The
question of risk versus benefit enters into the medical
decision as to what modality to apply in a particular case.
Ultrasonic methods are attractive for diagnosis because
they are non-invasive (in the sense of being externally
applied), and they are apparently safe. An extensive body
of literature on biological effects of ultrasound exists
including many review articles (e.g., Refs 5, 6, 7 and
8). It is sufficient to say that at the present time,
there are no data available from any clinical source that
indicate that presently used levels of diagnostic ultrasound are unsafe for humans (Ref 5).

The levels in routine use with several conventional
B-scan instruments were measured by Hill (Ref 9). His
results are shown in Figure 23, along with a curve from

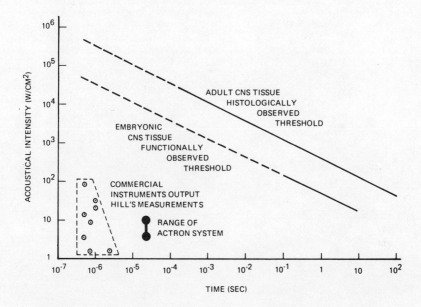

Figure 23.   Ultrasound Damage Thresholds (From Refs 7 & 9)

Dunn et. al. (Ref 7) showing threshold power levels for
histological effects in central nervous system tissue
(CNS). The peak powers used in diagnostic ultrasound are
several orders of magnitude below Dunn's curve for central
nervous system tissue of the mouse, which has been shown
to be one of the more sensitive tissues. The peak power
available from the present system is also shown in Figure 23.
While it is admittedly a different pulse length than the
conventional equipment, it too is much less than the
threshold curve. Everyone will agree that less power is
better, if it can do the same examination. The reason for
this discussion is to point out that we felt justified and
conservative in using the power levels employed. The
nature of the dose administered is well controlled in this
system, whereas in a conventional B-scan, the human operator
may dwell for a long period over one part of the anatomy.
Furthermore, even if a uniform scan motion of 5 cm/sec is
employed with a typical B-scan system, the average volume
dose with many B-scans spaced 1 cm apart is greater than
that from the holographic system in a typical run.

RESEARCH IN ULTRASOUND IMAGE GENERATION: A COMPUTERIZED ULTRASOUND PROCESSING, ACQUISITION, AND DISPLAY (CUPAD) SYSTEM[*]

A. Goldstein, J. Ophir, and A. W. Templeton

Department of Diagnostic Radiology
University of Kansas Medical Center
Kansas City, Kansas 66103

The information available to the physician from clinical ultrasound examinations is limited. Current research in our laboratory has identified some of the limitations found in presently available clinical imaging systems and has suggested ways of making significant improvements. In this paper the structure of our research system (CUPAD) is presented. Some of the current goals of this research are noted, and research schemes designed to achieve some of these goals outlined. A brief description of our acquisition hardware and processing software is provided and initial CUPAD scans discussed.

THE DISPLAY-THE WEAK LINK IN THE IMAGING SYSTEM

The clinical utility of ultrasound images is limited by the information content in the presently used storage tube display. Figure 1 shows a representative ultrasound scan displayed on a storage tube and recorded on polaroid film. Since the storage tube display is binary, only echo presence or absence information may be displayed. White represents echo presence. In this image a fetal head may be identified. In addition, amniotic fluid, the placenta, and fetal parts can also be identified.

[*](Supported by NSF Grant GJ-41682)

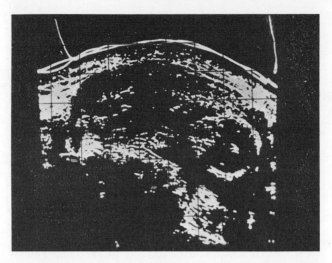

Figure 1: Storage tube display of an ultrasound scan
showing a longitudinal cross section of a pregnant uterus.
White areas indicate echo presence. The fetal head,
placenta, fetal parts, and amniotic fluid can be identified.

        The low information content in these images limits
interpretation to physicians experienced in ultrasound;
restricts diagnostic criteria to changes in organ shape
or size, the presence or absence of internal echoes within
an abnormality, the borders of a lesion, and echo attenuation; and also makes the detection of system malfunctioning
very difficult.

        The display is the weak link in the system. This is
demonstrated in Figure 2 which is a diagrammatical representation of the information flow in a pulse-echo ultrasound imaging system. Data is acquired from the patient
and transferred through the acquisition, processing, and
display sections of the imaging machine. The physician
views the display in making his diagnosis. Although
presently available clinical imaging systems acquire and
process large amounts of ultrasonic data, the limited
information presented on the storage tube display places
restrictions on diagnostic information.

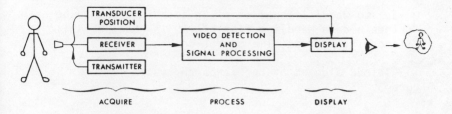

Figure 2: Information flow in a pulse-echo ultrasound imaging system.

RESEARCH GOALS

Our objective is to build an imaging system with an improved display, which will not be the weakest link in the information flow, and to use the system to learn how best to present ultrasonic data to the physician for improved medical diagnosis. The experimental questions thus become (1) what ultrasonic data should be acquired from the patient, and (2) what is the most effective method to acquire, process, and display the data for the physician.

In designing our research plan, the following decisions were made:

1. The imaging system should be developed with flexibility of data handling as a major criteria and the system should be used as a research tool to learn how to obtain the most effective clinical image.

2. A digital computer should be a central component of the system because of its memory and computational ability along with the flexibility and ease in implementation of software control. (Although we feel the computer is essential in a research system, it might not be necessary, desirable, or even practical in a final clinical imaging system.)

3. The system should be of modular design for ease in changing data handling schemes.

4. The contact B-scan should be the mode of data acquisition due to its wide clinical acceptance, and

5.   In acquisition a sampling technique should be used
to maintain data handling flexibility in research while
still permitting real time operation.

A block diagram of our research system is shown in
Figure 3.

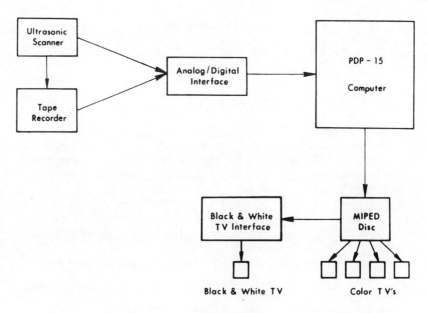

Figure 3:   Block diagram of CUPAD

The ultrasound data is acquired by the ultrasonic
scanner and digitized by a specially designed interface.
The data is then transferred to a PDP-15 computer (64K of
core, two 250K-word fixed head disks, and two 10.2 million
word disk pack drivers), in which it is processed and
stored. At selectable intervals during the scan, the data
is transferred to our Medical Image Processing, Enhancement, and Display (MIPED) system (NSF Grant GJ-28779) for
display on color or black and white T.V.s. MIPED contains
a 72-channel fixed head disk which refreshes the T.V.
images. The images are displayed in a 128 X 128 format

with data coded in 16 color levels or 16 shades of gray.
Three levels of magnification on the ultrasonic scanner
allow for a choice of resolution cell sizes of 1.8 mm on a
side, 1.2 mm on a side, or 0.67 mm on a side.

Acquisition may be varied by the control parameters on
the ultrasonic scanner or by a change of hardware function
in the interface. Processing may be varied in the computer
by software control. Display parameters may be varied on
MIPED or by software control in the computer.

In order to objectively evaluate different data
handling schemes, it is essential that the same raw analog
ultrasonic data be used. A tape recorder is used for this
purpose (Honeywell, model 96, 2 MHz bandwidth). The data
from actual clinical examinations is recorded from the
ultrasound scanner on the tape to produce an archive of
analog scan data. At a later time a stored scan is played
back into the interface to simulate a real scan.

A great deal of information is contained in the received echoes. The electrical signals obtained from the
transducer allow independent or simultaneous measure of
echo amplitude, frequency, phase, or transmission time
delay. Echo amplitude has been chosen as the first parameter to investigate (1) since it is easily obtained,
(2) since it is a potentially important diagnostic parameter, and (3) since an amplitude coded image allows an
indirect measure of another potentially important diagnostic parameter - tissue attenuation.

ACQUISITION HARDWARE

The geometry of acquisition in the contact B-scan is
shown in Figure 4. The transducer is held by an operator
and moved over the patient's skin surface in a random
fashion. Both the displayed image and the patient's
cross sectional anatomy are divided into corresponding
matrices of resolution cells. In the display, each resolution cell is represented by a single color level or shade
of gray. We are thus effectively assuming that all the
echo amplitude data contained in one resolution cell within
a patient is uniform. Therefore, in the acquisition
sampling scheme, each sample should correspond to the data
in only one resolution cell.

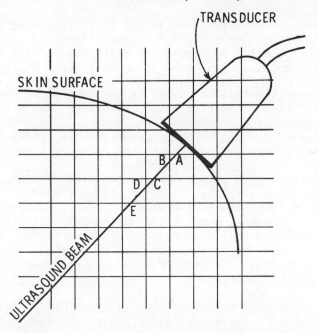

Figure 4: Acquisition geometry in a contact B-scan. The patient's skin surface and the penetrating ultrasound beam are shown. The patient's cross sectional anatomy and the displayed image are divided into corresponding matrices of square resolution cells.

One sample is obtained for each transmission pulse giving an acquisition rate of 1 KHz. The sampling logic is designed to sequentially sample each resolution cell in the beam path, beginning with the one closest to the transducer and ending with the last cell contained in the image. The random motion of the transducer in the contact B-scan presents two difficulties: (1) the correct digital registration of the sampled resolution cell, and (2) the proper normalization for the varying beam lengths in each resolution cell. The interface hardware has been designed to eliminate these problems.

The solution of the registration and normalization problems are shown in Figure 5 which is a block diagram of the interface hardware. The ultrasonic scanner was

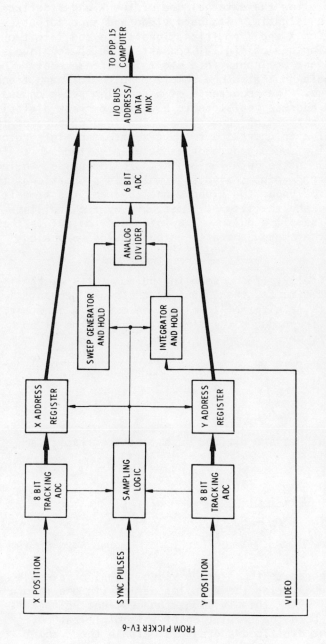

Figure 5: Block diagram of hardware interface. Bold arrows indicate digital data flow. Intermediate arrows indicate analog data flow. Thin arrows are control signals.

modified to allow the external use of the X and Y deflection position signals. Standard video and sync outputs are used. The X and Y position signals are fed into two 8-bit tracking analog/digital converters which continuously convert the analog inputs into digital address codes. The sampling logic generates one sampling pulse for each transmission pulse. The occurrence of a sampling pulse causes the X and Y address registers to hold the current digital output of the tracking converters. The leading edge of the sampling pulse triggers the sweep generator and integrator circuits. The sweep generator produces a voltage which is proportional to the duration of the sampling pulse, while the integrator produces a voltage which is proportional to the total area under the video signal between the time it is triggered and the time it is stopped by the trailing edge of the sampling pulse. The integrator output is then divided by the sweep generator output, using an analog divider.

Analytically, if the sampling pulse starts at t_A and stops at time t_B, the output of the sweep generator after time t_B is $k(t_B - t_A)$ volts, where k is a constant. The integrator produces the time integral of the video signal, $v(t)$, or

$$k \int_{t_A}^{t_B} v(t)dt.$$

These two outputs are divided such that the result is

$$\frac{1}{t_B - t_A} \int_{t_A}^{t_B} v(t)dt.$$

This corresponds to the average video signal or average echo amplitude during the sampling period (in one resolution cell). After the division, the quotient voltage is digitized. Upon the completion of this digitization, and while the digitized address is held in the X and Y address registers, a flag is raised and an I/O program

interrupt is requested. At this point the software assumes
control of the I/O data transfer from the interface output
address/data multiplexer to the computer.

The correct X-Y address of the echo origin and the
properly normalized video are therefore always obtained,
independent of the position, angulation, and motion of the
transducer.

PROCESSING SOFTWARE

Data flow in the computer and all other software functions are controlled by the user program, Figure 6. This
includes the control of the interface handler programs, the
preliminary data storage, the preprocessing programs, and
the assembly of the data buffer as well as the preparation
for display.

The handler is the lowest level program which communi-
cates directly with the hardware interface to facilitate
the correct transfer of data into one of two 512 word line
buffers. Upon command from the user program, the handler
instructs the hardware multiplexer to first transfer the
sample address and then the digitized video. These two
pieces of data are stored in two successive computer words
in the line buffer. The data is sequentially stored in one
line buffer until the full wavetrain has been sampled, and
then the data from the next wavetrain is stored in the
other line buffer. While the second wavetrain is being
acquired, the first wavetrain is being processed and transferred to the 16K data buffer. The switching of line
buffers enables the computer to synchronize the acquisition
and processing such that all incoming data is properly
handled. Since the data in the line buffers is stored in
a sequence corresponding to its distance from the transducer, it lends itself to an additional data handling step
(called preprocessing) in which calculations such as
correcting for tissue attenuation, beam pattern corrections,
and detection and removal of reverberations are performed.
From the line buffers the data is transferred to the data
buffer which is a 16K portion of core containing one 18-bit
computer word for each resolution cell. The appropriate
processing software to load the image buffer is chosen by
the operator.

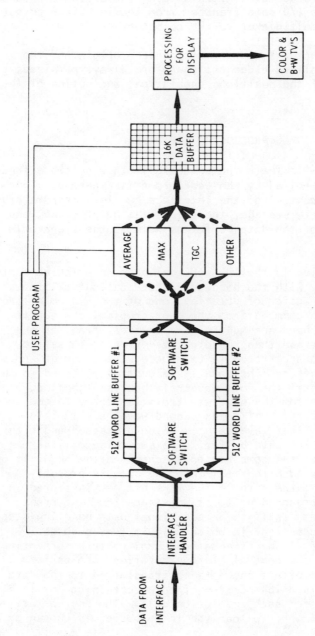

Figure 6: Software structure and data flow in computer. Bold arrows indicate major data flow paths. Broken arrows are alternative paths. Thin lines indicate software control by the user program.

If the maximum processing method is chosen, the newly
acquired amplitude sample is compared to the previously
acquired sample and the larger stored. Thus, we are
building an image of the maximum echo amplitude sampled in
each resolution call during the scan. Maximum processing
is designed to correctly display large amplitude, specularly reflected echoes such as from organ interfaces.

If the operator chooses the averaging processing method,
then the new amplitude sample is added to a running sum of
all sampled amplitudes in the appropriate word. A count
of the number of samples (also kept in the word) is incre-
mented. When the data is processed for display, the running
sum is divided by the sample size to obtain the average
amplitude sampled during the scan. Average processing is
designed to correctly display low amplitude, diffusely
scattered echoes such as from intraorgan structures.

The processing for a display step is performed at
fixed operator selected intervals. Acquisition is interrupted, the full 16K data buffer processed, word packed
and transferred to the MIPED disk to update the image.
CUPAD thus retains the interactive real time nature of the
ultrasound examination with the operator viewing the up-
dated image continuously through the scanning process.

At the end of a scan the operator can either erase
or save the image. If he saves the image, the full 16K
data buffer is moved to mass storage. The advantage of
storing the complete data buffer is that a retrieved
image may be further processed or post-processed with
feature extraction hardware. A disadvantage is the amount
of mass storage required. A maximum of 622 images can be
stored on a 10.2 million word disk pack.

INITIAL CUPAD SCANS

CUPAD scans are presently being obtained with 4-bit
input digital resolution and a magnification corresponding
to 1.8 mm resolution cells. Figure 7 is a CUPAD image of
the same anatomy shown in Figure 1. The gray scale shown
on the top is short due to limited film latitude. Average
processing was used. The placenta, fetal parts, amniotic
fluid, and the anterior uterine wall may be identified.

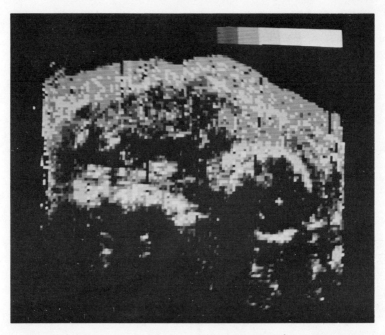

Figure 7. Black and white echo amplitude coded CUPAD image
of the same anatomy shown in Figure 1. Average processing
was used. White corresponds to the largest amplitude echo
and lower amplitude echoes are displayed by correspondingly
darker shades of gray. The gray bar is short due to
limited film latitude. The fetal head, placenta, amniotic
fluid, fetal parts, and anterior uterine wall may be
identified.

Figure 8 shows another scan of the same anatomy
processed by the maximum algorithm. Placenta, amniotic
fluid, and fetal parts can again be identified. Also,
the anterior uterine wall is more sharply defined. The
white spots are artifacts added to the image when it was
transferred to the MIPED disk.

About one year ago second generation gray scale
clinical imaging systems were developed by manufacturers.

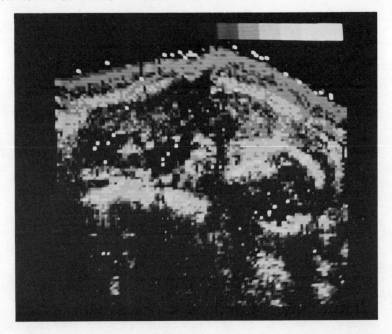

Figure 8: Black and white echo amplitude coded CUPAD
image of the same anatomy as Figure 1. Maximum processing
was used. The fetal head, placenta, amniotic fluid, and
fetal parts may be identified. There is good definition
of the anterior uterine wall. The white spots on this
image are artifacts added to the image when it was transferred to the MIPED disk for display.

Echo amplitude data is displayed on a black and white T.V.
coded in up to eight shades of gray. A scan converter
tube is used as memory. The images have higher resolution
and a faster acquisition rate than CUPAD images (which
typically require one-half to two minutes to generate).
Since the scan converter tube only allows the equivalent
of maximum processing, we will not use this equipment in
our present research other than to generate reference
images.

PLANNED RESEARCH

We are presently updating CUPAD to accommodate both quantitative and qualitative research with echo amplitude data. The specific improvements include:

1. Increasing the input digital resolution to 8-bits and

2. Redesigning the acquisition so that CUPAD can be calibrated to acquire, process, and store a 60 db dynamic range of echo amplitude data standardized with respect to a perfect reflector. Any portion of this 60 db dynamic range can be chosen for display.

The digital computer gives CUPAD the capability of rapid quantitative assessment of ultrasonic data in a clinical situation. We will perform experiments to obtain standardized quantitative echo amplitude data from image areas of interest. These data will be important for two reasons:

1. A knowledge of the magnitude of echo amplitudes will allow us to properly acquire and display them in images and build a proper data base for data handling studies, and

2. We will be able to evaluate the feasibility of using quantitative echo amplitude information to differentiate between normal and diseased tissue.

ALGEBRAIC RECONSTRUCTION OF SPATIAL DISTRIBUTIONS OF ACOUSTIC VELOCITIES IN TISSUE FROM THEIR TIME-OF-FLIGHT PROFILES

J. F. Greenleaf, S. A. Johnson,
W. F. Samayoa, and F. A. Duck
Biophysical Sciences Unit
Department of Physiology and Biophysics
Mayo Foundation, Rochester, Minnesota 55901

ABSTRACT

Two-dimensional distributions of acoustic velocities were measured in transverse sections
through intact isolated organs, using reconstruction techniques. Profiles of time-of-flight (TOF)
of 10 MHz pulses through the specimen were obtained by rectilinearly scanning two opposing
transducers along either side of the specimen in
the plane of interest. The received pulses were
digitized at a rate of one 8-bit sample-per-10
nanosec, for 512 samples and were analyzed with a
computer algorithm which calculated the TOF of the
pulse to within ± 10 nanosec. Typically, 256
measurements of TOF were made in each profile scan
for each of 37 angles of view separated by 5°.
TOF's through tissue, normalized by TOF through
water, were used to calculate velocity within the
specimen, using an algebraic reconstruction technique (ART). Images obtained represented acoustic velocities in individual cross sections within
the tissue specimen with a resolution of 64 by
64 elements (< 2 mm square). The disadvantage
of TOF reconstruction is that transmission scanning is required. Advantages over B- and C-scan

imaging are: 1) dynamic changes in gain of
receiver are not required, 2) attenuation occurs
on only one traversal through tissue, and
3) the absolute value of an important acoustic
parameter (velocity) is determined, which may
have significant diagnostic value.

INTRODUCTION

It has been known for many years that three-
dimensional functions can be determined from
their two-dimensional projections obtained by
line integrals along paths through the three-
dimensional function (1). Obtaining the solution
for three-dimensional distributions of functions
from their two-dimensional projections is popu-
larly called "reconstruction". Two-dimensional
projections or shadows of objects used in biomedical imaging (related through line integrals
to three-dimensional distributions of scalers)
can be obtained in many ways but in general
involve transirradiation of the object space with
some form of energy. Reconstruction methods have
been developed using many forms of energy such as
light (7), x-radiation(4,6), electrons (3), and,
more recently, sound (8). Reconstruction involves
the solution of large numbers of equations either
by Fourier transform techniques (2) or by iterative algebraic techniques of various kinds (9).

This laboratory has recently reported the
reconstruction of the distribution of acoustic
attenuation coefficients within excised tissues
using the algebraic reconstruction technique or
ART (5). These attempts at reconstruction of
acoustic impedances from projections of acoustic
energy through tissue met with difficulties
caused by refraction of the acoustic waves within
the tissue. This problem invalidates the
assumption that the line integrals through the
three-dimensional functions, which represent the
points on the two-dimensional function, or shadow,
are along straight lines. In order to meet this
basic problem, we have developed a technique of

measuring the propagation delay time of acoustic
pulses through tissue. Since the earliest arriving pulse can be detected with relative ease
and accuracy using high-speed A-to-D conversion
techniques with specialized digital computer
algorithms and since the arrival time of pulses
is unaffected by reflection of portions of the
energy within the pulse, problems of refraction
and reflection can be minimized by using propagation delay time profiles for reconstruction
rather than intensity profiles. Refraction effects are minimized since the earliest arrival
time is most probably that of the straightest ray
path.

The purpose of this paper is to describe
initial results of the application of reconstruction techniques to the determination of acoustic
indices of refraction within isolated tissues
calculated from profiles of propagation delay
time obtained by scanning transducers on either
side of the tissue of interest. In addition, the
effect of refraction on the resulting reconstructions will be evaluated and discussed.

METHODS AND PROCEDURE

Data were obtained by rectilinearly scanning
diametrically opposing transducers on each side
of the tissue being investigated (Figure 1).
Propagation time through the tissue was measured
by digitizing the received pulses at a rate of
one sample per ten nanoseconds with an amplitude
resolution of 8 bits. The transducers were 5 or
10 MHz Panametrics transducers having approximately 2 MHz bandwiths and were scanned using a
modified computer-controlled Picker Magna IV
scintiscanner. The receiving transducer was
connected to a Panametrics PR 5050 receiver which
was connected to a high speed A/D converter (BTR)
(Biomation Transient Recorder Model 8100, Biomation Industries, Palo Alto, California) capable
of digitizing at rates of up to 100 samples
per microsecond for 2048 8-bit samples.

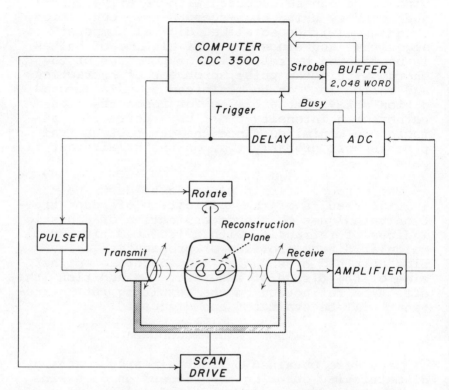

Figure 1   Schematic of computer controlled data
            acquisition system which obtains digi-
            tized ultrasonic pulses from B and C
            scans. Pulser and A/D converter are
            triggered by computer at 256 equispaced
            points along scan length of ~12 cm.
            Received signal is digitized into up to
            2048 eight-bit samples at rates of up
            to 100 per μsecond. After digitization,
            data are strobed into computer and upon
            completion of scan, the sample is rota-
            ted 5° and the scan process is repeated.

The transmitting transducer was connected to the
transmit port of the Panametrics PR 5050. The
samples were stored in the buffer memory of the
A/D converter until they were strobed into the
computer. In general, polystyrene converging
lenses of focal length 4 cm and diameter 2 cm
were used on the transducers. The length of the
scan was approximately 12 cm during which 256
pulses were propagated between the transmitter
and receiver and were digitized at a rate of 100
pulses per microsecond for a period of 256 to 512
samples. Details of A-to-D conversion programs
using this system are described elsewhere (6,8).
The pulse repetition rate was generated by the
internal clock of the computer which sent delay
and digitize commands to the BTR and at the same
time triggered the transmitter thus achieving
accurate temporal synchrony between the beginning
of A-to-D conversion by the BTR and the transmit·
pulse generated by the pulser. After an appropriate delay allowing the pulse to traverse the
tissue and arrive at the receive transducer, the
BTR began to digitize the 256 to 512 samples of
the signal. Subsequently the data samples in the
BTR were strobed into the computer one at a time.

The equations governing the propagation time
between the transmitter and the receiver can be
derived with the help of Figure 2 which demonstrates the geometry of the reconstruction problem.
If we let τ_w be the delay across a fluid-filled
gap, we note that the time of delay = $\tau_w = \frac{D}{v_w}$ where
D is the total gap length and v_w is the acoustic
velocity in fluid. The following equation represents the arrival time of the pulse as it traverses
ray path j through the tissue,

$$\tau_j = \sum_{i=1}^{N} \frac{1}{v_{ti}} L_{ij} + \frac{1}{v_w} (D - \sum_{i=1}^{N} L_{ij}), \qquad (1)$$

where D is the distance between the transducers,
N is the total number of pixels, v_{ti} is velocity
in tissue in region i and L_{ij} is length of ray
path j in region i.

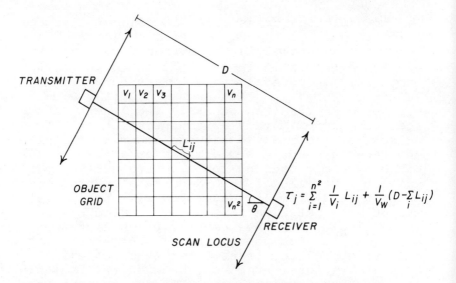

<u>Figure 2</u> Geometry of ultrasound transmission for
          algebraic reconstruction. Grid is
          fixed to coordinates of object. Velocity
          is assumed to be constant and isotropic
          within each pixel. L_{ij} can be varied
          according to path or ray which need not
          be straight. Equation represents delay
          of pulse through reconstruction region
          of grid.

     In general we measure the difference between
the arrival time through fluid and the arrival
time through tissue giving

$$\tau_w - \tau_j = \sum_{i=1}^{N} \left(\frac{1}{v_w} - \frac{1}{v_{ti}} \right) L_{ij}, \quad j = 1, 2, \ldots, M \tag{2}$$

which is the algebraic equation representing the
reconstruction problem. The L_{ij} are known from
geometric considerations and the quantities
$\left(\frac{1}{v_w} - \frac{1}{v_{ti}} \right)$ are to be calculated at each value of i

and are relative refraction indices representing the change in propagation time through pixel i in the reconstruction image relative to the immersing fluid. One can see that with enough measurements (i.e., equations) of acoustic propagation delay Equation 1 can be solved for the N unknowns. In the images to be shown in this paper, the plots represent the value of the change in propagation time $(\frac{1}{v_w} - \frac{1}{v_{ti}})$ and not the direct value of the velocity in the tissues. Therefore, image points which are brighter than the background will represent velocities faster than the velocity of the immersion fluid and image points darker than the background will represent values which are slower than that of the immersion fluid.

EXPERIMENTAL PROCEDURE

Tissues and organ samples were immersed in saline and scanned in the manner described previously. After each scan, the sample was rotated 5° about the axis vertical to the plane of the scan and the scan was repeated. This sequence was repeated through 35 rotational steps. This yielded 35 x 256 measured values, each representing the propagation delay of acoustic energy through the tissue for a unique ray path. If we solve for a 64 x 64 square array of velocity values, the system of equations is "over determined", which decreases its sensitivity to noise. The algebraic reconstruction technique (ART) is capable of solving this system of equations and obtaining solutions in a manner which is relatively stable and immune to noise (5).

Figure 3 illustrates the digitized signals in A-scan representation, for ultrasonic pulses which have traveled through saline (upper panel) and two different lengths of tissue. Note that the early arrival of the pulse is easily seen in the second and third panels. The gain of the receiving system has been fixed at a high value causing the A/D converter to clip, but without affecting the ability to determine arrival times.

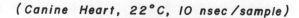

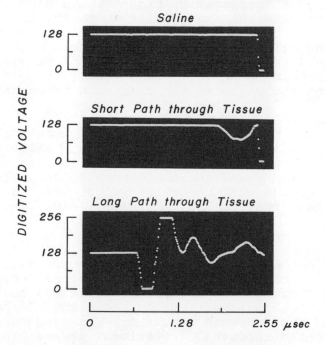

Figure 3   Digitized ultrasound pulses after trans-
            mission through saline and two different
            path lengths of tissue. Received signal
            was digitized to 256 eight-bit samples.
            Arrival time of pulse is first indica-
            tion of negative excursion of signal.
            Arrival times through various tissues
            such as cardiac muscle are usually
            earlier than for saline-filled gap.

    Data obtained by the scanning and digitization
procedure for one scan across a canine heart and
recorded on digital magnetic tape are shown in
Figure 4. This image represents 256 samples
along the time axis from left to right, each
sample 10 nanoseconds from its neighbor, while
256 pulses are plotted along the scan travel from

top to bottom. The brightened line indicates
the line for which the A-scan is represented in
the lower panel of Figure 3. One can see that
the arrival time of the acoustic energy through
the saline path is extremely constant, demonstrating
the stability of the synchrony between the pulsing
system and the delay counter in the BTR A-to-D
converter. One also notes the relative smoothness
of the profiles indicating that the pulse rate of
256 pulses per scan was adequate to characterize
the propagation delay times representing the
particular tissue under study (a canine heart
immersed in saline).

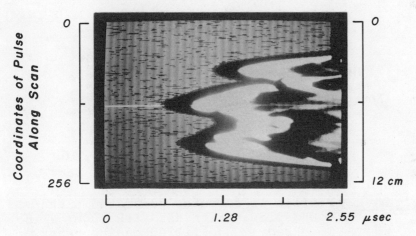

(Canine Heart, 22° C, 10 nsec / sample)

Figure 4   Digitized ultrasonic pulses: arrival
time (left to right) versus position
along scan traversal (top to bottom).
Grey level of image is proportional to
signal intensity (black < 0, white > 0
volts). Arrival time through saline
(top few lines) is later than arrival
times through tissue. Brightened line
indicates line depicted in A-scan format
in bottom panel of Figure 3. Arrival
time was defined as earliest indication
of negative excursion of signal.

A computer algorithm was written which scanned
each line depicted in Figure 4 and calculated the
point at which the signal rose above the background noise (defined as two standard deviations
of the noise in the first twelve samples representing the baseline). The algorithm then searched
backwards down the first pulse to that point
which represented the zero crossover (baseline
average) and took that value as the arrival time.
Application of this algorithm to the profile
depicted in Figure 4 results in a line profile
represented in Figure 5. The profiles are sometimes filtered with three-or five-point filters
and their dc value is adjusted so that propagation
times through the immersing liquid is zero. This
results in reconstructed images in which the

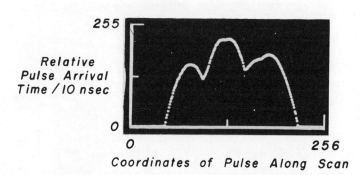

Figure 5   Profile of computer generated ultra-
            sound propagation delay through saline
            immersed, saline filled canine heart.
            Arrival times of 256 pulses are plotted
            as calculated by computer algorithm
            from digitized pulses represented
            in Figure 4. Arrival times are within
            ±10 nanoseconds - the digitizing rate
            of the A/D converter - and have a
            dynamic range of about 2.5 μseconds.
            Profile demonstrates high signal to
            noise ratios of data which are used in
            reconstruction.

background is usually black and represents propagation time equal to that through the liquid in which the tissue is immersed. Thirty-six such profiles are submitted to the reconstruction program, each profile representing the propagation arrival times for a scan at each of 36 angles of view separated by 5°.

RESULTS

In order to test the fidelity and resolution of the reconstruction technique, we first chose excised canine hearts obtained from other experiments as tissue subjects since they are easy to obtain and they contain interesting features within the ventricles. However, we do not want to imply that reconstruction of the intact heart in vivo is possible since ultrasound at these frequencies cannot traverse the lungs. The right panels of Figure 6 are optical photographs of the surface of a sliced canine heart illustrating the right and left ventricles, the papillary muscles, and the endocardial and epicardial surfaces for two separate transverse levels separated by 2 cm. The left panels of Figure 6 represent reconstructions of refractive indices through the level at which the heart was sliced as indicated in the respective photographs. The intensity of these images is proportional to the value $(\frac{1}{v_w} - \frac{1}{v_{ti}})$ or the difference in propagation times for the individual elements within the reconstructed area. Notice the fidelity of reconstruction of the ventricular papillary muscles, the endocardial surface of the right ventricle, and the epicardial surface. Since the heart was frozen before it was sliced, some slight changes in geometry probably occurred, such as alteration of the size of the ventricular cavities. However, one can easily note the similarity of the epicardial surfaces of the reconstructions and the photographs.

(Water - Filled, 22° C)

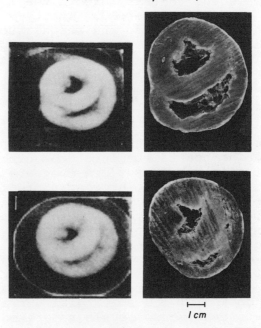

I cm

Figure 6   Reconstructions of relative propagation
delays within canine heart (left) compared to photographs of sections through
corresponding levels (right). Diametrically opposing transducers were
rectilinearly scanned on either side of
the tissue in the plane to be reconstructed. Pulses were propagated through
the tissue at each of 256 equally spaced
points along the 12 cm scan, and were
digitized with a magnitude resolution
of 8 bits and a temporal resolution of
±10 nanoseconds. Resulting propagation
delay profile was used to reconstruct
relative propagation delays within each
of the 64 x 64 array of pixels making
up the image. Papillary, epicardial
and endocardial surfaces are represented
with good fidelity. Increase in size
of ventricles may have occurred during
tissue preparation for photography.

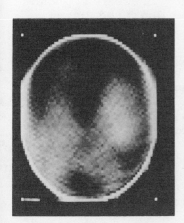
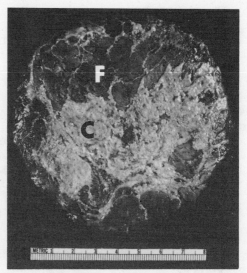

Figure 7   Reconstruction of relative propagation
delays within excised human breast
tissue (left) compared to photograph of
section through corresponding level
(right). Scan data were obtained in a
manner similar to that described in
Figure 6. Areas denoted by F are fatty
tissue of low acoustic velocity while
areas denoted by C are connective tissue
and are of higher velocity. The general
outline of the various regions depicted
by the reconstruction seems accurate;
however detail is missing.
The difficulty of slicing the tissue
at the precise level of the reconstruction along with blurring effect caused
by width of the acoustic beam probably
caused loss of comparative detail.

Figure 7 ,right, represents a photograph of a
slice through a human female breast obtained after
subcutaneous mastectomy. The regions indicated
by F are fatty tissue, while the regions indicated
by C are connective tissue. The reconstruction
through this level of the breast is shown in
the left panel of Figure 7 and indicates that
differences in acoustic velocity between fat and
connective tissue can easily be depicted using
this technique. It was apparent in obtaining
these data that attenuation of ultrasound intensity
within the breast was not a significant problem
and that very large masses of human breast tissue
could be scanned using this technique. The attenuation characteristics of skin should be such
that the intact breast could be scanned with
little additional loss of energy.

The accuracy of the assumption that the ray
paths through tissue are straight lines can be
studied to a first order approximation with a
simple model of propagation through a right
cylinder.

Figure 8 indicates the geometry of the model
used for analysis. The top half of the cylinder
illustrates the results of ray tracing the acoustic energy which is generated by the transducer-
lens assembly used for the canine heart reconstructions reported previously. The index of
refraction of the cylinder relative to saline was
assumed to be constant at .94, about that of
cardiac tissue. Note the overall beam is bent
several degrees as it traverses the cylinder
from right to left. However, the rays actually
detected by the receiver traverse only a narrow
region of the cylinder and thus represent propagation times in a region similar to the assumed
straight line path between transmitter and receiver. The thickness of the traversed region is
somewhat thicker when the transducers are at the
ends of a diameter of the cylinder, i.e., near
its center.

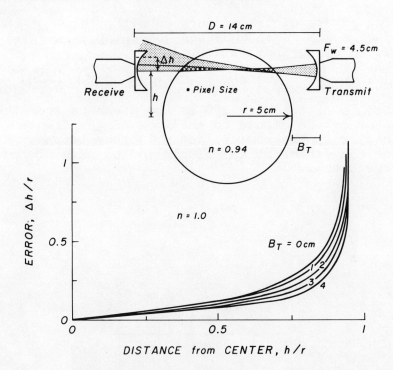

Figure 8   Geometry of cylindrical model used for
analysis of acoustic refraction. Transducer lens assembly is identical to that
used in canine heart reconstructions.
Beam is assumed to be made up of individual rays. Central ray of beam is
refracted amount Δh depending on distance Δh/r from center of cylinder and
depending on distance from transmitter
to cylinder Bt. F_w = focal length in
water. Gap length D was fixed for error
analysis.

The previous reconstructions reported in this
paper were done for a distance between transmitter and cylinder, Bt = 2 cm. Note that the
error is less when the receiver is closest to the
cylinder. A simpler model geometry is illustrated
in Figure 9 which describes the ray path from a
point source to a point receiver. The geometry
closely resembles that which might be expected

using arrays of transducers on either side of the
object space. The equation governing this case,
derived using Snell's law, is

$$\Delta h = (D - \sqrt{r^2 - h^2}) \, \text{Tan} \left[\text{Sin}^{-1}(h/r) - \text{Sin}^{-1}(\frac{n_2 h}{n_1 r}) \right],$$

where Δh is the error, D is the distance from the
center of the circle to the face of the transducer, r is the radius, and h is the distance
from the center of the circle to the axis of the
transducers.

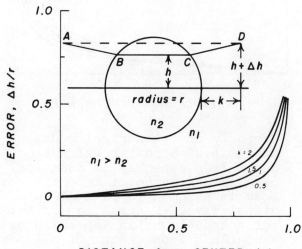

DISTANCE from CENTER, h/r

Figure 9   Geometry of cylindrical model used for
           analysis of acoustic refraction for
           point source and receiver. Error Δh
           is a function of distance of transmit-
           receive axis from center of cylinder
           and of distance of transducers from
           cylinder. For k < 0.5 most errors are
           less than 5%. At approximately
           h/r = 0.9 total reflection occurs.
           Arrays of transducers on either side of
           the object are modeled by this geometry.

Here the ray that traverses from transmitter
to receiver through the cylinder follows a line
B-C which is parallel to but displaced an amount
Δh from the straight line path A-D. The plots of
$\Delta h/r$ indicate that the transmitter and receiver
should be as close as possible to the cylinder in
order to minimize this effect. It should be
possible in imaging either breast or abdominal
cross sections to place the transducers much
closer to the object than 0.5 r. In addition,
one should chose an immersing fluid which would
minimize the effect. In the case of the breast,
however, it seems difficult to chose an immersing
liquid which will match even the average velocity
in the breast since such properties have been
reported to vary widely from patient to patient
(10).

DISCUSSION

Arrival times of acoustic energy propagated by
transmission through a specimen at many angles of
view have been used to reconstruct the refractive
indices within selected excised tissue samples.
The use of arrival time of the pulse seems to
minimize the undesirable effects of refraction.
Fidelity of reconstruction of geometry was qualitatively shown to be good although accuracy
of calculated refraction indices was not studied.
Our previous attempts at reconstructing attenuation coefficients were met with difficulties
caused by refraction (8). Accurate reconstruction of the refraction index may lead to a method
of correcting for refraction and allow the calculation of both acoustic impedance and attenuation
coefficients. Impressive results have recently
been obtained by Jellins et al (11) who improved
B-scan images of the breast with very simple
velocity corrections. This success indicates
that application of refractive index corrections
to compound B-scan images may also be very useful.

The effect of refraction on the assumption
that the ray paths are straight lines joining the
transducers seems minimal, at least for homogeneous

cylinders, and for 64 x 64 pixel reconstructions.
However, it is clear that high resolution images
of attenuation coefficients, acoustic impedences
or refractive indices within heterogeneous tissues
will require ray path corrections to be incorporated into the reconstruction process. It is
not currently known whether the first reconstructions obtained by assuming straight ray paths is
close enough to the final answer to allow an
iterative solution to be implemented in which the
refractive indices of the first image are used
to calculate the ray paths to be used in the
second reconstruction and so on. Such an iterative process - if it converges - should allow
higher resolution images to be obtained.

Computer implementation of the algorithms
required to 1) determine arrival times of the
pulses and 2) reconstruct the refraction indices
from the propagation time profiles currently
requires about 20-30 minutes per cross section on
a CDC 3500 computer. The calculation of arrival
times from digitized pulses could be implemented
on a special purpose device to decrease the total
calculation time by about an order of magnitude
(to 3-5 minutes per section).

The use of advanced techniques in ultrasonic
imaging combined with high-speed, high-fidelity
signal processing equipment may allow quantitative
imaging of basic tissue properties within intact
animals and man. Reconstruction techniques
generate quantitative values representing basic
properties of tissue rather than qualitative
values which represent geometric distributions of
acoustic interfaces such as those obtained with
B-scans or transmission or reflection C-scans.
The advantages of transmission reconstruction
techniques over B- and C-scan imaging are 1)
dynamic changes and gain of the receiver are not
required since the signal can be clipped, 2)
attenuation occurs on only one traversal through
the tissue, and 3) the absolute value of an
important acoustic parameter (velocity) is determined which may have significant diagnostic value.

ACKNOWLEDGMENT

The authors wish to thank Dr. Earl H. Wood for his encouragement in this project and Dr. Richard A. Robb for assistance in programming the BTR. We also thank Mrs. Jean Frank and co-workers for their secretarial and graphic assistance. We are also indebted to Dr. Arnold L. Brown and associates of the Department of Pathology and Anatomy for acquiring and preparing some of the samples of human tissue.

This research supported by grants N01-HT-4-2904, HL04664, and RR-7 from the National Institutes of Health, United States Public Health Service; AF F44620-71-C-0069 from the United States Air Force.

REFERENCES

1. Gordon, R., and G. T. Herman:
    Reconstruction of pictures from their projections.
    Comm ACM 14(12):759-768, 1971.

2. Bracewell, R. N., and J. A. Roberts:
    Aerial smoothing in radio astronomy.
    Aust J Phys 7(4):615-640, 1954.

3. DeRosier, D. J., and A. Klug:
    Reconstructions of three-dimensional structures from electron micrographs.
    Nature 217:130-134, 1968.

4. Gordon, R., R. Bender, and G. T. Herman:
    Algebraic reconstruction techniques (ART) for three-dimensional electron microscopy and x-ray photography.
    J Theor Biol 29:471481, 1970.

5. Herman, G. T., and S. Rowland:
    Resolution in ART: An experimental investigation of the resolving power of an alge-

braic picture reconstruction technique.
J Theor Biol 33:213-223, 1971.

6. Robb, R. A., S. A. Johnson, J. F. Greenleaf,
   M. A. Wondrow, and E. H. Wood:
   An operator-interactive computer-controlled
   system for high fidelity digitization and
   analysis of biomedical images.
   Proceedings of the Society of Photo-
   Optical Instrumentation Engineers,
   August 27-29, 1973, Volume 40, pp 11-26.

7. Sweeny, D.:
   Interferometric measurement of three-dimen-
   sional temperature fields.
   Mechanical Engineering Ph.D. Thesis,
   University of Michigan, p 125.

8. Greenleaf, J. F., S. A. Johnson, S. L. Lee,
   G. T. Herman, and E. H. Wood:
   Algebraic reconstruction of spatial dis-
   tributions of acoustic absorption within
   tissue from their two-dimensional acoustic
   projections.
   Acoustic Holography Vol 5, Plenum Press,
   New York, 1974, pp 591-603.

9. Gordon, R., and G. T. Herman:
   Three-dimensional reconstruction from projections: a review of algorithms.
   International Review of Cytology
   38:111-151, 1974.

10. Kossoff, G., E. K. Fry, and L. Tellins:
    Average velocity of ultrasound in the human
    female breast.
    The Journal of Acoustical Society of
    America 53(6):1730-1736, 1973.

11. Jellins, L., and G. Kossoff:
    Velocity compensation in water-coupled
    breast echography.
    Ultrasonics 11(5):223-226, September, 1973.

CARDIOVASCULAR DIAGNOSIS WITH REAL TIME ULTRASOUND IMAGING

O. T. von Ramm, F. L. Thurstone, J. Kisslo

Departments of Biomedical Engineering and
Medicine
Duke University, Durham, N.C.

In 1954, Edler and Hertz first used ultrasound to
identify cardiac structures. Since that time, this non-
invasive technique has become a useful diagnostic tool
which is currently enjoying a phase of rapid technological
advancement. New developments have primarily been aimed
towards the generation of two-dimensional tomographic
images of cardiac structures in an effort to overcome the
inherent limitations of a one-dimensional imaging technique,
the time-motion (T-M) display mode. Currently, this
technique is the standard method of clinical cardiac
examinations.

Efforts directed towards the generation of two-dimensional ultrasound B-mode images have followed several
avenues. King (1) and Kikuchi and Okuyana (2) utilized an
EKG triggered compound B-mode method to obtain single
images at predetermined times of the cardiac cycle. A
similar approach taken by Gramiak (3) entailed the storage
of both the EKG and simultaneous B-mode data in a computer
so that single frame ultrasound images at various times of
the cardiac cycle could subsequently be reconstructed.
More recently attempts have been aimed at increasing data
acquisition rates to the point where two-dimensional images
can be generated in real time. Bom, et al (4), and King
(5) have described multitransducer imaging systems which
produce a simple linear scan of the cardiac anatomy whereas
Henry (6) and Eggleton (7) utilized mechanical motion of a
single transducer to provide a real time sector scan of the
heart.

The operation of the real time imaging system described here relies on phased array principles to steer and focus the ultrasound beam in the near field of a linear array of ultrasound transducers.* Underlying the development of this system were several design considerations specifically directed towards its application in clinical cardiology. These include image frame rates commensurate with cardiac dynamics, a field of view sufficiently large to allow image formation of most or all of the left ventricle and the utilization of a relatively small hand held transducer which would allow some freedom in imaging various planes of the cardiac anatomy through the restricted acoustic windows of the thorax. In addition, efforts were directed towards improving the final image quality by maintaining high azimuthal and range resolution throughout the field of view and by logarithmically compressing the received echo signals from each element of the array.

This non-linear processing aside from improving image resolution, depth of field and reducing sidelobe response compresses the dynamic range of echo information so that regions in the tissues characterized by diffuse or specular reflectors can be displayed simultaneously on a typical CRT monitor.

The imaging system produces real time tomographic images in a circular sector format diagrammatically illustrated in Figure 1. Note that the tomographic plane is parallel to the long axis of the transducer array. The maximum sector angle of 60° is also indicated. In the final image, each complete 60° sector scan is composed of up to 256 individual B-mode line images. The current linear array utilizes 16 ultrasound transducers during transmit and receive. The pulse generated by each of these elements consists of approximately 4 to 6 half cycles with a center frequency of 1.8 MHz. At the site of contact, the array measures 24 by 14 mm where the latter dimension corresponds to the length of each element in elevation which is the dimension normal to the tomographic plane.

---

* For a more complete description of the system operation, see Thurstone and von Ramm (8).

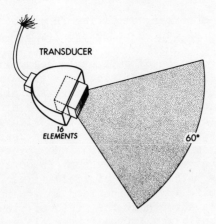

Figure 1:   The sector scan.

The high data rates necessary to image dynamic cardiac
structures are achieved by timing the transmitted pulses
from the individual elements of the array in a manner such
that the effective azimuthal orientation of the insonification pulse can be altered from one transmit operation to
the next. For example, if the 16 elements of the array are
excited with electrical pulses in a linear time sequence,
the resulting acoustic plane wave will propagate at an azimuth
angle from the direction normal to the array. By providing
excitation pulses with a spherical time relationship, the
transmitted beam may be focused at a specified range. In
normal operation, a spherical time relationship is combined
with linear timing to produce a beam focused at any azimuth
angle.

During reception, echoes arising from targets in the
direction of the transmitted pulse arrive at the transducer
elements at different times. This necessitates a delay of
the received signals so that the effective orientation of
the array during reception corresponds to the orientation
during transmission. In this system, accurate phasing is
accomplished via the use of switchable wide band lumped
constant analog delay lines which pass the acoustic information with different delay times for each element. In
this way, only echoes from a desired azimuthal orientation

will arrive at a summing amplifier in phase. After summation, the signal can be processed for a B-mode display.

The incorporation of switchable delay lines in the
electrical configuration of the receiver permits the receiver to be focused at a given range and allows tracking
of this focus in synchrony with the range of returning
echoes. This technique in conjunction with a focused transmit beam is responsible for the improved resolution of this
system. The azimuthal resolution (6db points) varies from
less than 2 mm at a minimum range of 35 mm to 4 mm at a
range of 170 mm. This depth of field (i.e. 35 - 170 mm) is
generally adequate for applications in adult cardiology.
Range resolution throughout the field of view is approximately 1.5 mm while resolution in the elevation dimension
corresponds to values predicted for an unfocused slit aperture (6db beam width - 8 mm at 110 mm range).

Flexibility is an important aspect of this imaging
system. All transmit timing and receive information is
stored in the core memory of a PDP 11-20 computer so that
the phase at each element position within the array aperture can be independently controlled within the maximum
delay envelope of 7.875 microseconds. In this way, modifications to imaging parameters such as focal points, etc,

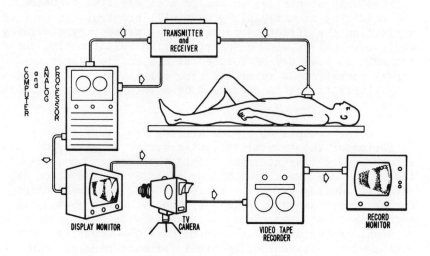

Figure 2:   Imaging system configuration.

merely require the storage of appropriate data within the core memory. On the other hand, scan format changes such as total number of lines per frame, sector angle, etc., are easily implemented through minor software changes in the system operating routines.

The architecture of the prototype imaging system currently under clinical evaluation is diagrammatically illustrated in Figure 2. The computer in association with the appropriate timing logic controls the entire scan sequence. In this figure, the switchable delay lines controlled by the computer are referred to as the "Analog Processor." The primary display oscilloscope, a Hewlett-Packard 1311 A, is viewed directly by a television camera so that sequences of particular interest can be recorded on video tape for later playback or frame by frame analysis. Normally, the acoustic image frame rates are synchronized to standard television rates of 30 frames per second. However, at the discretion of the clinician, the system may be used in an asynchronous mode to obtain other frame rates. During the synchronous

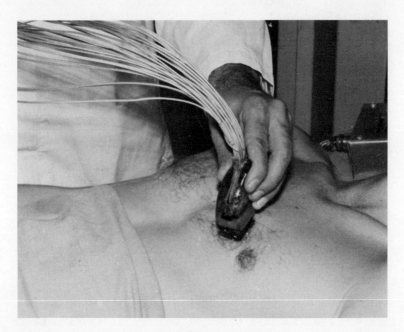

Figure 3: Linear array applied to the chest of a patient oriented to image the long axis of the left ventricle.

operation, the total number of acoustic scan lines is determined by the maximum range to be imaged. For a typical
maximum range of 15 cm each tomographic image frame is composed of 160 individual B-mode lines. In the asynchronous
mode, 256 line images can be obtained at the rate of 20 per
second for a full 60° sector and a similar maximum range.
Naturally, much higher frame rates are possible in this
mode by simply decreasing the total number of lines per
frame. (E.g. 64 line images can be generated at the rate
of 80 per second.)

Figure 3 illustrates the application of the linear
array to the chest of a patient. Here, the orientation of
the transducer is suitable to allow imaging of the left
ventricle in its long axis. The tomographic plane thus
imaged is indicated in Figure 4 and is labelled "Longitudinal." By rotating the transducer 90° from the orientation

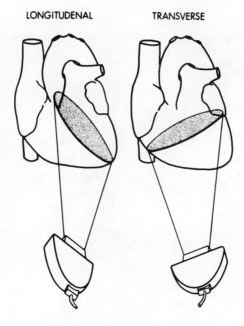

LONGITUDENAL   TRANSVERSE

Figure 4: Two standard orientations of the transducer to
image the long and short or transverse axis of the left
ventricle.

shown, various cross sections through the short axis of the left ventricle may be obtained. This standard clinical view is labelled "Transverse" in the latter figure.

An example of a tomographic image of the long axis of the left ventricle produced by this system is shown in Figure 5. This picture was produced by stop framing the video tape recorder and photographing the television monitor.

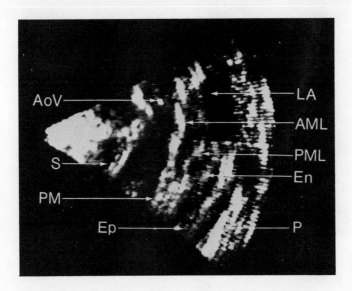

Figure 5: Long axis of left ventricle.
AoV = aortic valve cusps, S = interventricular septum, PM = papillary muscles, Ep = epicardium, P = pericardium, En = endocardium, PML = posterior mitral leaflet, AML = anterior mitral leaflet, LA = left atrium.

In this view, the patient's chest wall is on the left side, the posterior heart wall on the right. The aortic root is seen at the top of the scan while the left ventricular apex is toward the bottom. The large left ventricle is in diastole and the aortic cusps can be seen in the closed position whereas the anterior and posterior mitral leaflets are in the open position. The left atrium is behind the aortic root while the right ventricle is anterior to the septum. The internal limits of the left ventricle are well

defined by the interventricular septum and posterior wall
endocardium, myocardium, and epicardium. A pericardial
effusion is also visualized. Note the chordae tendinae
attached to both mitral leaflets and the papillary muscles.

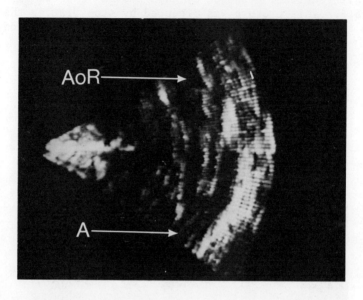

Figure 6: Long axis of left ventricle from aortic root
(AoR) to apex (A).

Figure 6 is an acoustic image of the left ventricle in long
axis from the aortic root to the apex. The dilated right
ventricle of this patient with an atrial septal defect
sufficiently displaced the left ventricle in range to permit
visualization of the entire left ventricular cavity in the
long axis. Again, the ventricle is in diastole. Note the
anterior and posterior mitral apparati, the interventricular
septum, root of the aorta, and the small size of the left
atrium typical of late diastole.

A series of scans through the short axis of the left
ventricle is shown in Figure 7. In panel "a" the section
is at the level of the mitral leaflets and shows the
anterior and posterior mitral leaflets open in diastole.
Panel "b" shows a short axis section through the left

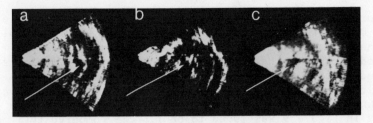

Figure 7: Serial cross sections through the short axis
of the left ventricle.   a) at the level of the mitral
orifice indicated by the arrow.   b) at the level of the
papillary muscles - arrow points to the left ventricular
cavity.   c) at the apex - arrow points to the narrowed
left ventricular cavity.

ventricular cavity at the level of the papillary muscles.
Panel "c" is a section through the left ventricular apex
and reveals a narrowed left ventricular cavity at its tip.
Short axis cross sections of the left ventricle at the level
of the papillary muscles in a patient with pulmonary hypertension is shown in Figure 8.   Panel "a" shows the
ventricle in diastole whereas "b" shows the ventricle in
systole.   Arrows indicate the left ventricular cavity.
Note the flattened appearance of the ventricle during both
phases of the cardiac cycle, a finding typical in patients
with pulmonary hypertension.

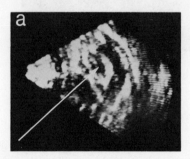
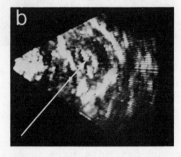

Figure 8: Short axis cross section of the LV at the level
of the papillary muscles of a patient with pulmonary hypertension - arrows point at ventricular cavity.   a) ventricle
in diastole.   b) ventricle in systole.

Figure 9 shows a short axis cross section at the level of the mitral valve orifice in a patient with severe mitral stenosis. The reduced mitral valve area available for blood flow from the left atrium to the left ventricle is immediately apparent when the orifice in this figure, indicated by the arrow, is compared to the mitral orifice shown in Figure 7a. In fact, by planimetry the orifice area in Figure 9 was found to be 1.5 cm^2 compared to 5 cm^2 for the orifice of Figure 7a.

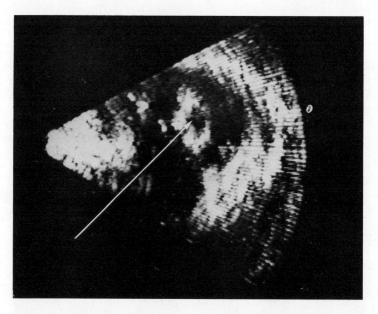

Figure 9: Short axis cross section at the mitral orifice of a patient with mitral stenosis. The arrow points at orifice.

During initial clinical trials of this imaging system, over 100 patients have been examined. High quality images of the aortic root, interventricular septum, left atrium, mitral and tricuspid leaflets, papillary muscles and posterior heart wall have been obtained with regularity. There are, however, several limitations in the performance of the current system. Among these are the ring-down

artifacts produced by the array which obliterate the first
3 to 4 cm of the final image, a total sector arc of 60°
which is sufficient to image the entire short axis cross
section of the left ventricle but is often inadequate in
displaying the entire long axis of the same ventricle as
well as the low center frequency of the linear array.
Current modifications and developments are oriented towards
overcoming these limitations in an effort to produce even
more diagnostically useful images which will permit
accurate quantification of various cardiac geometries and
functions.

REFERENCES

1. King D.L.: Cardiac ultrasonography. Proceedings of
    the 15th Annual Meeting of the American Institute of
    Ultrasound in Medicine, October 12-15, 1970, E-7, Case
    Western Reserve University, Cleveland, Ohio.

2. Kikuchi Y. and Okuyama D.:   Ultrasono-cardiotomography.
    JEE, #47, October, 1970.

3. Gramiak R., Waag R.C., Simon W.: Cine ultrasound
    cardiography. Radiology 107:175, 1973.

4. Bom N., Lancee C.T., Honkoop J., and Hugenholtz P.G.:
    Ultrasonic viewer for cross-sectional analyses of
    moving cardiac structures. Biomedical Engineering,
    November, 1971.

5. King D.L.: Real-time cross-sectional ultrasonic imaging
    of the heart using a linear array, multielement trans-
    ducer. JCU 1:196, 1973.

6. Henry W., Griffith J.: Cardiac scanning. Proceedings
    of the 19th Annual Meeting of the American Institute
    of Ultrasound in Medicine, October 5-10, 1974, Seattle,
    Washington.

7. Eggleton R.C., Dillon J.C., Feigenbaum H.D., Johnston
    K.W., and Chang S.: Visualization of cardiac dynamics
    with real time, B-mode, ultrasonic scanner. JCU 2:228,
    1974.

8. Thurstone F.L., von Ramm O.T.: A new ultrasound technique
    employing two-dimensional, electronic beam steering.
    Acoustical Holography , volume 5. P.S. Green, Editor,
    Plenum Press, 1974.

     This work was supported in part by USPHS grants HL
14228, HL 12715, HL 17670, and HS 01613.

SAMPLED APERTURE TECHNIQUES APPLIED TO

B-MODE ECHOENCEPHALOGRAPHY

D. J. Phillips, S. W. Smith, O. T. von Ramm,
F. L. Thurstone
Department of Biomedical Engineering
Duke University
Durham, North Carolina 27706

Whereas A-mode echoencephalography has been found diagnostically useful, B-mode displays for similar object volumes have not yet provided the additional information that might be expected from a two-dimensional presentation. Poor image quality results from a number of causes stemming from the presence of skull bone (White, 1967; 1968). Large differences in specific acoustic impedance between skull bone and soft tissue combined with irregularities of skull bone surfaces result in unfavorable reflections and reverberations of acoustic energy. Severe attenuation of ultrasound at diagnostically useful frequencies further reduces system sensitivity and target acquisition. Additionally, significant differences between acoustic velocities in bone and tissue result in beam degradation and registration errors in the B-mode image. With one possible exception (Somer, 1969), the nature and extent of these aberrations have thus far limited practical utilization of B-mode scanners in clinical examinations of head structure.

Perhaps the most significant aberrations are caused by variations in skull thickness over conventional transducer apertures. Unfortunately, important imaging parameters such as lateral resolution and target sensitivity would be compromised by a reduction in the aperture dimension. The present study uses an electronically steered and focussed multi-element transducer array (Thurstone and von Ramm, 1974), in an attempt to minimize acoustic aberrations caused by skull thickness variations. Implementation of a multi-element array is an intriguing possibility in that

the transducer aperture size is maintained while decreasing
skull thickness variations over an individual element.

Unlike a simple transducer with a fixed lens the
sampled aperture approach permits utilization of numerous
signal processing techniques which might improve target
resolution and picture quality when imaging through an
aberrating medium. Two distinct approaches presently under
investigation are: (1) a phase compensation technique
based upon an a priori knowledge of the phase character of
the aberrating medium, and (2) a signal processing scheme
which is effective for one class of phase aberrations
likely to be encountered. Fundamental considerations and
differences of each approach will now be discussed.

PHASE COMPENSATION

In many applications the use of discrete elements
filling a relatively large effective aperture presents
major advantages for a linear array over a simple transducer of comparable dimension. Independent control over
each element in transmit and receive provides a method to
produce any desired wavefront response within the limitations imposed by element size and spacing with respect to
the wavelength of energy used. Phase compensation has
previously been described for imaging through an aberrating medium (Goodman, 1966) and has recently been proposed for an astronomical application (Muller & Buffington,
1974).

Phase ambiguities or time shifts are introduced when
acoustic energy propagates through a section of skull
varying in thickness. Figure 1 shows two elements of a
transducer array simultaneously transmitting an acoustic
pulse through a section of skull. Due to the large dis-
parity in acoustic velocity between that in skull and that
in soft tissue or water, differences in skull thickness
produce relative changes in the phases of the acoustic
wavelets emerging from the inner table of the skull. The
phase difference between wavefronts from any two elements
is given by the product of the relative change in skull
thickness times the difference of the reciprocal acoustic
velocities of the two media.

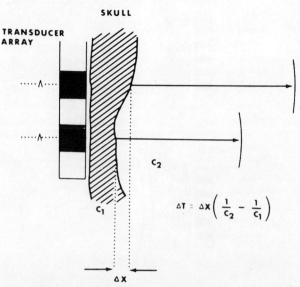

Figure 1. Phase Aberrations Introduced by a Section of
Skull Varying in Thickness.

Figure 2 illustrates the basic principle of the phase
compensation technique. Five representative elements of
the transducer array are shown with a section of skull
placed in front of them. Part A shows the phase aberrated
wavefronts emerging from a variable thickness skull when
the elements are phased to produce a focussed wavefront in
a medium of constant acoustic velocity. In B, an acoustic
wavefront provided by another source passes from the inner
table of the skull to the transducer elements. Knowledge
of the spatial character of the incident wavefront and the
arrival times at each transducer element allows for the
determination of the relative changes in skull thickness in
front of the elements comprising the aperture. When these
relative phase variations are incorporated into the transmit timing as shown in C, the emergent wavefronts exhibit
a restored phase character similar to that originally
intended.

As a first order approximation, the skull bone is
modeled as an attenuating acoustic lens positioned in front
of the transducer array. To ascertain the validity of this

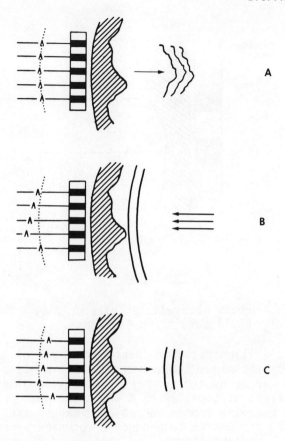

Figure 2.   Basic Principles of Phase Compensation Technique.

model and the usefulness of the phase compensation technique, feasibility studies of acoustic transmission through
skull flaps were performed in a water tank environment.
Two different skull flaps were used in the experiments.
One is from a 50-year-old adult removed from the frontal-
temporal region, and the other is from the temporal region
of a 5-year-old child. Both flaps were frozen upon removal
from the individuals and were thawed before experimental
implementation and continually stored in water. A diagram
of the experimental arrangement is shown in Figure 3. The
multi-element array measures 24 mm in length and consists
of 16, equally-spaced, transducer elements. Each element

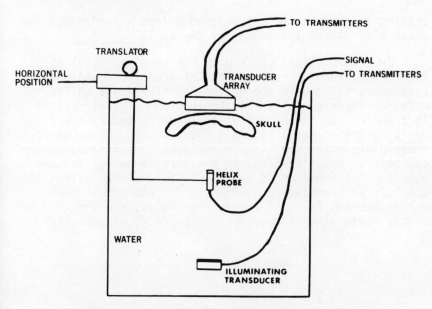

Figure 3.   Diagram of Experimental Set Up.

is 0.35 mm in width and 14 mm in elevation. The center
frequency was experimentally measured to be 1.8 MHz. The
linear array is placed just under the surface of the water
with the skull flap positioned in close proximity. Independent control of transmit phasing for each element was
provided by continuously variable, mono-stable multivibra-
tors whose outputs were directly coupled into the solid
state transmitter-pulsers. To record the acoustic field
pattern a small transducer is desirable in order to pre-
serve an omnidirectional response and to resolve small
spatial variations of acoustic pressure. A Helix transducer probe measuring 0.46 mm in diameter was used for this
purpose. The acoustic pressure was displayed on a cathode
ray tube and recorded on photographic film as a function of
the lateral position of the translated field probe. A
simple piston transducer, 12.5 mm in diameter served as an
illuminating transducer which provided a smooth acoustic
wavefront incident at the inner table of the skull. The
relative arrival times of acoustic energy relate changes in
skull thickness in front of the transducer aperture. These
relative arrival times are incorporated into the transmit

phasing to produce the compensated delays which reduce
phase aberrations introduced by the skull.

As a theoretical check with the proposed model of
skull bone, computer simulations of acoustic field pressure
were constructed with use of the relative phase shifts
obtained experimentally when the section of skull placed
in front of the linear array was interrogated by the illuminating transducer. Figure 4 shows two of the acoustic
field plots which are used to evaluate the ultrasound beam
character. Relative acoustic pressure is normalized and
plotted as a function of lateral position. The extent of
the lateral translation is 25 mm. The element transmit
delays are set for a 10 cm Gaussian focus, and the experimental field plot is recorded in the focal plane. The

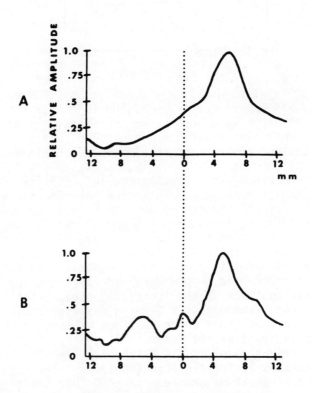

Figure 4.   Transmission Through a Skull Flap and the
            Corresponding Computer Simulation.

upper plot shows the acoustic pressure at a range of 10 cm
when the adult skull is placed in front of the linear
array. Of particular interest is an azimuthal shift of
the main lobe to the right and the broad main lobe beamwidth. The lower plot shows the computer simulation where
the experimentally measured compensated delays were incorporated into the transmit timing to construct the acoustic
field pattern. In each simulation a transmit pulse character was specified along with an attenuation factor for
each element to more accurately represent the experimental
situation. If the proposed model of the skull is valid as
a first order approximation, then the character of the
simulation should correlate closely with the acoustic
field plot obtained experimentally. The computer simulation shows an azimuthal shift of 4.9 mm to the right compared with the 5.0 mm shift in the experimental plot. The
6 db beamwidths in both plots are noted to be 7.0 mm. The
overall geometrical similarities lend considerable support
to the proposed first order model of the skull.

The following experiment is presented to illustrate
preliminary findings utilizing the phase compensation. As
shown in Figure 5, three field plots of acoustic pressure
are recorded. Part A shows the control field plot
recorded at a range of 10 cm when the transducer elements
are phased for a 10 cm focus and the skull removed. The
6 db beamwidth is experimentally measured to be 4.5 mm
compared with a calculated 4.3 mm beamwidth for the diffraction limited aperture. Part B shows the beam plot
with the adult skull positioned between the array and
field probe. The 6 db beamwidth is increased to 14.0 mm,
and a main lobe refraction of 3.0 mm to the right is also
apparent. The relative phase changes as a function of
skull position in front of the linear array were measured,
and the phase compensation is shown in part C. Although
the acoustic beam is attenuated due to a single transmission through skull bone, the phase character due to
variations in skull thickness is restored. The refracted
lobe was returned to within 0.5 mm of the axis defined in
the control, and the 6 db beamwidth was significantly
improved and measured to be 4.5 cm. Although the foregoing
studies were performed for the transmit mode in a direction
normal to the face of the transducer, it was hoped that
similar improvements could be realized for electronically
steering to an azimuth angle and for focussing the acoustic
energy in the receive mode.

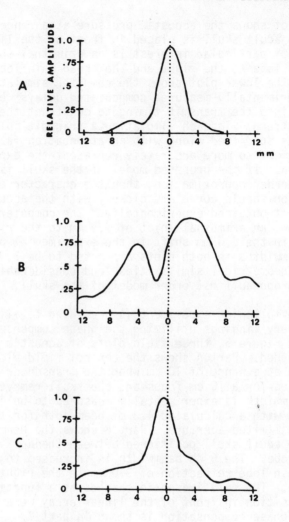

Figure 5. Acoustic Field Plots for the Phase Compensation
           Through an Adult Skull Flap.

    Implementation of the compensation technique in a
clinical environment will be more complex since additional
procedures and techniques must be utilized if the data is
to accurately represent relative changes in skull thickness. Of basic importance is the necessity of producing

a smooth acoustic wavefront incident at the inner table, within a live, intact head.

Shown in Figure 6 is a horizontal slice through an entire head. The illuminating transducer is placed on one side of the head and the linear array is placed on the opposite side. For a typical trans-skull diameter of 14 cm, an aperture size for the illuminating transducer can be chosen so that regardless of the skull variation on the near side of the skull, the ultrasound wavefront at the far side will be smooth. Although the acoustic wavefront from the illuminating transducer is irregular after passage through the near section of skull, by the time it reaches the opposite side of the head and propagates further into the far field, the wavefront must be regular for a carefully chosen aperture size for the illuminating transducer. The phase compensation technique is of little value without the ability to produce a smooth acoustic wavefront within the head.

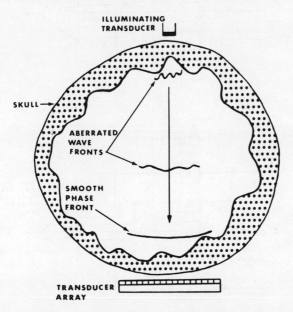

Figure 6.   Schematic Diagram of the Phase Compensation Technique Implemented on an Intact Skull.

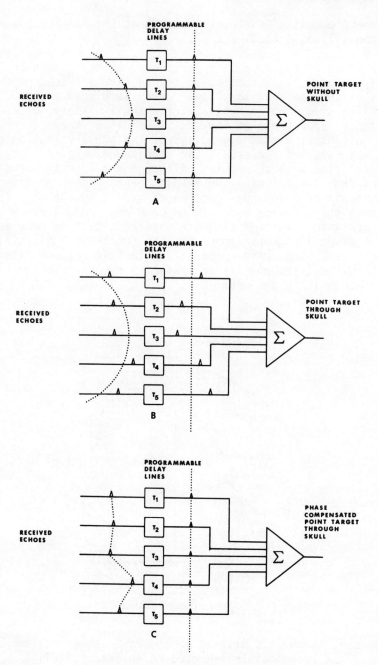

Figure 7.   Phase Compensation Technique in the Receive Mode.

The next step in the analysis was to perform phase compensation studies for the transmit-receive mode in the water tank environment. The experimental procedure was similar to that described for the transmit studies. Part A in Figure 7 shows the received echoes for five representative channels from an acoustic point source in a non-aberrating medium. The returning echoes arrive at individual elements of the array according to the spherical character of the incident wavefront. Since the programmable delay lines are sequenced to provide a focus for all points throughout the object volume of interest, the electronic pulses corresponding to the received echoes emerge from the delay lines at identical times as shown to the right. In part B, a skull sample was placed in front of the linear array. Due to the aberrating nature of the skull the emerging wavefront no longer resembles a diverging spherical wave. Since the delay times are sequenced for a focus in a homogeneous medium it is not surprising that the outputs from the delay lines are unable to provide phase coherence for an echo returning from a point target when the skull is interposed. In part C the received echoes from the same point source through skull are restored to phase coherence through interactive alignment of the output from each delay line to that in a reference channel. This procedure removes the timing errors caused when imaging through skull of varying thickness or composition.

This technique was performed in a water tank environment through an adult section of skull. The problem of a relatively high acoustic attenuation through skull was seen to be a major problem when using a center acoustic frequency of 1.8 MHz. Although it was easy to establish phase coherence of the returning echoes through iterations of the programmable delay lines, the frequency of received echoes was significantly reduced. This problem became more acute in the transmit-receive mode than in the transmit mode described earlier because the present situation involves two passes through the skull. Depending upon skull thickness the frequency dependent attenuation significantly reduces the center frequency of the returning echoes. These findings indicate that a lower acoustic frequency is desirable to both increase the signal to noise ratio and to decrease the variation in acoustic frequency of the received echoes. In addition an increased aperture

size of the linear array is desirable to restore the resolution loss from an increased wavelength.

    Because of a relatively large aperture dimension the
need for some processing technique over that of a linear
summation can be appreciated from thickness studies of our
skull samples. Figure 8 shows the root mean square skull
phase variation vs. aperture size at 1.8 MHz and 1 MHz.
The data was measured at ten different locations on two
skull samples. The RMS phase variation is a monotonically
increasing function. The standard deviation similarly
increases due to the large variabilities from point to
point on the skull samples and the fewer sample combinations at large apertures. After the phase variation
reaches some limit, say 1 radian (Fried, 1965) increasing
the aperture size does not significantly increase the
resolution capability using conventional linear summation.
Based on our data, for transducer sizes greater than 6 mm
at 1.8 MHz and greater than 18 mm at 1 MHz some form of
compensation or signal processing is needed to approach
diffraction limited resolution.

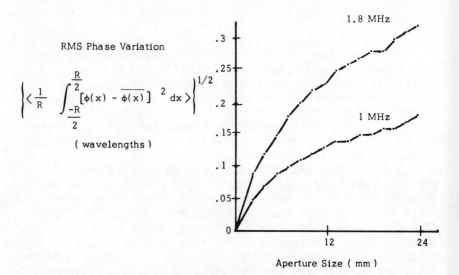

Figure 8.   Root Mean Square Phase Variation Due to the
            Skull vs. Aperture Size for 1.8 MHz and 1 MHz.

SIGNAL PROCESSOR

As an alternative to the phase compensation technique, methods of receive processing have been considered which would not require the on-line measurement of skull phase variation but which might be insensitive to such an aberration. The skull thickness itself and, hence, the phase variation across a one-dimensional transducer can be written as an nth order polynomial:

$$\phi(y) = A + By + Cy^2 + Dy^3 + Ey^4 + \ldots$$

where $\phi(y)$ is the relative phase at some position y on the surface of the skull. That is, the sum of a mean phase shift, a linear phase variation and higher order terms. On the average the magnitude of these coefficients will decrease with higher orders; i.e., the finer grain variations will have smaller amplitudes than, for instance, the average slope of the skull across the transducer.

The mean thickness and the slope of the skull make up the first two terms of the phase polynomial. These do not degrade the resolution of an A-mode line or an individual line of a B-scan image. However, for a linear B-scan or a compound B-scan, these terms can cause distortions and misregistrations in both the range and lateral dimensions. An electronic sector scanner as described by Somer (1968) or Thurstone and von Ramm (1974) produces a B-scan image through a single fixed spot on the skull. Besides the obvious advantages of real time, a sector scanner image is shifted uniformly in range and azimuth by these first terms but suffers no distortion.

Second order and higher order phase variations across the aperture do degrade the lateral resolution of a single A-mode line and every type of B-scanner. Coherent summation, linear or non-linear, yields a good combination of resolution, dynamic range and signal to noise ratio in the absence of phase aberration. However with aberration the fidelity of images is seriously degraded to such an extent that the effective aperture size is reduced to that dimension over which the phase variation is less than $\lambda/8$ (Goodman, 1966). Incoherent summation is the least sensitive to phase aberration (Cox, 1974), but exhibits quite poor signal to noise ratio and hence dynamic range. The

solution would be to find some combination of coherent and
incoherent processing to minimize the various disadvantages
of each. One possibility is to assume that there are many
areas on the skull where the phase aberration will closely
approximate a linear phase shift plus some function which
is symmetric about the center of the linear array, i.e.,

$$\phi(y) = A + By + Cy^2 + Ey^4 + \ldots$$

In these areas of the skull some types of symmetric processors will exhibit a better response than a coherent
summation or an incoherent sum.

In Figure 9 we show a block diagram for a 6 element
array which compares normal linear summation with the
operations of one symmetric non-linear receive processor
under current consideration. Linear summation is just a
simple sum of the signal outputs from each element in the
sampled aperture. In the lower diagram, however, the out-

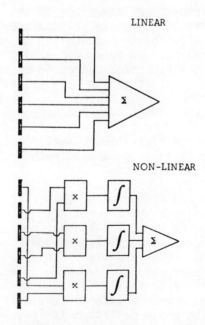

Figure 9. Block Diagram of Linear Summation versus Non-
           Linear Processing for Six Element Transducer
           Array.

puts from symmetric elements are multiplied together, integrated in time and then summed with the results of the same operations from other symmetric pairs. If the elements of a given pair are in phase, with or without the presence of an aberration, the receive point spread function (PSF) for that pair will be unshifted.

Figure 10 compares the PSF of a linear summation with that of our non-linear processor for an unaberrated system. One can note the narrow maximum and more numerous side lobes typical of a multiplicative processor. For any aberration which can be described as an even function whose origin is at the center of the sampled aperture, the point spread function will be unchanged from this control. Thus, the system is insensitive to a cosine function, a parabola, etc. The system still responds to a linear variation with a simple azimuthal shift, so that there is no loss of resolution for any phase aberration which can be described by a combination of mean thickness, linear variation, quadratic variation and other even functions.

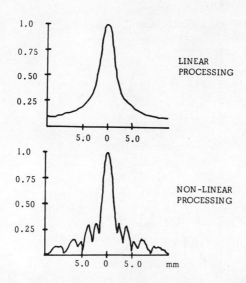

Figure 10. Calculated Point Spread Function for Linear Summation versus Non-Linear Processor with No Aberration.

One should note that the coherent multiplication and integration of symmetric elements is twice as sensitive to small deviations from a symmetric phase variation as a symmetric summation. However, the multiplication and a short pulse length decreases the background level associated with the incoherent aspect of the processor much more rapidly than a symmetric summation.

As an example of the processor's operation, Figure 11 compares the calculated point spread functions in receive mode at 2 MHz for a linear summation with no aberration, a linear summation with data from an adult skull, and the even function processor with the same data. The improvement in 6 db beamwidth from 18 mm to 6 mm is significant.

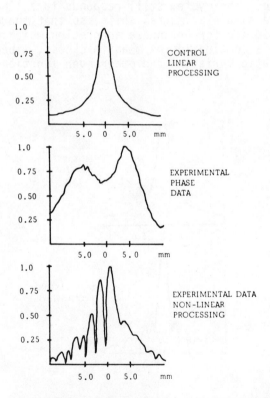

Figure 11.   Calculated Point Spread Functions for Linear-
             Summation Control versus Linear Summation with
             Aberration, versus Non-Linear Processing with
             Aberration.

Naturally, the skull phase variation cannot always be described as a slope plus an even function, but this type of receive processing should improve the lateral resolution at a significant number of locations on the skull surface.

In summary, two basically different approaches have been described which could have practical usefulness in ultrasonic investigations of brain tissue. It is difficult to predict how much improvement is necessary in lateral resolution and target dynamic range before B-mode images of cephalic structure routinely provide information of diagnostic value. The improvements presented here are encouraging and may lead to new techniques that enable the obtainment of acoustic images of cephalic structure having acceptable resolution for diagnostic evaluation.

References:

1. Cox, H., "Line Array Performance When the Signal Coherence is Spatially Dependent." Journal of the Optical Society of America, 54: 1743, 1973.

2. Fried, D.L., "Statistics of a Geometrical Representation of a Deformed Wavefront." Journal of the Optical Society of America, 55: 1427, 1965.

3. Goodman, J.W., Huntley, W.H., Jr., Jackson, D.W., Lehman, M., "Wavefront-Reconstruction Imaging Through Random Media." App. Phys. Lett., 8: 311, 1966.

4. Somer, J.C., "Electronic Sector Scanning with Ultrasonic Beams." Proceedings of the First World Congress on Ultrasound Diagnostics in Medicine, 1969, June 2-7, Vienna, Austria.

5. Somer, J.C., "Electronic Sector Scanning for Ultrasonic Diagnosis." Ultrasonics, 6: 153, 1968.

6. Thurstone, F.L., von Ramm, O.T., "Electronic Beam Scanning for Ultrasonic Imaging." Proc. II World Congress on Ultrasonic Diagnostics in Medicine, Rotterdam, the Netherlands, Excerpta Medica, Amsterdam, 43, 1973.

7.   Thurstone, F.L., von Ramm, O.T., "A New Ultrasound
     Imaging Technique Employing Two-Dimensional Electronic
     Beam Steering." Acoustical Holography, 5, Plenum
     Press, New York, pp. 249-259, 1974.

8.   White, D.N., Clark, J.M., White, M.N., "Studies in
     Ultrasonic Echoencephalography - VII: General
     Principles of Recording Information In Ultrasonic
     B- and C-Scanning, and the Effects of Scatter, Reflec-
     tion, and Refraction by Cadaver Skull on this Infor-
     mation." Med. and Biol. Engrg., 5: 3-14, 1967.

9.   White, D.N., Clark, J.M., Chesebrough, J.N., White,
     M.N., Campbell, J.K., "Effect of Skull in Degrading
     the Display of Echoencephalographic B and C Scans."
     Journal of the Acoustical Society of America, 44:
     1339-1345, 1968.

Acknowledgements:

     We would like to thank Dr. Ralph Barnes and Dr. Pat
McGraw, Bowman Gray School of Medicine, Winston-Salem,
N.C., for their assistance in obtaining the skull samples.

     This work was supported by USPHS grants HL 12715,
HL 14228, HS 01613, and by the Food and Drug Administration, Bureau of Radiological Health.

HIGH-RESOLUTION B-SCAN SYSTEMS USING A CIRCULAR ARRAY

Albert Macovski and Stephen J. Norton

Electrical Engineering Department
Stanford University
Stanford, California 94305

INTRODUCTION

The desirable characteristics of an improved cross-sectional or B-scan imaging system include diffraction-limited lateral resolution in both dimensions at all depths, large-aperture receiver sensitivity, and real-time imaging. No B-scan system, in use or published, meets all of these desirable traits. Some array systems are proposed in this paper which attempt to meet all of these characteristics in a relatively simple structure.

Most commercial B-scan systems, using manually scanned single transducers operating in the near field region, have none of these characteristics. A number of new approaches have recently emerged which supply some of the desired characteristics. Somer [1,2] introduced a real-time system using phased-array techniques to generate a sector scan. This system is not dynamically focused and has near-field resolution patterns in both dimensions. Kossoff [3] has achieved the lateral resolution and sensitivity of a large aperture by using an array of concentric transducer rings. In the receive mode the array is dynamically focused by connecting the rings together with time-varying delay networks. The system operates on axis with mechanical scan used to form an image. It is thus not real time but provides the desired resolution characteristics. Thurstone [4] has reported on a real-time sector scan system for cardiology. The array was dynamically focused and deflected in one

dimension using computer controlled delay elements. In
the orthogonal dimension the system used near field patterns
so that the focusing characteristic was realized in one
lateral dimension.

CIRCULAR ARRAYS

The primary problem is that of achieving focusing in
both dimensions over a large depth range. One method of
achieving the desired resolution is the matrix transducer
array shown in Fig. 1. Here a controlled delay element is
connected to each transducer element to compensate for the
difference in propagation time between a reflecting point
in the object and each element. Thus if the propagation
time to each element n is r_n/c, a corresponding delay

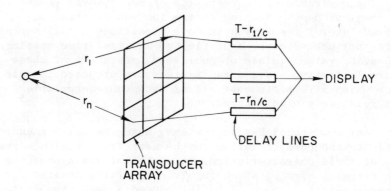

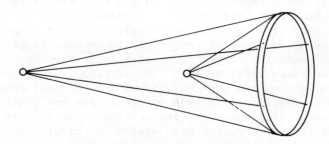

Fig. 1 Matrix and Ring Array Configurations

element of $T - r_n/c$ results in a uniform delay of T from the reflecting point through all paths to the collecting point. This configuration, at the present state of technology, is difficult to implement because of the large number of time-varying delay structures which have to be controlled in a relatively complex manner. The circular array shown at the bottom of Fig. 1 is a relatively simple structure which can provide many of the desired characteristics. Circular arrays have inherently desirable properties in that their Fresnel transform and Fourier transform have the same amplitude characteristic as shown in the appendix. Thus the same field patterns are maintained at essentially all depths except for a small region very close to the array. A uniformly driven annulus has an amplitude pattern at a depth z given by $J_0\left(\dfrac{kRr}{z}\right)$ shown in Fig. 2 where $k = \dfrac{2\pi}{\lambda}$ is the wavenumber, R is the radius of the ring and r is the radial coordinate. As with all graphs in this paper, the ordinate is the relative response and the abscissa is the argument of the function. This response is obtained using the Fourier Bessel transform. It assumes, as will be done throughout this paper, that the system is operated with pulse durations which approach steady-state or sinusoidal behavior. In addition it is assumed that the ring elements are sufficiently narrow in the radial direction so that the element patterns themselves are relatively isotropic and do not significantly affect the array pattern. As shown in Fig. 2, the sidelobe pattern is excessive. If used in both the transmit and receive mode this response is squared providing an overall response given by $J_0^2\left(\dfrac{kRr}{z}\right)$. This pattern has reduced sidelobes which however, are still greater than desired. In addition to causing false responses, the integrated effect of the energy in the sidelobes can limit the dynamic range and thus obscure important subtle echoes.

An interesting system was proposed and demonstrated by Burckhardt [5], which corrected the side lobe problem. Two scans are taken at each lateral position. Firstly a line is addressed with the ring array uniformly weighted on both the transmit and receive modes resulting in the overall amplitude response $J_0^2\left(\dfrac{kRr}{z}\right)$. Then the ring is weighted in

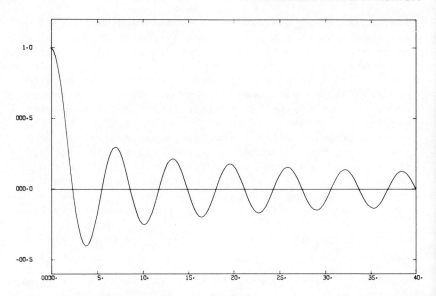

Fig. 2   Field Pattern of a Uniformly Driven Ring Array, $J_0^2\left(\dfrac{kRr}{z}\right)$

phase by $e^{j2\theta}$ on transmit and $e^{-j2\theta}$ on receive providing an overall amplitude response of $J_2^2\left(\dfrac{kRr}{z}\right)$. When these are subtracted to provide $J_0^2\left(\dfrac{kRr}{z}\right) - J_2^2\left(\dfrac{kRr}{z}\right)$, most of the sidelobes cancel resulting in an overall response which approaches that of a diffraction-limited full aperture. This approach is excellent for manual B-scans although it presents problems with real-time considerations and is quite complex. The requirement for two scans at each line in the object halves the number of scan lines which can be obtained in a given time interval. Since the number of lines is already significantly limited by the round trip time in the body, this further limitation is quite severe. In addition, the storing of the complex amplitude information for an entire line is a relatively complicated task.

In this paper the desired responses are approached or achieved in a single scan thus providing real-time performance without storage requirements. In the transmit mode the transducer elements around the ring are weighted by

$\cos^2\theta$. The resultant amplitude pattern in the Fresnel and Fraunhoffer regions as shown in the appendix is given by

$$\mu(r,\phi,z) = \frac{1}{2}\left[J_0\left(\frac{kRr}{z}\right) - J_2\left(\frac{kRr}{z}\right)\cos 2\phi\right].$$

Using the Bessel function identity

$$J_{n-1}(x) + J_{n+1}(x) = \frac{2nJ_n(x)}{x}$$

the response can be rewritten as

$$\mu(r,\phi,z) = \frac{J_1\left(\frac{kRr}{z}\right)}{\left(\frac{kRr}{z}\right)} - J_2\left(\frac{kRr}{z}\right)\cos^2\phi.$$

When placed in this form the response in the y direction ($\phi = 90^\circ, 270^\circ$) is seen to be identical to that of the desired full aperture focused system. In the x direction the response is $J_0\left(\frac{kRr}{z}\right) - J_2\left(\frac{kRr}{z}\right)$ which has significant side lobe content. Thus the weighting with $\cos^2\theta$ in the transmit mode achieves the desired response at all depths in one lateral dimension.

To achieve the same response in the other lateral dimension the array is modified in the receive mode to a $\sin^2\theta$ weighting. This provides a response given by

$$\mu(r,\phi,z) = \frac{1}{2}\left[J_0\left(\frac{kRr}{z}\right) + J_2\left(\frac{kRr}{z}\right)\cos^2\phi\right]$$

$$= \frac{J_1\left(\frac{kRr}{z}\right)}{\left(\frac{kRr}{z}\right)} - J_2\left(\frac{kRr}{z}\right)\sin^2\phi.$$

The overall response representing the product of the two array configurations is given by

$$\mu_0(r,\theta,z) = J_0^2\left(\frac{kRr}{z}\right) - J_2^2\left(\frac{kRr}{z}\right)\cos^2 2\theta.$$

This overall response, as shown in Fig. 3, is equal to $J_0^2\left(\frac{kRr}{z}\right) - J_2^2\left(\frac{kRr}{z}\right)$ along the vertical and horizontal axes.

This is the same response achieved by Burckhardt with his
two scan system. Along the diagonals, however, the response
is again $J_0^2\left(\dfrac{kRr}{z}\right)$ with its somewhat excessive side lobe
amplitude.

The side-lobe reduction can be improved by transmitting
and receiving with rings of different sizes, R_1 and R_2.
This has the added practical advantage of minimizing the
switching requirements in that the transmitter and receiver
angular weightings to their respective rings can remain
fixed. In this configuration the side-lobe patterns at
45° are minimized since they occur at different radii for
the two ring patterns. Fig. 4 shows the normalized integrated sidelobe energy along the diagonals for various
ratios of R_1/R_2. The integration was done from a radius
outside the main central lobe to a radius far removed from
the central pattern. As is shown in Fig. 4, the sidelobe
energy minimizes at ring diameter ratios of about 1.2 and
2.0. The diagonal response is considerably improved over

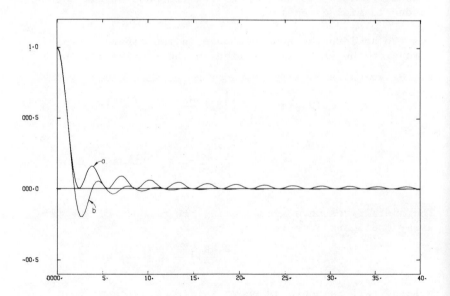

Fig. 3   a) Field Pattern Along the Diagonals.   b) Field Pattern
         Along the Vertical and Horizontal Axes.

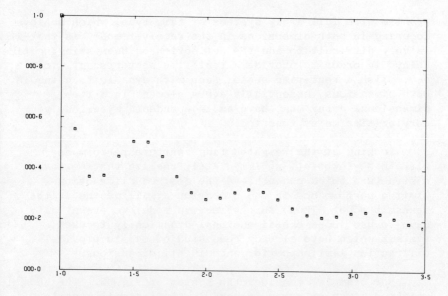

Fig. 4   Relative Integrated Side Lobe Energy Along the
         Diagonal Direction Versus Ratio of Transmitter and
         Receiver Ring Diameters.

the single annulus case.   Further study is planned to
optimize this configuration.

The annular configuration studied can provide real-
time performance with adequate lateral resolution.   They do
not, however, have the receiver sensitivity of a system
having a full receiving aperture.

SYSTEMS WITH DYNAMIC FOCUSING

To further improve the performance, dynamic focusing
is considered.   It must be recalled that dynamic focusing
can only be achieved in the receiving mode.   The Kossoff
array system [3] with its concentric annular rings achieves
diffraction-limited resolution in all directions through
dynamic focusing in the receive mode.   The transmitter
pattern, using the same rings at fixed relative delays,
provides a relatively broad illumination source.   The

problem with full array systems of this type, which achieve
focusing in both dimensions in the receiver mode, is that it
is very difficult to add the deflection or beam-swinging time
delays in order to provide a real-time sector scan. To
accomplish a real-time sector scan with dynamic focusing in
both dimensions, essentially every element in a two-
dimensional array must have an independent, dynamically-
varying time delay function.

Looking at the weighted ring aperture performance
with this viewpoint, we immediately realize that we have
a structure which can achieve the desired diffraction-
limited performance in one dimension at all depths in the
transmit mode. Thus, in the receive mode, we essentially
have to use those one-dimensional dynamically focused
systems which have already been studied [4] to provide
diffraction limited performance in all dimensions.

For example, consider the system of Fig. 5. The $\cos^2\theta$
weighted circular array is used in the transmit mode to
achieve the desired resolution in the y direction. In
the receive mode a dynamically focused line array is used
which provides the required time delays to keep the array
focused on the reflections from the propagating pulse. Thus
the overall response at any plane z is given by

$$\mu_0(r,\theta,z) = \left[\frac{J_1\left(\frac{kRr}{z}\right)}{\left(\frac{kRr}{z}\right)} - J_2\left(\frac{kRr}{z}\right)(x/r)^2 \right] \operatorname{sinc}\left(\frac{2Rx}{\lambda z}\right)$$

where $\operatorname{sinc}\left(\frac{2Rx}{\lambda z}\right)$ is the response of a cylindrical lens of
width 2R focused at plane z. This response provides
full-aperture diffraction-limited resolution in both dimensions since the undesired side lobe portion of the illuminating ring array is attenuated by the sinc function in
the x direction.

To achieve a dynamically focused array the time delays
must be varied in accordance with the distance from the
reflecting point in the object to each point along the array.
For objects along the axis this time τ is given by

$$\tau = \frac{1}{c}\sqrt{z^2 + x^2} \quad .$$

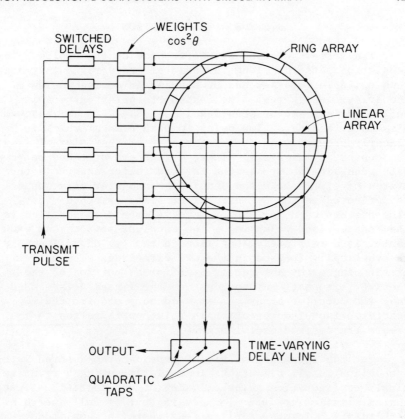

Fig. 5   Sector Scan System Using a Weighted Ring Array
         Transmitter and a Dynamically-Focused Line Array
         Receiver

For $z \gg x$ $\tau \cong \dfrac{z}{c} + \dfrac{x^2}{2zc}$.

Thus, using the paraxial approximation, each pair of
elements on either side of the linear array is connected to
a quadratically tapped delay line whose delay is varied
inversely with z or propagation time. A CCD (Charge
Coupled Delay Line) is an ideal candidate for this tapped
delay line. To achieve the desired delay variation the
clock frequency is linearly increased with time as the sonic
pulse propagates causing the time delay to vary inversely
with z as required.

SYSTEMS WITH DEFLECTION

The systems discussed thusfar provide the desired
resolution patterns along the axis of the array. They are
thus suitable for A-scans or B-scans involving motion of
the array. For a real-time B-scan the array pattern is
deflected in a sector scan fashion [1,2,4]. This is accomplished by adding a time delay to each element which is
linearly proportional to the distance of that element from
the center of the array. These delays are switched between
each scan line to create the desired sector scan. In the
system of Fig. 5 only the annular transmitter array beam
is deflected in the y direction. Thus switched delay
elements are used for each pair of elements in addition to
the $\cos^2\Theta$ fixed weighting. The receiver pattern, at each
depth, is a relatively long line in the y direction which
is dynamically focused in the x direction. The pattern
of the ring array intersects a different portion of the
receiver line pattern dependent on the linear delays used.
Thus the receiver array system need only provide the
quadratic focusing operation in this configuration to
create the desired sector scan.

The configuration of Fig. 5 provides the desired lateral
resolution in all dimensions in real-time using a relatively
simple configuration. The only one of our initial desirable
characteristics it fails to meet is that of full aperture
sensitivity. In many applications where the amplitude
of subtle echoes may be in the noise level, a full aperture
receiver is desired to maximize the collection efficiency.
In Fig. 6, with relatively small additional complexity, the
system is made to provide full aperture sensitivity. The
linear receiver array is transformed into an array of longer
vertical transducers which fill the aperture. The field
pattern of this receiver array, is primarily a near field
type of pattern in the y direction, and is dynamically
focused in the x direction. Since this receiver pattern
is not long enough for an entire sector scan in the y
direction, both the transmitter and receiver arrays are
deflected in the x direction rather than the y direction
as shown in Fig. 5. Thus the additional complexity
required in the full array system is that the receiver
requires an array of delay lines for deflection in addition

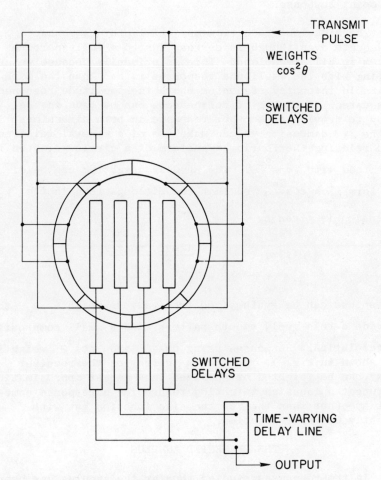

Fig. 6 Sector Scan System Using a Weighted Ring Array
Transmitter and a Linear Array of Line Segments
for the Receiver.

to the tapped time-varying line for dynamic focusing. The
resolution patterns of this system is essentially the same
as that of the system of Fig. 5 since the pattern in the
y dimension of the receiver array does not contribute to
the point response.

The side-lobe reduction may still prove somewhat
inadequate even though it corresponds to a full-aperture
system in all dimensions. This is primarily because we are
dealing with the amplitude responses rather than the more
desirable intensity responses where the amplitude response
is squared. A number of refinements can be made in the
array to provide a smoother response in both dimensions.
In the y dimension we can make use of a general relationship relating the Fourier transform of a circle weighted in
amplitude with $\cos^{2n}\theta$. The angular weighting can be of
the form $\sum_{n} a_{n} \cos^{2n}\theta$ so that the field pattern in the y
dimension is given by

$$\mu(y,z) \;=\; \sum_{n} a_{n} \frac{(2n)!}{n!\,2^{n}} \frac{J_{n}\!\left(\dfrac{kRy}{z}\right)}{\left(\dfrac{kRy}{z}\right)^{n}}$$

A response can be synthesized from various weights a_{n} to
provide a relatively smooth pattern with a small compromise
in resolution. Responses using $\cos^{4}\theta$ and $\cos^{6}\theta$ weightings
are shown in Fig. 7. In the x direction the receiver
array can be weighted or apodized in a manner approximating
a truncated gaussian weighting to provide a response somewhat smoother than that of the sinc function but with
slightly poorer resolution.

THE SEGMENTED ANNULUS

In the responses studied thusfar the annulus has been
considered a uniform structure with the angular weighting
applied on a continuous basis. In an actual array a finite
number of segments are used which modifies the resultant
pattern. Assume a circular array of N arc segments each
of angular extent $\beta N/2\pi$ where $\beta \leq 1$ since $\beta = 1$ implies
a continuous ring. The resultant function of θ is

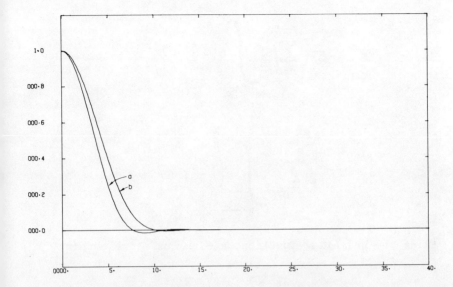

Fig. 7 Field Pattern of a Ring Array Weighted by
a) $\cos^4 \theta$ and b) $\cos^6 \theta$.

given by

$$U(\theta) = \sum_{n=0}^{N-1} g\left(n\,\frac{2\pi}{N}\right) \text{rect} \left[\frac{\theta - n\,\dfrac{2\pi}{N}}{\beta\,2\pi/N}\right]$$

where $g(\cdot)$ is the desired angular weighting function and
the rect function is unity between $\pm 1/2$ and zero otherwise. To find the field patterns we use a Fourier series
to expand the function of θ and then use the relationships developed in the appendix to find the Fourier transforms. $U(\theta)$ can be expanded into the form given by

$$U(\theta) = a_0 + \sum_{m=1}^{\infty} a_m \cos m\,\theta .$$

The values of a_0 and a_m are found by exchanging the
orders of summation and integration to obtain

$$a_0 = \frac{1}{2\pi} \sum_{n=0}^{N-1} \int_{n\frac{2\pi}{N} - \frac{\beta\pi}{N}}^{n\frac{2\pi}{N} + \frac{\beta\pi}{N}} g\left(n\frac{2\pi}{N}\right) d\Theta$$

and

$$a_m = \frac{1}{\pi} \sum_{n=0}^{N-1} \int_{n\frac{2\pi}{N} - \frac{\beta\pi}{N}}^{n\frac{2\pi}{N} + \frac{\beta\pi}{N}} g\left(n\frac{2\pi}{N}\right) \cos m\,\Theta\, d\,\Theta\,.$$

Using the weighting function $g(\Theta) = \cos^2\Theta$ we obtain

$$a_0 = \frac{1}{2\pi} \beta \frac{2\pi}{N} \sum_{n=0}^{N-1} \cos^2\left(n\frac{2\pi}{N}\right) = \beta/2$$

$$a_m = \frac{2}{\pi m} \sin\frac{m\pi\beta}{N} \sum_{n=0}^{N-1} \cos^2\left(n\frac{2\pi}{N}\right)\cos\left(mn\frac{2\pi}{N}\right)$$

$$= \frac{\beta}{2} \text{ sinc } \frac{m\beta}{N} \text{ for } m = 2,\ N\pm2, 2N\pm2\ldots\text{etc.},$$

$$\beta \text{ sinc } \frac{m\beta}{N} \text{ for } m = N, 2N,\ldots\text{etc.},$$

and zero for all other values of m,

where sinc $x = (\sin \pi x)/\pi x$.

The resultant angular pattern $U(\Theta)$ becomes

$$U(\Theta) = \frac{\beta}{2} + \frac{\beta}{2} \text{ sinc } \frac{2\beta}{N} \cos 2\Theta + \beta\sum_{n=1}^{\infty}\left\{ \frac{1}{2} \text{ sinc } \frac{(nN+2)\beta}{N} \cos(nN+2)\right.$$

$$\left. + \frac{1}{2} \text{ sinc } \frac{(nN-2)\beta}{N} \cos(nN-2)\Theta + \text{ sinc } n\beta \cos nN\Theta \right\}$$

If this is compared to the continuous distribution we see two major effects of segmentation. Firstly, the second harmonic component $\cos 2\theta$ is slightly diminished as compared to the constant term. This means that the desired $\dfrac{J_1(\cdot)}{(\cdot)}$ response will have an additional $J_0\left(\dfrac{kRr}{z}\right)$ term which is multiplied by $1 - \text{sinc}\,\dfrac{2\beta}{N}$. To minimize this component, and its associated sidelobes we use high N and/or small β. For reasonable values such as $\beta = .5$ and $N = 15$ (15 arc segments covering half of the total circumference) this additional J_0 factor has an amplitude less than $.01$.

The second major effect is that of the harmonics of N and their sidebands. These give rise to high order Bessel functions of the form $J_{nN}(\cdot)$, $J_{(nN+2)}(\cdot)$, and $J_{(nN-2)}(\cdot)$. In the region of the main lobe of the response where the arguments of these higher order Bessel functions are very small, they essentially do not contribute. The distant response where the arguments are very large, however, is similar for all orders so that these terms could conceivably contribute to distant sidelobes. For values of β close to unity, approximating a continuous ring, these components disappear due to the sinc function. Even for moderate values of β, they can be shown to be relatively negligible. For moderately high values of N, the sinc functions will be approximately equal resulting in the relationship

$$\mu_h \cong \sum_{n=1}^{\infty} \text{sinc}\, nN\beta\left[J_{nN}(\cdot) + \frac{1}{2}J_{n(N+2)}(\cdot) + \frac{1}{2}J_{n(N-2)}(\cdot)\right]$$

where μ_h is the field response due to the high orders along the high-resolution y axis. This can be subject to successive approximations as given by

$$\mu_n \cong \sum_{n=1}^{\infty} \frac{(nN+1)J_{nN+1}(\cdot)}{(\cdot)} + \frac{(nN-1)J_{nN-1}(\cdot)}{(\cdot)}$$

$$\cong \sum_{n=1}^{\infty} \frac{2(nN)^2 J_{nN}(\cdot)}{(\cdot)^2}$$

Despite the multiplying factor, the square of the argument in the denominator causes this function to drop off considerably in regions distant from the main response.

It is important to note that systems using a segmented
annulus spaced uniformly in angle have no "grating lobe"
response. These grating lobes normally restrict the
angular deviation of sector-scan systems using uniformly
spaced linear arrays. Thus, as in Figs. 5 and 6, if a
circle array is used as the transmitter it will minimize
the grating lobe problem. The region of the receiver grating lobe will receive reduced energy from the annulus
transmitter and thus have a diminished response. This
feature is of considerable help in the design of the
linear receiver array.

COMPUTER RESULTS

Using a 3cm diameter circle of 60 point segments,
the magnitudes of the patterns were computed for a wavelength of .75mm and a depth z of 12cm. Fig. 8 shows
the calculated responses along the axes and on the diagonals
for a $\cos^2\theta$ weighting on transmit and a $\sin^2\theta$ weighting

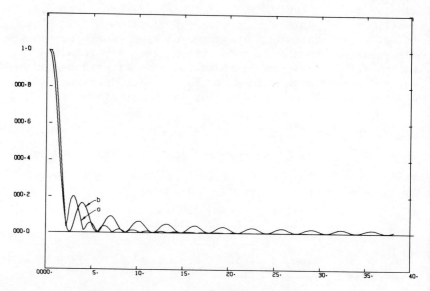

Fig. 8   Computed Field Patterns of the Magnitude of the
Response a) Along the Vertical and Horizontal Axes
and b) Along the Diagonals.

on receive.   Fig. 9 shows the calculated response along
the x and y axes using the system shown in Figs. 5 and 6
with the line or rectangular array having 30 segments.
The responses are shown for the same dimensions as that
of Fig. 8 using a $\cos^2\theta$ weighting on the transmitter
annulus.   To indicate the type of sidelobe reduction
obtainable on the y axis the response due to a $\cos^2\theta$
and $\cos^4\theta$ weighting is shown in Fig. 10.   The effect of
deflection is shown in Fig. 11.   The response of the annular
transmitter array weighted with $\cos^2\theta$ is shown with linear
delays applied to the elements corresponding to 10° deflection.

This work was partially supported under grant
5 P01 GMI 7940-03 from the National Institute of General
Medical Sciences.

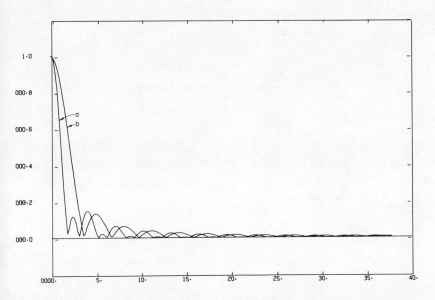

Fig. 9   Computed Field Pattern Magnitudes of the Circle-
Line Array Along the a) x Axis and b) y Axis.

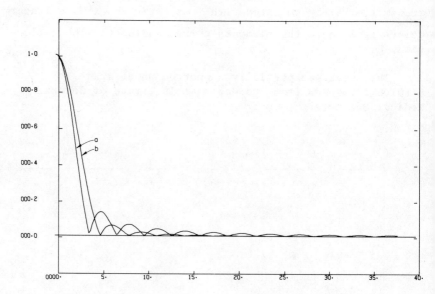

Fig. 10 Computed Field Pattern Magnitudes Using a Ring
         Array with a) $\cos^2\theta$ Weighting and b) $\cos^4\theta$
         Weighting.

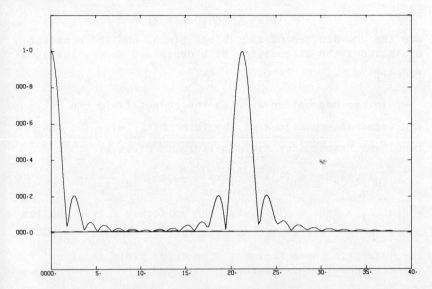

Fig. 11   Computed Field Pattern Magnitudes of the Ring
Array System in the Undeflected State, and Using
Linearly Varying Delays to Obtain 10° Deflection.

APPENDIX

A function separable in polar coordinates can be expressed
as summations of functions of the form $g(r)e^{jm\Theta}$. The
Fourier transform of this function is given by

$$\mathcal{F}[g(r)e^{jn\Theta}] = (-j)^n e^{jn\phi} \mathcal{H}_n[g(r)]$$

where $\mathcal{H}_n[\cdot]$ is the Hankel transform of order n which is
given by

$$\mathcal{H}_n[g(r)] = 2\pi \int_0^\infty rg(r)J_n(2\pi rp)dr \ .$$

The far field or Fraunhoffer region in scalar diffraction theory[6] is given by

$$U(r,\phi) = c_0 \exp\left(j \frac{\pi}{\lambda z} r^2\right) \mathcal{F}[U(r_1,\Theta)]$$

where r, ϕ are the image coordinates at depth z; r_1, θ are the coordinates of the object plane, and the constant c_0 includes the attenuation with depth and a constant phase factor.

In the case of an annulus the object field can be expressed as summations of the form $\delta(r_1 - R)e^{jn\theta}$. The resultant Fraunhoffer field pattern is given by

$$U_n(r, \phi) \;=\; c_1(-j)^n e^{jn\phi} \exp\left(j \frac{\pi}{\lambda z} r^2\right) J_n\left(\frac{kRr}{z}\right)$$

where $c_1 = 2\pi Rc_0$. Thus, for the uniformly driven annulus where $n = 0$, the amplitude distribution becomes $J_0\left(\frac{kRr}{z}\right)$. The quadratic phase factor in the image field is of no significance in this system.

In the Fresnel region, closer to the array, the field amplitude is given by

$$U(r, \phi) \;=\; c_0 \exp\left(j\frac{\pi}{\lambda z} r^2\right) \mathcal{F}[U(r_1, \theta)\exp(j \frac{k}{2z} r_1^2)]$$

which is identical to the Fraunhoffer region except for the quadratic phase factor. Again using object field components of the form $\delta(r_1 - R)e^{jn\theta}$ the Fresnel field pattern is given by

$$U_n(r, \phi) \;=\; c_1(-j)^n e^{jn\phi} \exp\left(j \frac{\pi}{\lambda z} r^2\right) \exp\left(j\frac{\pi}{\lambda z} R^2\right) J_n\left(\frac{kRr}{z}\right)$$

Thus the amplitude components are identical in the Fresnel and Fraunhoffer regions so that the desired patterns are maintained at all depths of interest.

Ignoring the phase factors and normalizing the constant c_1 to unity the effect of weighting with $\cos^2\theta$ can be evaluated by expanding it into $e^{jn\theta}$ components.

$$\cos^2\theta \;=\; \frac{1}{2} + \frac{1}{4} e^{j2\theta} + \frac{1}{4} e^{-j2\theta}$$

The resultant amplitude μ of the field components are given by

$$\mu(r,\phi,z) \;=\; \frac{1}{2} J_0\!\left(\frac{kRr}{z}\right) - \frac{1}{4} e^{j2\phi} J_2\!\left(\frac{kRr}{z}\right) - \frac{1}{4} e^{-j2\phi} J_2\!\left(\frac{kRr}{z}\right)$$

$$\;=\; \frac{1}{2}\left[J_0\!\left(\frac{kRr}{z}\right) - J_2\!\left(\frac{kRr}{z}\right) \cos 2\phi \right].$$

Using the Bessel function recursion identity

$$J_{n-1}(x) + J_{n+1}(x) \;=\; \frac{2n J_n(x)}{x} \quad,$$

as was pointed out in the body of the paper, allows the field pattern to be expressed as a diffraction limited pattern with side lobes occurring at specific angular regions.

It is instructive to investigate the generalized weighting of the form $\cos^{2n}\theta$. This periodic function can be expanded into components of the form $\cos 2n\theta$ using the expression

$$\cos^{2n}\theta \;=\; \frac{1}{2^{2n}}\left\{ \sum_{R=0}^{n-1} 2 \binom{2n}{k} \cos 2(n-k)\theta + \binom{2n}{n} \right\}$$

The resultant field pattern due to an annulus weighted by $\cos^{2n}\theta$ can be studied by first evaluating the amplitude of the field pattern due to $\cos 2m\theta$ which is given by

$$\mu(r,\phi,z) \;=\; (-1)^m J_m \frac{kRr}{z} \cos 2m\phi$$

As a simple illustration we evaluate the field pattern with a weighting of $\cos^4\theta$ where

$$\cos^4\theta \;=\; \frac{3}{8} + \frac{1}{2}\cos 2\theta + \frac{1}{8}\cos 4\theta$$

The amplitude of the resultant field pattern is given by

$$\mu(r,\phi,z) \;=\; \frac{3}{8} J_0(\cdot) - \frac{1}{2} J_2(\cdot)\cos 2\phi + \frac{1}{8} J_4(\cdot)\cos 4\phi$$

Using the previously given recursion relationship we have

$$\mu(r,\pi/2,z) \;=\; \frac{3}{4}\left[\frac{J_1(\cdot)}{(\cdot)} \;+\; \frac{J_3(\cdot)}{(\cdot)}\right]$$

$$=\; 3\,\frac{J_2(\cdot)}{(\cdot)^2}$$

We can generalize this result to the case of an annular ring weighted by $\cos^{2n}\theta$ where the field at $\phi = \pi/2$ is given by

$$\mu(r,\pi/2,z) \;=\; \frac{(2n)!}{n!\,2^n}\,\frac{J_n\!\left(\frac{kRr}{z}\right)}{\left(\frac{kRr}{z}\right)^n}$$

where $\mu(r,\pi/2,z)$ is equivalent to $\mu(y,z)$.

This pattern becomes smoother with increasing n until it becomes a constant as n approaches ∞. On the x axis, with increasing n, the pattern exhibits increasing side lobes until it becomes a cosine wave as n approaches ∞.

REFERENCES

1. J.C. Somer, "Electronic Sector Scanning with Ultrasonic Beams," Proc. of the First World Congress on Ultracound Diagnostics in Medicine, 1969, June 2-7, Vienna Austria.

2. J.C. Somer, "New Processing Techniques for Instantaneous Cross-Sectional Echo-Pictures and for Improving Angular Resolution by Smaller Beams," Proc. of the Fourth Congress of the Internat'l. Society for Ultrasonic Diagnosis in Opthalmology, 1971, May 6-9, Paris, France.

3. G. Kossoff, Meeting of the American Association for Ultrasound in Medicine, Seattle, Wash., October, 1974.

4. F.L. Thurstone and O.T. van Ramm, "A New Ultrasound Imaging Technique Employing Two-Dimensional Electronic Beam Steering," Acoustical Holography, Vol. 5, Plenum Press.

5. C.B. Burckhardt, P.A. Grandchamp and H. Hoffman, "Methods for Increasing the Lateral Resolution of B-Scan," Acoustical Holography, Vol. 5, Plenum Press.

6. J.W. Goodman, "Introduction to Fourier Optics," McGraw-Hill.

PROGRESS IN ANNULAR-ARRAY IMAGING

David Vilkomerson and Bernard Hurley

RCA Laboratories

Princeton, N. J. 08540

Abstract

An improved algorithm is presented for computing an image from the amplitude and phase on an annular array. The old algorithm is discussed first and the new one developed from it. Simulation data shows the new algorithm to be 40 times faster than the old for 128 x 128 element images.

I. Introduction

We present here an improved method of computing an image from the amplitude and phase information collected on an annulus. The theoretical foundations, historical development, and important restrictions on applicability of such annular-aperture imaging are treated in a paper given at the previous symposium[1].

We will first review the algorithm used for image formation previously and then develop the new algorithm. The emphasis here will be on the calculating procedure. The theory of the imaging will be only sketched in, and for details of the mathematical basis for the algorithm the reader will be referred to Reference 2. The improvement in efficiency is demonstrated for a simulated imaging situation.

Though the particular improvement we have made is specifically for the annular-array imaging system, we hope that the approach will be of interest and value to others doing ultrasonic imaging by computer holography.

II. The Previous Method

Annular-array imaging can be thought of as a series of computational steps whereby the intensity of the (assumed) point sources that are considered to form the object and image are calculated from the amplitude and phase on an annular-array. Referring to Fig.1, we calculate the intensity of a point p' at x', y', z in four steps.

Step 1: Aiming the Array

The annular array is "aimed" at the particular point p' by adjusting the phase of the received signal at each element of the array to compensate for the different distances from the point p' to the elements. (This gives the effect of the point source being on-axis at infinity.) The necessary adjustment to aim at point p' is multiplication of the signal by the conjugate of the signal that would be received from a point source at p', i.e.

$$h_{p'}(n) \triangleq S(n) \,\big|_{\text{point source at } p'} \qquad (1)$$

where $S(n)$ is the signal as a function of element number n.

Therefore to <u>aim</u> the array we form a complex amplitude on the array of

$$A_{p'}(n) = S(n)\, h_{p'}^{*}(n) \qquad (2)$$

If and only if $S(n) = h_{p'}(n)$, $A_{p'}(n) = 1$ for all n.

Step 2: Calculating the Zero-order Intensity

Summing the amplitudes $A_{p'}(n)$ and squaring the absolute magnitude gives the zero-order intensity:

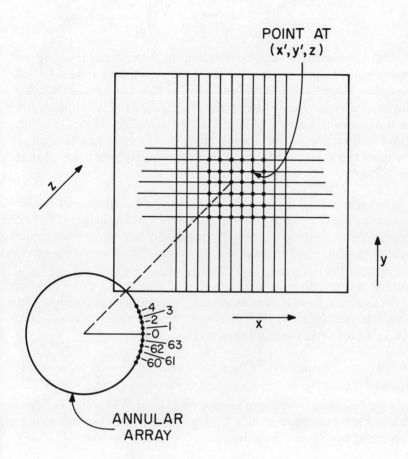

Fig. 1. The geometrical arrangement of the receiving
         annular array, shown here with 64 elements,
         and the object plane consisting of points on
         a cartesian grid.

$$I_{p'0} = \left| \sum_{n=0}^{N-1} A_{p'}(n) \right|^2 \qquad (3)$$

This operation can be viewed as measuring the energy of a signal that has passed through a matched filter[3]. A matched filter's impulse response is the conjugate of the signal for which it is matched; the "signal" considered here is that from the point p', and the integration of eq. 3 over the annulus gives the energy from the signal that has passed through the matched filter formed by multiplying the signal by $h_p^*(n)$.

The matched filter can be shown to be the ideal filter against Gaussian noise[3]. However, signals similar to the desired one can still come through. In the case we are considering, that means that a point source nearby to p' will contribute to the $I_{p'0}$ intensity calculated in eq.3. The contribution from such a nearby point will not be as great as from p' because h_p^* will not be exactly conjugate to such a signal; the resultant $A_{p'}(n)$ will vary around the annulus, but its integral will be non-zero.

It can be shown[4] that the contribution of point source of unit intensity not "aimed at" will vary as $J_0^2\left(\frac{2\pi p}{D}r\right)$, shown in Fig.2, where p is the radius of the annulus, D the distance between the object plane and the annulus, and r the distance from the point.

Unfortunately the J_0^2 response obtained from the annulus does not fall quickly enough to provide usable images in any but the simplest situations[5]. The $\Lambda_1^2 = (J_1(x)/x)^2$ response also shown in Fig. 2 is the response of a full disc aperture, such as a lens of the same diameter as the annulus, for the same imaging conditions. Note that a source of one-tenth unit intensity can be "resolved" at an argument of Z = 4 with the Λ_1^2 response while it would be completely obscured for the J_0^2 response. Moreover, the energy received in the sidelobes can surpass the energy in the central lobe.

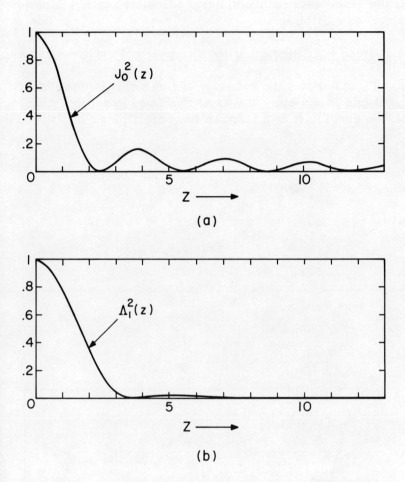

Fig.2. The normalized response of (a) an annulus and (b) a full disc of the same radius to a point source as a function of distance from the axis $(Z = \frac{\pi p}{D} \cdot r$ where p is the radius of the aperture, D the distance to the plane of the point and r the distance from the axis to the point).

This poor imaging response can be corrected, however; to the zero-order point intensity calculated in eq. 3 can be added correction terms.

Step. 3: Calculating the Correction Terms

Fig. 3 shows the behavior of the higher-order Bessel functions J_1^2, J_2^2 etc. It was shown in [1] that the signal on the annulus from a point in the object plane is

$$S = e^{i z \cos \varphi} \tag{4}$$

where $z = 2\pi p r/D$

so that the Bessel relation [6]

$$J_n(z) = \frac{i^{-n}}{2\pi} \int_0^{2\pi} e^{i z \cos \varphi} e^{in\varphi} d\varphi \tag{5}$$

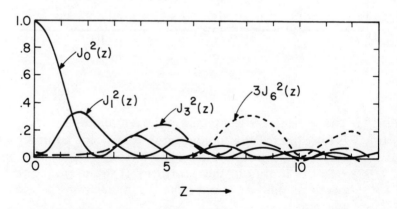

Fig.3. The Bessel functions of order 0, 1, 3, and 6 obtained from the annular array by adding progressive phase shifts to the aimed signal.

shows the if a progressive phase shift $n\varphi$ is added to the signal as a function of position around the annulus a higher-order Bessel function will result. These higher-order Bessel functions are <u>larger</u> away from the aimed-at point; this suggests subtracting the higher-order intensities $I_{p'\ell}$ calculated from

$$I_{p'\ell} = \left| \sum_{n=0}^{N-1} A_{p'}(n) \, e^{i\ell n \cdot \frac{2\pi}{N}} \right|^2 \qquad (6)$$

These correction terms can reduce the influence of sources away from the aimed-at point.

Step 4: Calculating the Correct Intensity

The intensity for p' can be calculated to the same precision as using a full disc aperture, i.e. the effective response curve can be made equal to the Λ_1^2-function, by

$$I_{p'} = \sum_{k=0}^{L} t_k \, I_{p'k} \qquad (7)$$

with the $I_{p'k}$'s of eq. 6. The values of t_k needed for the correction procedure, and the method of their calculation, are discussed in [1].

When these four steps for the point p' are finished, the next point on the object, i.e. at $x'+1$, y, z, is taken as p' and the four steps repeated to compute <u>its</u> intensity. In this way, a point at a time, the intensity of the whole object is calculated.

III. The New Method

The crucial difference between the old and new method is the change from rectangular to cylindrical coordinates. The new way of describing the object is shown in Fig. 4.

The transformation to cylindrical coordinates in itself brings a certain convenience to the calculation; however, it is the cylindrical symmetry and the periodicity of the signals on the annulus that allows the use of the discrete Fourier transform. It is the use of this transform that permits a real advantage in computational speed to be obtained.

The finite Fourier transform is excellently described in [2]. The definition is that a sequence of N complex numbers $A(n)$, $n = 0, 1, 2, \ldots N-1$ has finite Fourier transform of

$$\overline{A}\,(j) = \frac{1}{N}\sum_{n=0}^{N-1} A\,(n)\,e^{-i\,\frac{2\pi}{N}\,nj} \tag{8a}$$

and the original sequence of N numbers can be recovered from $\overline{A}\,(j)$ by the inverse

$$A\,(n) = \sum_{j=0}^{N-1} \overline{A}\,(j)\,e^{i\,\frac{2\pi}{N}\,nj} \tag{8b}$$

It should be noted that in eq. 8a and eq. 8b the sequences are multiplied by multiples of $e^{i\theta}$ where $\theta = \frac{2\pi}{N}$; this immediately establishes the periodicity of the transform as N, and allows the sequences to be thought of as occupying points on a circle, each point being separated in arc from the last by $2\pi/N$.

We also see that in this interpretation of the finite Fourier transform the arrangement of points in the image and the elements on the receiving annulus echo the arrangement of points in the sequences of the finite Fourier transform, and it will be seen that the calculations can be done in this domain easily. Using the finite Fourier transform in less ideal situations is difficult[7].

The finite Fourier transform is of interest to us because there exists an algorithm to calculate it at high speed, the

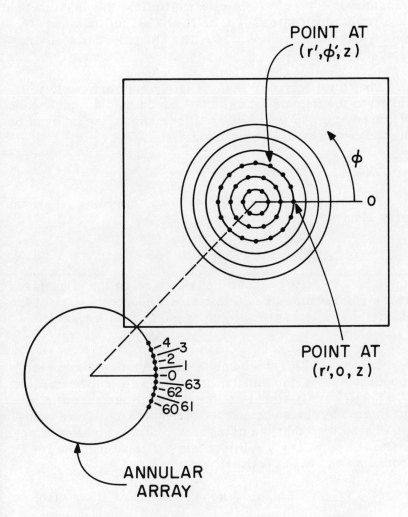

POINT AT
(r', ϕ', z)

ϕ

0

POINT AT
$(r', 0, z)$

4
3
2
1
0
63
62
60
61

ANNULAR
ARRAY

Fig.4. The receiving annulus and the object plane
consisting of points on a plane on a polar grid.

Fast Fourier Transform (FFT)[8] . This algorithm permits the computation of a sequence of N length to the finite Fourier transform in $\sim N \log_2 N$ complex multiplies; the usual method requires N^2 multiplies. For an N of 256, for instance, N $\log_2 N$ is 2048; N^2 is 65,536. The FFT produces a saving of over 30 to one.

The finite Fourier transform is not only extremely well suited to the form of our calculations but has a rapid means of computation as well. We will see how it can be used on the four steps of the new method.

Step 1': Aiming the Array

Fig. 4 shows a point at $r',0$ being "aimed at", i.e. the array elements are multiplied so that

$$A_{r'0} (n) = S(n) h^*_{r',0} (n) \qquad (9)$$

where $h_{r',0}(n)$ is the signal that would be on the elements from a point source at $r',0$ in correspondence to eq.1 of the old method. [Note that $h_{r'0}(n)$ is even, i.e. $h_{r'0}(n) = h_{r'0}(-n)$].

Looking at Fig. 4 we see that the rotational symmetry about the axis of the annular array means a point source at r',φ , will have a signal on the elements like that at $r',0$ but rotated by the same number of degrees φ . If we renumber the elements so that the new "zero" annular element is opposite the point at r',φ, we obtain the same impulse response as eq. 8. Specifically,

$$A_{r',\varphi}(n) = S(n) h^*_{r',\varphi} (n) \qquad (10)$$

$$= S(n) h^*_{r',0} (m-n)$$

where

$$m = \frac{\varphi}{2\pi} \cdot N$$

and

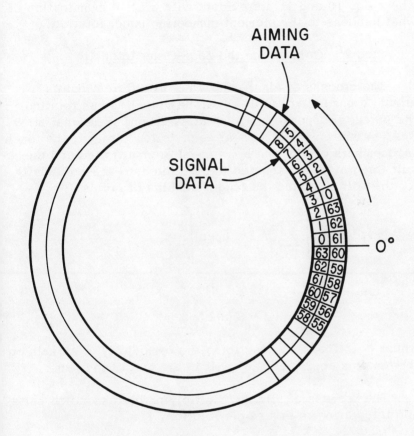

Fig.5. A schematic representation of the procedure for aiming the annular array: for each new point at the aimed-at radius, the aiming data is "clicked" around one position, and the touching elements of signal and aiming data multiplied and the result summed.

$$\text{If } (m-n) < 0, \, h(m-n) = h[N + (m-n)] \tag{11}$$

where eq. 10 and 11 are evident with a little examination of what happens to the element numbering under rotation.

Step 2': Calculating the Zero-Order Intensity

The process of aiming the array at different points around a particular radius r' can be visualized as rotating the impulse response $h_{r',0}(n)$ array around the signal array, $S(n)$, as shown in Fig. 5. At each "click" around, the touching members of $S(n)$ and $h_{r',0}^{*}(m-n)$ are multiplied and the result integrated and squared to give the zero-order intensity for the points around the object annulus of radius r':

$$I_{r'}(m) = \left| \sum_{n=0}^{N-1} A_{r',0}(m-n) \right|^2 \tag{12}$$

where

$$A_{r',0}(n) = S(n) \, h_{r',0}^{*}(n) \tag{13}$$

where eq. 10 and 11 describing the periodicity of signals on the annulus apply, as they will in the rest of the paper.

The process of sliding, multiplying corresponding parts, and integrating is that of convolution, i.e.

$$\sum_{n=0}^{N-1} A_{r'0}(m-n) = \sum_{n=0}^{N-1} S(n) \, h_{r',0}^{*}(m-n)$$

$$\approx \int_{0}^{T} S(\tau) h^{*}(t-\tau) d\tau \approx S(t) * h^{*}(t) \tag{14}$$

We note that the output function is periodic with period N as are the two functions convolved together.

The exact analogue of convolution for discrete arrays is

$$\frac{1}{N} \sum_{n=0}^{N-1} A_1(n) A_2(m-n)$$

and it can be shown[9] that

$$\frac{1}{N} \sum_{n=0}^{N-1} A_1(n) A_2(m-n) \longleftrightarrow \overline{A}_1(j)\overline{A}_2(j)$$

just as the well-known result in the continuous Fourier transform that convolution in the time domain is equivalent to multiplication in the Fourier domain.

With this tool we can rewrite eq. 12.

$$I_{r'\,0}(m) = \left| \sum_{n=0}^{N-1} A_{r',0}(m-n) \right|^2 ; \qquad (12)$$

because

$$\sum_{n=0}^{N-1} A_{r',0}(m-n) = \sum_{n=0}^{N-1} S(n) h^*_{r',0}(m-n) \qquad (15)$$

$$I_{r'0}(m) = \left| G_{r'\,0}(m) \right|^2$$

where

$$G_{r'0}(m) \longleftrightarrow G_{r'0}(j) = \overline{S}(j)\,\overline{h}^*_{r',0}(j) \qquad (16)$$

This says that the N multiplications $\overline{S}(j)\,\overline{h}^*_{r',0}(j)$ and a finite Fourier transform produce a set of N zero-order intensity values corresponding to a N-point annulus in the image.

In the old method, eq. (3) for obtaining the zero-order intensity of a point

$$I_{p'0} = \left| \sum_{n=0}^{N-1} S(n)h_{p'}^*(n) \right|^2$$

required the same N complex multiplications. We see that once in the Fourier domain we obtain an annulus of N image points for the same number of multiplications needed for one point in the old method. It is this advantage, N to 1, that more than makes up for the calculation of the finite Fourier transform when the FFT is used.

Step 3': Calculating the Correction Terms

As discussed, the correction terms are generated by adding a progressive phase shift to the "aimed-at-signal" $A(n)$, i.e. for a single point p'[10].

$$I_{p'\ell} = \left| \sum_{n=0}^{N-1} S(n)h_{p'}^*(n) e^{i \frac{2\pi}{N} \cdot \ell n} \right|^2 \tag{6}$$

What we wish is a similar form using our annular form:

$$I_{p'\ell}(m) = \left| \sum_{n=0}^{N-1} S(n)h_{r',0}^*(m-n) e^{i \frac{2\pi}{N} \ell n} \right|^2 \tag{17}$$

We have seen in Step 2' that for $\ell = 0$ this can be re-cast as a finite Fourier transform. Examination of the finite Fourier transform shows that[11]

$$e^{i \frac{2\pi}{N} mn} F(n) \longleftrightarrow \bar{F}(j - m) \tag{18}$$

which is the analog of the familiar Fourier transform rule that

$$e^{i\omega_o t} f(t) \longleftrightarrow F(\omega - \omega_o) \tag{19}$$

Therefore eq. (17) can be rewritten in terms of the finite Fourier transform as

$$I_{r'\ell}(n) = \left| G_{r'\ell}(n) \right|^2$$

where

$$G_{r'\ell}(n) \longleftrightarrow \overline{G}_{r'\ell}(j) = \overline{S}(j)\overline{h}_{r'}^{*}(j-\ell) \qquad (20)$$

Again comparison of the number of multiplications needed by the old method and the new, eq.6 and 20 respectively, shows that N multiplications provide correction terms for <u>one</u> image point in the old method and <u>N</u> annular image points in the new method. As the Fourier transforms are already computed for calculation of the zero-order intensity, this is pure gain. The greater the number of correction terms needed, the greater the saving in computation. (In [1] it is shown that the number of correction terms and the desired field of view are proportional.)

Step 4′: Calculation of Correct Intensity

This step in the new method does not differ from step 4 of the old method: sum the properly weighted intensities to form the correct intensity.

$$I_{r'}(n) = \sum_{k=0}^{L} t_k I_{r'k}(n) \qquad (21)$$

When the new method is used in place of the old for images of reasonable size, significant savings in computation time are obtained. Because the saving in computation depends upon N, the number of points in the annulus considered, the larger the image the greater the saving. In fact for very small images, the additional computer time needed to get the FFT procedure going offsets any gain from using the new method. The relation between image size and time-saving has not yet been thoroughly explored. However, some idea of the advantage of the new method may be seen from a simulated image calculated using a Univac 70/46 general-purpose computer:

2048 point image - old method = 1 hr.
"   - new method = 90 seconds.

IV. Discussion

The new method discussed fits into the general philosophy of employing an annular-array: place more emphasis upon computing elements, and less on receiving elements because, unlike receiving elements, the computing elements are readily available. The new method depends heavily on the finite Fourier transform and the FFT. There has been a tremendous interest in both the software and hardware in this field [12] of which we intend to take advantage.

Image-restoration procedures must be analyzed in terms of their sensitivity to noise [13]. We are doing image-restoration in that we are correcting the poor image of the annular aperture. We have tested the sensitivity of the annular-array imaging system by simulating signals with varying degrees of noise and examining the computed image. We have found that to within a factor of three the noise has a direct and linear effect on image signal/noise ratio.

The sensitivity of the annular array is reduced from that of a fully populated array of the same diameter. This is because of the "array gain" of N in signal/noise ratio between the S/N ratio at each element of the array and of the total array. This arises, as shown in Fig. 6, from the (assumed) coherent detection method employed: the signal vectors from each element sum to a total vector N times as long as each individual signal while the noise vectors from each element are like random steps in a random walk so their total length is $\sqrt{N}$ as long as each noise signal. The signal/noise

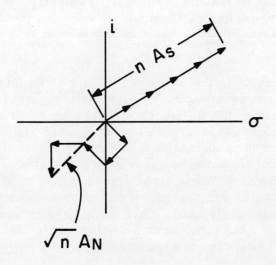

Fig.6. For a N-element array using coherent detection the signal vectors of individual amplitude A_s will sum in a straight line of length NA_s; the noise vectors of amplitude A_n will sum to a vector of length $\sqrt{N}\,A_n$.

power-ratio of the array is the ratio of vector lengths squared, or N times the S/N ratio of an element.

If a square array of 128 x 128 elements is replaced by an annular array of 256 elements, the array gain goes down by 64, and even if there is no loss in signal/noise ratio due to the image-restoring, the annular array will require 64 times the signal intensity to obtain the same image as the full array.

For imaging in the human body, this 18 dB loss in sensitivity is not as damaging as might be thought. For the 20 cm propagation path to the center of the body and back, .25 mm wavelength ultrasound will be attenuated 160 dB[14]; if the wavelength is increased to .28 mm, because the absorption is proportional to frequency, the attenuation drops to 142 dB. The same image signal/noise ratio can be attained with the annular array as with a full array if the slightly larger wavelength is used. This leads to lower resolution, by about 15%, but this seems a small price to pay for the simplified fabrication of the annular array.

V. Conclusions

By using the finite Fourier transform and the FFT, a reduction in the computation time for forming an image from an annular array can be achieved. This reduction is between 10 and 100 for images of the size of interest for diagnostic ultrasonic imaging.

The authors wish to thank Reuben S. Mezrich for valuable discussions on these topics.

References

1. Vilkomerson, D., "Acoustic Imaging with Thin Annular
   Apertures", Acoustical Holography Vol. 5, P. Green,
   Ed., Plenum Press, N.Y. (1974).

2. Cooley, J. W., Lewis, P., Welch, P., "The Finite
   Fourier Transform", IEEE AU-17, 77 (1969).

3. See, for example, Mason, S., Zimmerman, H.,
   Section 7.11 in "Electronic Circuits, Signals, and
   Systems", Wiley & Sons, Inc., New York (1960).

4. Vilkomerson, D., Op. cit., page 287.

5. McLean, D. J., "The Improvement of Images Obtained
   with Annular Apertures", Proc. Roy Soc. A 263,543
   (1961).

6. Jahnke, E. and Ende, F., "Tables of Functions",
   Fourth Ed., page 149, Dover Publications, N.Y. (1945).

7. Bergland, G. D., "A Guided Tour of the Fast Fourier
   Transform", IEEE Spectrum, July (1969).

8. Cooley, J. W. and Tukey, J. W., "An Algorithm for the
   Machine Calculation of Complex Fourier Series",
   Mathematics of Computation 19, 297 (1965).

9. Cooley, J. W., Lewis, P., and Welch, P., op. cit.,
   Section IV.

10. We note that we can recognize eq. 6 and 16 as measur-
    ing the power spectrum of A(n), as the multiplication by
    the phase shift is equivalent to forming the finite
    Fourier transform as seen from comparison with eq. 8a.

11. Cooley, J. W., Lewis, P., and Welch, P., op. cit.,
    Section III, Theorem 4.

12. See, for instance, Part 2 of "Digital Signal Processing", Rabiner, L. and Rader, C., Eds., IEEE Press, N. Y. (1972).

13. Barnes, C. W., "Object Restoration in a Diffraction-Limited Imaging System", J. Opt. Soc. Am. 56, 575, (1966).

14. Vilkomerson, D., "Analysis of Various Ultrasonic Holographic Imaging Methods for Medical Diagnosis", Acoustical Holography Vol. 4, G. Wade, Ed., Plenum Press, N. Y. (1972).

SYSTEM FOR VISUALIZING AND MEASURING ULTRASONIC WAVEFRONTS*

R.S. Mezrich, K.F. Etzold, D.H.R. Vilkomerson

RCA Laboratories

Princeton, N. J. 08540

Abstract

A system for the measurement and visualization of ultra-sonic wavefronts is described that features high sensitivity and acoustic wavelength limited resolution over apertures up to 15 cm in diameter and at frequencies up to 10 MHz. It operates by interferometrically detecting the motion of a thin, acoustically transparent metallized pellicle as the ultrasonic wave passes through it. The basic arrangement is that of the Michelson interferometer with the addition of an open loop method to stabilize the response and a deflection system in one leg of the interferometer to scan the pellicle.

I. Introduction

We report on an instrument for the measurement and visualization of ultrasonic waves that has application in dosimetry, in the study of ultrasonic interactions with biologic tissue, in the calibration of ultrasonic transducers and the visualization of their radiation patterns, and as an aid to the design of more sophisticated ultrasonic imaging devices. Although primarily designed as a laboratory instrument for the study of ultrasonic waves[1] it has sufficient sensitivity to be

---

* Some of this material is taken from an article in the December 1974 issue of RCA REVIEW by permission of the editor.

useful in some clinical applications, especially the observation of such external organs as breasts and limbs.

As we shall discuss more fully below, its main features are accuracy, high resolution, a large acoustic aperture, a very large angular response, a broad frequency response, a wide dynamic range, and a simple, cheap arrangement.

Presently the system is being used for the examination of tissue sections for pathology studies, for tissue visualization, and for basic measurements of acoustic wave propagation through and reflection from biologic tissue. It is also being used by us and, with a unit we recently delivered, by the Bureau of Radiological Health (BRH) to study and measure transducer characteristics and the radiation patterns they produce.

In the following we shall illustrate several of these applications and give examples of the system operation. We first describe the basic principles of operation of the system, which we call the ultrasonovision.

II.   Basic Principle of Operation

The basic arrangement of the ultrasonovision system is that of the optical Michelson interferometer, schematically illustrated in Fig. 1. The key elements of the system are a light source (laser), a beam-splitter, a reference mirror which is external to the sound field, and a thin flexible mirror through which the acoustic wave passes. In our experiments the flexible mirror is a thin (~ 6 micron) metallized plastic film (pellicle) which is in the fluid through which the acoustic wave propagates.

The pellicle is so thin that it is essentially transparent to the ultrasonic wave passing through it for frequencies as high as 10 MHz and for angles of incidence from 0° to 40°. Being transparent means that the pellicle motion, or displacement, is almost exactly equal to the displacement amplitude of the acoustic wave.

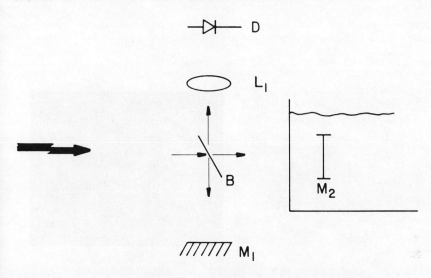

Fig. 1. Basic arrangement of system. M_2 is pellicle in water tank, M_1 is reference mirror.

The essential idea of the system is to accurately meas-ure the displacement amplitude of the acoustic wave by in-terferometrically measuring the motion of the pellicle through which the wave passes. By scanning the laser beam over the pellicle the displacement amplitude at each point of the ultra-sonic field is measured; an image of the acoustic field is generated by processing the signal derived from the inter-ferometer as the laser beam scans to brightness-modulate a synchronously scanned cathode ray tube.

The image size of the system will be limited by the pel-licle aperture (up to 150 mm diameter) and the ultimate lateral resolution by the size of the laser beam. (The practical limit is the acoustic wavelength.) The acoustical numerical aper-ture, the sensitivity, and the dynamic range are determined by the pellicle response (discussed later) and the character-istics of the optical interferometer.

Some results of imaging and measurement of acoustic wavefronts attained with the ultrasonovision are shown in Figs. 2-5.

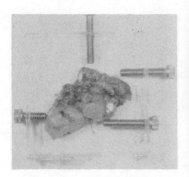 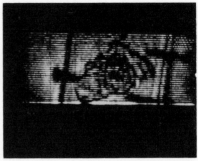

Fig. 2. Optical Image (left) and Acoustic Image (right) of
breast tissue with malignant mass. Screw tips inserted to
touch mass. Dark areas in acoustic image are malignant
structures.

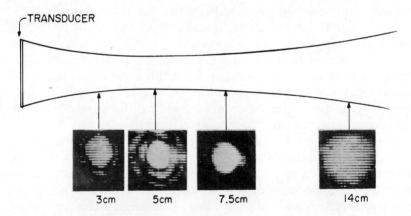

Fig. 3. Radiation pattern from ultrasonic transducer, at
different distances from face of transducer.

Fig. 2 shows an optical and acoustic image, to the same scale, of excised breast tissue with a malignant tumor. The acoustic frequency was 3.0 MHz. The tumor can be clearly seen in both images, black in the acoustic image, light in the optic image. As will be discussed below, the ultrasonic transmittance through each part of the tissue can be measured at the same time that the tissue is visualized. It has been found by such measurements that some types of malignant tissue have a lower acoustic transmittance than benign structures, which in turn have a lower transmittance than normal tissue.

The image of Fig. 2 was obtained with the aid of an acoustic polystyrene lens which imaged the object onto the pellicle.

Fig. 3 shows the field pattern, with no acoustic lens used, of a "focused" acoustic transducer operating at 2.25 MHz. The nominal focal length of the transducer was 7.5 cm. The figures show the field pattern as the transducer was moved to different distances from the pellicle, and it is seen that the actual focus is at 5 cm. The Bessel rings (or side lobes) are clearly seen at focus.

As an example of the utility of ultrasonovision in examining and checking transducers, Fig. 4 shows images from a clinical diagnostic transducer that was still in active use at the time of the experiments, although the user had noticed "something was wrong".

From the acoustic image of the transducer surface it can be seen that sound is only being emitted from some regions around the edges. The pattern in the far field is uniform (as could be predicted).

The cause for the improper sound emission was that the protective coating over the transducer had delaminated – which was not evident by just looking at the surface.

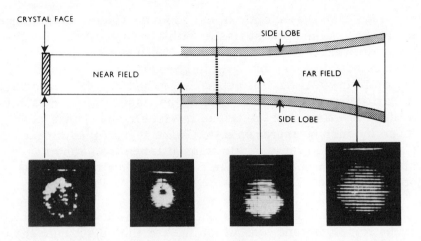

Fig. 4. Acoustic image of transducer surface and radiation pattern of damaged, but still used, diagnostic transducer. Bright spots on surface (left) indicate the only radiating regions of transducer.

As all transducers can be expected to change their characteristics due to the rigors of constant clinical use, sometimes drastically without the user even being aware of the changes, it is clear that the ultrasonovision should prove useful in performing rapid and simple periodic checks to maintain the performance of diagnostic tissues.

Fig. 5 is a further example of acoustic imaging of soft tissue, in this case an adult hand. Due to limitation in the ultrasonic beam that insonified the hand, the complete picture is a collage of small sections. The lower picture is a detail of the region of the palm below the index finger, showing what is probably a bifurcation in the blood vessel leading to the first two fingers.

The important components of the ultrasonovision are the pellicle, the optical arrangement of the interferometer, and the electronics system (which include the means for detection

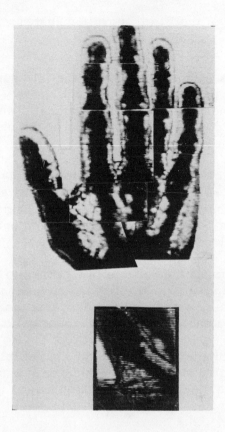

Fig. 5. Acoustic image of adult hand. Lower picture shows detail of region of palm near index finger, with a bifurcation of blood vessel visible.

and display of the acoustic image). Important considerations are the sensitivity (or signal to noise), the need for (and the means used to obtain) stability in the interferometer, and the effects of component characteristics used in a practical system.

In the following section we give a detailed analysis of these and other matters.

III. Analysis of the Ultrasonovision System

A. Pellicle

The basis for the operation of the ultrasonovision is that the motion of pellicle, at every point, is very nearly equal to the displacement amplitude of the acoustic wave passing through it. By measuring the motion of the pellicle the acoustic wave is measured.

The pellicle is a thin (~6μ) metallized plastic film that is suspended in the water through which the acoustic wave passes. At the present time pellicles as large as 6 in. in diameter have been used and the measured optical flatness has been of the order of 1 wave/in.

The motion of the pellicle in response to an acoustic wave has been calculated following an analysis of Brekhovski.[2] The result is that the angular response is flat over acoustic beam angles of ±40° and to acoustic frequencies at least as high as 10 MHz. The ratio of the motion of the pellicle (perpendicular to the plane of the pellicle) on the optically illuminated side to the displacement amplitude of the acoustic wave (again in the direction normal to the pellicle) is .99.

These results were verified experimentally by measuring the waves transmitted through and reflected from the pellicle. As an example, Fig. 6 shows a graph of the relative sensitivity as a function of the angle of incidence (θ) of the acoustic beam. Also plotted is the theoretically expected response. Note that in both cases the response falls as cos (θ), since the normal (perpendicular) component of the incident wave is proportional to cos (θ). The agreement between the expected and experimental results is good over the angular range 0° to ± 40°.

B. Basic Interferometer

The operation of the interferometer (Fig. 1) may be understood by the following analysis. Let an acoustic wave of

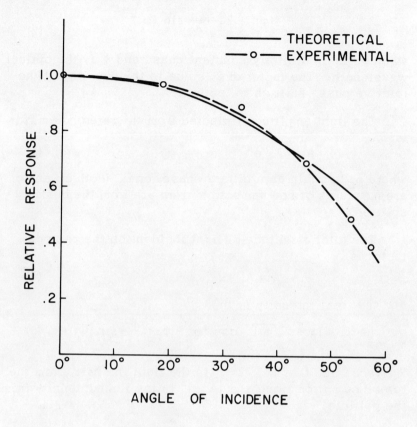

Fig. 6. Pellicle response, theoretical and experimental, as a function of angle of incidence of acoustic beam.

displacement amplitude Δ pass through the pellicle. The deviation of the water, and hence the pellicle at that point is

$$\mathcal{E} = \Delta \sin \omega t \qquad (1)$$

When a light beam is reflected by the pellicle its phase will be changed in proportion to the pellicle motion and the amplitude of the reflected optical wave becomes

$$A_p = A_{po} e^{j(\varphi p + 2\Delta \frac{2\pi}{\lambda} \sin(\omega t))} \qquad (2)$$

when φ_p is an arbitrary constant phase and λ is the optical wavelength. The factor of 2 is due to the doubling of the relative phase shift on reflection.*

The light amplitude reflected from the reference mirror is

$$A_r = A_{ro} e^{j\varphi r} \qquad (3)$$

where φ_r is again an arbitrary phase term. (Both φ_p and φ_r are measures of how far both mirrors are from the beam splitter.)

The total amplitude of light incident on the photodetector is

$$A = A_r + A_p \qquad (4)$$

and the total intensity is then

$$I = I_r + I_p + 2\sqrt{I_r I_p} \cos \left(\varphi_p - \varphi_r + \frac{4\pi\Delta}{\lambda}\sin(wt)\right) \qquad (5)$$

We recognize that $|A_{ro}|^2 = I_r$ is the light intensity from the reference mirror alone and $|A_{po}|^2$ is the light intensity from the pellicle. [†]

---

* There is an additional phase shift, due to the acousto-optic interaction of the light with the ultrasonic wave passing through the water after it passes the pellicle, which should be added to the phase term in eq. 3. To avoid complicating the analysis we do not include this effect here, but discuss it fully in Appendix I.

† It is assumed that the two optical waves are coherent and overlap completely - conditions that can be readily met. If this condition is not met a factor $K(|K| \leq 1)$ multiplies the expression $\sqrt{I_r I_p} \cos [\varphi_p - \varphi_r + (4\pi\Delta/\lambda) \sin(\omega_s t)]$.

The photodiode generates a current (i_s) proportional to the incident light intensity $i_s = \eta I$ where η, the quantum efficiency, is of the order of .3 amps/watt (for red light incident on a silicon photodiode). A high-pass filter is used to suppress the low frequency terms. The signal after the filter is

$$i_s = 2\eta\sqrt{I_r I_p}\, \cos\left(\varphi_p - \varphi_r + 4\pi\,\frac{\Delta}{\lambda}\sin(\omega t)\right) \tag{6}$$

Assume that $\varphi_p - \varphi_r = 90°$; the general case where the relative phase shift can have arbitrary values will be considered below. Then

$$i_s = 2\eta\sqrt{I_r I_p}\, \sin\left(4\pi\,\frac{\Delta}{\lambda}\sin(\omega t)\right) \tag{7}$$

If $\Delta/\lambda \ll 1$, which, for the frequency range $.5 - 10$ MHz holds for acoustic intensities less than 1 watt/cm^2, the current, to a good approximation becomes

$$i_s = 8\pi\eta\,\frac{\Delta}{\lambda}\sqrt{I_r I_p}\, \sin\omega t \tag{8}$$

This signal is directly proportional to the displacement amplitude of the acoustic wave.

Once the light intensities incident on the photodiode from the reference mirror and pellicle are measured, which can be done with great precision in practice, the electrical signal is an accurate measure of the displacement amplitude, and hence wave intensity, at all frequencies and all intensities. The system can be accurately calibrated.

As a check, the system was calibrated against a U. S. Navy-Underwater Sound Reference Laboratory transducer at a frequency of 1.0 MHz. The quantity $\eta\sqrt{I_1 I_2}$ was found by measuring the maximum and minimum values of the interference ($= \eta\left[I_r + I_p \pm 2\sqrt{I_r I_p}\right]$); this includes the overlap factor K of the previous footnote. After taking into account the effect due to the acousto-optic interaction (see Appendix I) the agreement between the system and the Navy transducer was within 5%.

C. Sensitivity and Dynamic Range

The sensitivity and dynamic range are limited by noise. The major components of noise are thermal noise, generated in the load resistor and shot noise. The shot noise is due to the steady component of the light incident on the photodiode ($I_r + I_p$ in eq. 5), and is dominant for light levels at the photodiode greater than 0.1 mw. We assume shot noise to be dominant in the following.

The shot noise current is

$$i_{ns}^2 = 2 \, e\eta \, I \, F \tag{7}$$

where $e = 1.6 \times 10^{-19}$ is the electronic charge, η is the photodiode quantum efficiency, I is the incident light ($\approx I_p + I_r$) and F the system bandwidth.

The expression for the signal to noise becomes

$$\frac{i_s^2}{i_n^2} = \eta \, \frac{(8\pi)^2}{4eF} \, \frac{\Delta^2}{\lambda^2} \, \frac{I_r \, I_p}{I_r + I_p} \tag{10}$$

Since the acoustic intensity is proportional to Δ^2, the minimum detectable acoustic intensity decreases linearly with the available light.

Experimentally, with $I_r \approx 5 I_p = 2.5$ mw, $\eta = 1/3$, $\lambda = 6328$Å and $F = 10$ MHz a minimum displacement amplitude of .1Å was measured. The bandwidth used in this experiment was far greater than necessary, since, as we will see below, the effective bandwidth of the entire system is only ~50 Khz. This would imply a true minimum detectable displacement, using the parameters above, of .007Å.

The ultimate limit on the sensitivity appears to be given by the maximum dissipation allowed in the photodiode, which for commercially available silicon photodiodes, is about one watt. By eq. 10 we can estimate the ultimate sensitivity of the ultrasonovision to be of the order of 10^{-10} watts/cm^2 at 1.5 MHz with a 1 MHz bandwidth. While this is not as sensitive as can be achieved with piezoelectric detectors, the sensitivity is adequate for in vitro studies of biologic tissue and for the in vivo examination of the appendages and external organs such as arms, legs, and breasts.[3]

The maximum level of acoustic intensity which can be measured accurately is determined by eq. 7. At high values of intensity the condition $\Delta/\lambda << 1$ no longer obtains and nonlinear effects, notably harmonics, appear. Eq. 7 may be expanded in a Fourier-Bessel series expansion as[4]

$$i_s = 2\eta\sqrt{I_r I_p} \left\{ 2\sum J_2 n+1 \left(\frac{4\pi\Delta}{\lambda} \right) \sin((2n+1)\omega t) \right\} \qquad (11)$$

For small values of the argument $J_1(x) \approx \frac{x}{2}$, which gives the expression of eq. 8. As the argument increases the approximation becomes less valid. We may arbitrarily set the upper limit of validity at $x = 1/4$, where the approximation of eq. 8 is valid to better than 1%. Then

$$\Delta \leq \frac{\lambda}{16\pi} \approx 125 \text{ Å} \qquad (12)$$

At a frequency of 1.5×10^6 hz, this corresponds to an acoustic intensity of about one w/cm^2. Above the limit the ultrasonovision system does not fail to operate, but simply becomes less linear in its response.

The range of accurate measurement, then, extends from 10^{-10} w/cm^2 to 1 w/cm^2.

D. Scanning

In the previous discussion we described the measurement of the acoustic intensity at a single point of the pellicle illuminated by the light beam. The displacement amplitude, and hence acoustic intensity at every point of the pellicle may be measured by scanning the light beam over the pellicle. A method that maintains the Michelson arrangement and permits scanning the light beam over a large aperture with a minimum of large elements is achieved when the deflection system is put in one arm of the interferometer and the optics designed so that the light beam, at every poisition, is normally incident on the pellicle. The light beam is reflected from the pellicle and passes back through the deflection system to the beam splitter, where it recombines with the beam from the reference mirror. The advantage of this arrangement is that the beam splitter and reference mirror need be no larger than the light beam incident on the interferometer - on the order of one mm - while the scanned aperture may be much larger.

A practical arrangement for the deflector, which uses mirror galvanometers for the deflection elements, is shown in Fig. 7. Lenses L_2 and L_3 form a telescope that images the first galvanometer (G_1), used for horizontal deflection, into the second galvanometer (G_2) which deflects the beam

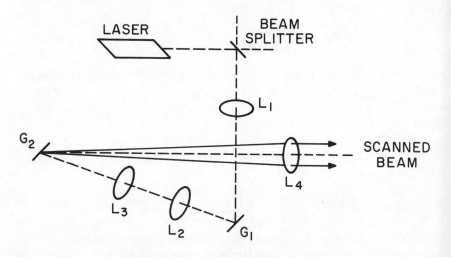

Fig. 7. Optical arrangement to provide uniform laser beam scanning over the pellicle. Key elements are vertical galvanometer (G_1), horizontal galvanometer (G_2) telescope lenses (L_2, L_3), and relay lenses (L_1, L_4).

vertically. Since galvanometer G_2 is at the front focal plane of lens L_4, the beams exiting from L_4 are parallel to the axis of L_4. As the galvanometers rotate the exit beam is displaced proportionately (by the factor $d = f_4 \theta$, where $\theta/2$ is the mirror rotation) and at every position will be normally incident on the pellicle. Lenses L_1, L_2, L_3 together form an equivalent lens, of focal length

$$f = \frac{f_1 \cdot f_3}{f_2} \tag{13}$$

with back focal plane at galvanometer G_2. The size of the beam at the pellicle is proportional to the ratio f_4 to this reflective focal length. With $f_1 = f_4$ and $f_2 = f_3$ the size of beam at the pellicle is approximately equal to the beam incident on L_1. The size of the beam could be reduced with appropriate changes in the focal lengths.

With this arrangement, and with the pellicle perpendicular to the axis of lens L_4, the beam reflected back through the system will coincide exactly in position, size, and angle with the incident beam. This is true regardless of the position of the scanned beam on the pellicle.

E.   Stability ("Wiggler")

In eq. 7 it was assumed that the relative phase difference, $\varphi_p - \varphi_r$, was 90°. The reason for this may be qualitatively seen in Fig. 8 which is a plot of eq. 6 (with $\Delta \approx 0$) as a function of the relative phase difference. (We define $\varphi_{pr} \triangleq \varphi_p - \varphi_r$.) At $\varphi_{pr} = 90^\circ$ the slope of the curve – the ratio of optical intensity change to phase difference change is maximum. At this point the response of the system, which is proportional to the slope, is also maximum. At other values of the relative phase the slope, and hence the response, is smaller and, for example, at $\varphi_{pr} = 0$ and $\varphi_{pr} = 180^\circ$ it is zero.

Practical arrangements of interferometers are susceptible to mechanical and thermal drift with the result that it is

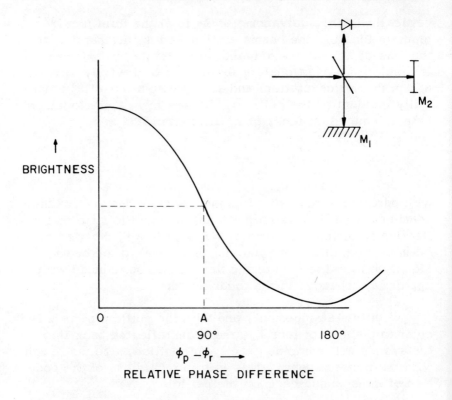

Fig. 8.   Interferometer response as a function of relative phase between pellicle and reference beam mirror.

difficult to maintain the condition $\varphi_{pr} = 90°$. Further, in large aperture systems such as this, it is difficult to achieve sufficient flatness, less than 1/8 wave, over the entire aperture. The result of this would be to cause such large variations in the output as to make a practical system useless. While feedback methods can be used to avoid these problems, we employ a simpler open loop method.

The method is to purposely vary the relative phase difference over at least 180° by vibrating, or "wiggling" the reference mirror. (Other methods of varying the phase are possible but wiggling the reference mirror is simplest.) The effect is that at least once per cycle of the wiggler (by wiggler we mean the vibrating mirror) the operating point of the system will pass through the point $\varphi_{pr} = 90^{\circ}$, despite large mechanical or thermal variations and to a great extent, regardless of lack of exact flatness of the pellicle.

Since the system response is maximum at $\varphi_{pr} = 90^{\circ}$, the peak value of the measured signal, will give the same value as if the system were always adjusted to $\varphi_{pr} = 90^{\circ}$.

As long as the net phase change in the system, including effects of external disturbances and the wiggler motion, is greater than 1/4 wave over the period of the wiggler, the system will be stable. It will show none of the effects that plague ordinary interferometers, such as drift due to thermal or mechanical changes or air currents. It is stable even in the presence of violent mechanical shock due to shaking of the table or due to water waves set up when samples are placed in the water tank.

There are several constraints on the range of the allowable wiggler frequencies.

The lower limit of the wiggler frequency is determined by the scan rate of the deflector: the wiggler frequency must be sufficiently high to allow at least one cycle of the wiggler during each spot (or resolution element) of the scan.

If the frame time of the system is T and the number of elements is N^2 then the wiggler frequency is constrained to

$$f_{\omega} > \frac{N^2}{T} \qquad (14)$$

As an example, with N = 100, T = .5 sec (i.e. 2 frames/sec). The lower limit of the wiggler frequency is 20 KHz.

There are three constraints on the upper limit of the wiggler frequency.

The first is due to the nonlinear relationship between phase changes in the reference leg and detected intensity (see eq. 6). This nonlinearity can cause higher harmonics of the wiggler phase change to appear in the signal bands, which can be comparable to the signal if the ratio of the acoustic signal frequency to the wiggler frequency is low (< 10). The second constraint is due to the use of a peak detector. For the response from the peak detector to match the actual peak value the relative phase of the interferometer should pass through 90° at the same moment that the acoustic signal is at its maximum value. The probability of this is maximized when the ratio of acoustic frequency to wiggler frequency is high. With ratios of the order of 15 the error due to this is ~1%. The final constraint is imposed by the use of pulsed acoustic signals. The period of the wiggler must be less than the duration of the acoustic pulse.

In practice wiggler frequencies between 25 KHz and 80 KHz are used.

While there are a number of ways to achieve the necessary phase modulation, a particularly simple method is to mount the reference mirror on a piezoelectric disc which is chosen to be resonant at the desired wiggler frequency. The peak displacement of the disc must be $\lambda/4$ or approximately 1500Å. If the thickness of the disc is greater than .015 cm this motion is well within the elastic limits of piezoelectric materials.[5]

F.   Complete System

The complete system, incorporating the deflection optics wiggler, signal display and visual display is shown in Fig. 9.

The pellicle is mounted in a large water tank, the front wall of which is made of reasonably good optical quality glass. Insonified objects may be acoustically imaged onto the pellicle, as shown on the figure, by acoustic lenses made of styrene.[6] For the determination of radiation patterns, for example from transducers, the lenses are removed.

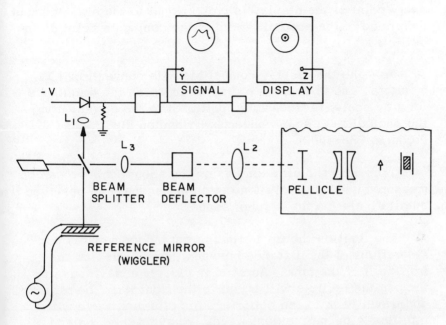

Fig. 9. Schematic of complete system, including oscilloscope and wiggler. Also shown is acoustic lens arrangement to image object onto pellicle.

The optical signal is detected by a silicon photodiode,
converted to a proportional current which is then filtered,
amplified and displayed on an oscilloscope (the signal
oscilloscope). Once the system has been calibrated, which
means that the light from the reference mirror and pellicle is
measured and the amplifier gain known, the voltage on the
oscilloscope is an accurate representation of the acoustic
displacement amplitude at every point on the pellicle. The
galvanometers may be stopped to accurately measure the intensity at a particular point or may be scanned over any desired portion of the pellicle.

The acoustic field is visualized by using the signal to
brightness modulate a second CRT. This CRT is synchronously scanned with the galvanometers so the brightness of

every point of the display is proportional to the displacement amplitude of the acoustic wave at the conjugate point on the pellicle.

To make the display compatible with conventional TV monitors, and to achieve a display in which the brightness (or gray scale) is linearly proportional to the current from the photodiode a scan converter (Princeton Electronic Products PEP-400R) is used.

Thus with the ultrasonovision the acoustic wave can be measured quantitatively (on the signal scope) and observed qualitatively (on the display scope).

The limit on the frame time is set by geometrical considerations of the distance between the pellicle and the front wall of the tank. As seen in Fig. 9, a wave passing through the pellicle will travel to the front wall, be reflected and travel back to the pellicle. The returning wave will interfere with the incident wave. To avoid this, pulsed acoustic waves are used and the duty cycle adjusted so the reflected wave and the following acoustic wave do not pass through the pellicle at the same time. This duty cycle is the main consideration in determining the frame rate of the ultrasonovision system.

The electronics for the ultrasonovision are schematically shown in Fig. 10.

The detection electronics has a receive gate synchronized to the pulsed acoustic wave so the signal detected and processed is due to the transmitted wave and not the reflected wave, or other reverberating waves in the tank.

The repetition rate of the pulses depends on the spacing between the pellicle and the tank wall while the duration of each pulse is determined by the scan rate. Typical values used are: a scan rate of 50 μ sec/mm, a pellicle to wall spacing of 60 mm and a duty cycle of $\frac{1}{7}$ with a period between pulses of 150 μ sec.

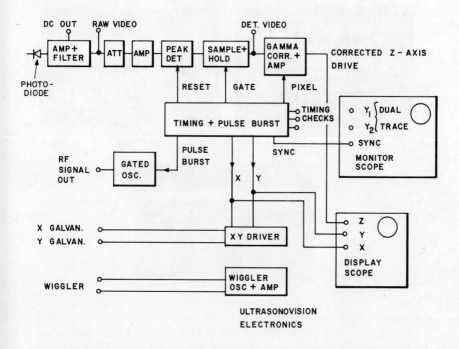

Fig. 10. Block diagram of system electronics.

The gate width, or sample time is made approximately equal to the acoustic pulse width (~ 20 μ sec in the example above). Since this is less than the time that any spot is illuminated by the scanner the detected signal is "held" electrically for the balance of the period. Note that this pulse width determines the actual bandwidth of the system – in this case 50 Khz.

By synchronizing the initiation and rate of the laser scanner with the acoustic pulses we can take advantage of the finite propagation time (1.5 mm/μ sec in water) with the result that the acoustic signal can be range gated. By proper synchronization of the scanner and adjustment of the detector gate, small volumes of the insonified object can be examined while signals from other parts of the object are rejected.

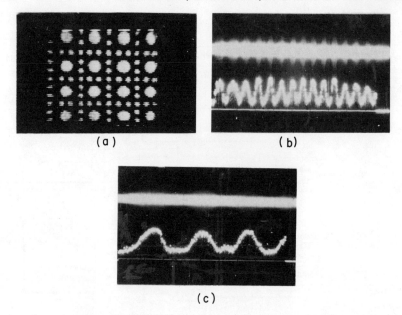

(a)   (b)

(c)

Fig. 11. Ultrasonic image of test object, aluminum sheet
with small (2.5 mm) and large (5 mm) holes. (b) shows signal
from one horizontal scan through line of small holes, and
(c) expanded view of (b), demonstrating the system resolution. The frequency is 3 MHz and an F:2 lens was used.

The primary attribute of the ultrasonovision system is
that it allows the quantitative and qualitative evaluation of
ultrasonic images with high resolution. Several examples
of imaging have been given above. As a further example,
including quantitative analysis, Fig. 11 shows an image of
a test object, which is a piece of aluminum caning consisting of large (5 mm) and small (2.5 mm) holes. The object
was insonified at 3.0 MHz and imaged onto the pellicle
with an F_2 acoustic lens. The image is well resolved.

The quantitative aspects of the system are shown
in parts b, c of the figure, which shows the electrical
signal from one scan through the image of the small
holes of the pattern. (Part c is an expanded view of b).

Once the light intensity in each leg of the interferometer is measured, the measurement of the electrical signal gives a rapid and accurate measure of the acoustic intensity at every and any point of the image.

Summary

The Ultrasonovision system is an accurate, high resolution system for the measurement and visualization of ultrasonic images. It is easily calibrated and accurate enough to serve as a primary standard in the measurement of ultrasonic intensity. It has a uniform frequency response, extending to at least 10 MHz. It has an angular response which is flat to $\pm 40^\circ$ and wavelength-limited   resolution. The sensitivity is linearly dependent on the intensity of the laser used (5 nanowatt/cm^2 with a 15 mw laser) and its response is linear over 8 orders of magnitude. Although primarily intended as a laboratory instrument for the study of ultrasonic interactions and propagation characteristics, with the use of large lasers its sensitivity is adequate for some clinical applications (e.g. the imaging of breasts and appendages).

These attributes, together with its relative simplicity of construction and use should ease and encourage the serious study of ultrasonic fields and increase the eventual utility of ultrasonic methods.

Acknowledgments

We gratefully acknowledge the advice and generous assistance (including supplying the tissues and the pathological reports) of members of the staff of Hahnemann Medical College and Hospital, in particular Dr. M. Haskin, Dr. C. Calderone, and Dr. B. Kingsley. We also want to thank M. Haran of the Bureau of Radiological Health (FDA) for first suggesting consideration of the acousto-optic effect. Last, we want to thank Dr. Jan Rajchman of RCA Laboratories for his constant support and encouragement.

Appendix I - Effect of Acousto-optic Interaction

In the ultrasonovision system the motion of the pellicle gives rise to a phase shift on the light beam reflected from the pellicle. Due to the arrangement used, with the pellicle suspended with fluid through which the acoustic wave passes, the light beam passes through the ultrasonic wave itself as the ultrasonic wave "emerges" from the pellicle. The acousto-optic interaction, in which the pressure of the acoustic wave causes a change in the index of refraction of the fluid leads to an additional phase shift on the light beam. In this appendix we consider the magnitude of this phase shift compared to that due to the motion of the pellicle alone.

The change in index of refraction due to a change in pressure in a fluid may be derived, to a good approximation, from the Lorentz-Lorenz condition[7,8] with the result that

$$\Delta n = \frac{(n^2 - 1)\ (n^2 + 2)}{6 n\, \rho_o\, c^2}\, P \qquad\qquad A1.1$$

where Δn is the change in the index of refraction, n is the static index of refraction, ρ_o is the static density, c is the acoustic velocity, and P is the "excess pressure" due to the acoustic wave.

The total phase shift due to the index change is found by summing over the interaction length (i.e. the path the light beam takes through the ultrasonic wave) or

$$\varphi_n = \frac{4\pi}{\lambda} \int_0^L \Delta n\, dx \qquad\qquad A1.2$$

where λ is the optical wavelength.

The expression for an acoustic traveling wave is

$$p = \sqrt{2\rho\, cI}\ \cos\left\{ \frac{2\pi}{\Lambda}\ (k - ct) \right\} \qquad\qquad A1.3$$

where Λ is the acoustic wavelength, I is the acoustic intensity and c again is the acoustic velocity.

Combining eq. A1.4, A1.2 and A1.1 the expression for the phase shift becomes

$$\varphi_n = M \frac{4\pi}{\lambda}\sqrt{2\rho_o c}\sqrt{I}\int \cos\left\{\frac{2\pi}{\Lambda}(x-ct)\right\}dx \qquad \text{A1.4}$$

where

$$M = \frac{(n^2 - 1)(n^2 + 2)}{6n\rho_o c^2} \qquad \text{A1.5}$$

If $t = 0$ is the time when the acoustic wave is just incident on the pellicle, the interaction length (L) is given by $L = ct$. The phase shift becomes

$$\varphi_n = \frac{4\pi}{\lambda}\sqrt{2\rho_o c}\sqrt{I}\frac{\Lambda}{2\pi}\sin wt \qquad \text{A1.6}$$

If the acoustic wave is a "gated sine wave", as is used in the ultrasonovision system, the duration of the signal due to the acousto-optic interaction is as long as the duration of the acoustic signal. Once the wave has passed through the pellicle there is no further change in the phase of the light beam, since the interaction length remains constant.

The phase shift due to the acousto-optic interaction may now be compared to that due to the pellicle motion.

The pressure of the traveling acoustic wave is, as before. The displacement amplitude is

$$\Delta = \frac{\sqrt{2\rho_o c I}}{\rho_o c^2}\frac{\Lambda}{2\pi}\sin\left\{\frac{2\pi}{\Lambda}(x - ct)\right\} \qquad \text{A1.7}$$

The phase shift due to the pellicle motion assuming the pellicle is at $x = 0$, is

$$\varphi_p = -\frac{4\pi}{\lambda}\Delta$$

$$\qquad \text{A1.8}$$

$$= -\frac{2\Lambda}{\lambda}\sqrt{\frac{2I}{\rho_o c}}\sin \omega t$$

The ratio of the magnitude of the phase shift due to the pellicle motion and acousto-optic interaction, from eqs. 6,7

$$\frac{|\varphi_p|}{|\varphi_n|} = \frac{1}{\rho_o c^2 M} \qquad \text{A1.9}$$

where M is as before. Taking values for water

$$N = 1.33$$

$$\rho_O = 1$$

$$c = 1.5 \times 10^5 \text{ cm/sec}$$

which means

$$M = 1.64 \times 10^{-11}$$

and

$$\frac{|\varphi_p|}{|\varphi_n|} = \frac{1}{2.25 \times 10^{10} \times 1.64 \times 10^{-11}} \qquad \text{A1.10}$$

$$= 2.7$$

Thus the phase shift due to the pellicle motion is 2.7 times as large as that due to the phase shift from the acousto-optic interaction.

The sign of φ_p is opposite that of φ_n, which means the phase shifts subtract, and so the pellicle motion as calculated by eq. 8 of the text should be increased (by about 30% in the case where water is the fluid).

We are indebted to M. Harran of the Bureau of Radiological Health, FDA, who first suggested the consideration of the acousto-optic interaction.

References

1. R. Mezrich, K. Etzold, D. Vilkomerson, RCA Review
    35, p. 483 (1974).

2. L. Brekhovskikh "Waves in Layered Media", p. 61,
    Academic Press, N. Y. (1960).

3. D. Vilkomerson, "Acoustical Holography", V.14,
    Plenum Press, N. Y. (1972).

4. A. Gray, G. Matthews, T. Macrobent, "Bessel
    Functions", p. 32, Macmillan Co., London (1952).

5. D. Berlincourt, D. Curran, H. Jaffe, "Physical
    Acoustics", Vol. 1A, p. 169, Academic Press, N.Y.
    (1964).

6. D. Folds, J.A.S.A. 53, p. 826 (1973).

7. Born, M. and E. Wolf, "Principles of Optics", p.593,
    Pergamon Press, London, 1965.

8. Heuter, T. and R. Bolt, "Sonics", p.352, J. Wiley &
    Sons, New York, 1955.

DIGITAL COMPUTER SIMULATION STUDY OF A REAL-TIME COLLECTION,

POST-PROCESSING SYNTHETIC FOCUSING ULTRASOUND CARDIAC CAMERA

S. A. Johnson, J. F. Greenleaf, F. A. Duck,
A. Chu, W. R. Samayoa, and B. K. Gilbert
Biophysical Sciences Unit
Department of Physiology and Biophysics
Mayo Foundation, Rochester, Minnesota 55901

ABSTRACT

High resolution ultrasound images (90% of Rayleigh limit
at all depths) were obtained by computer analysis of digitized
(at 10 and 20 points per microsecond, 8 bits per point)
signals detected with a 32 element lead zirconium titanate
3.0 MHz array. Every sixth element was used as a transmitter and pulsed 31 times while the signals from the remaining 31 elements were addressed sequentially by an analog
switch and their signals digitized and recorded. Each pixel
in the cross-sectional image was produced by calculating the
inner product of a specific window function and the ensemble
of digitized signals. Thus the array is mathematically
focused optimally for each pixel in the image. The acquisition and storage of all received signals allows subsequent
approximate calculation of spatial distributions of such
parameters as reflection, index of refraction, attenuation,
etc. The extension of these techniques through the use of a
fast analog and digital computer interface to obtain real-
time and stop-action imaging is treated. Examples of images
produced by these algorithms, the time requirements of various
algorithms as determined by the computation speeds of present
and anticipated digital and analog processing hardware is
presented. Supported in part by NIH research grants HT-4-2904,
RR-7, and HL-04664. Also supported in part by a contract from
the Office of Naval Research to E. M. Eyring, with whom S. A.
Johnson was a part-time research associate.

INTRODUCTION

A review of the literature dealing with ultrasonic and
acoustic imaging will reveal that a host of methods exist
for detecting, recording or producing images from scalar,
longitudinal pressure waves. Detection schemes sensitive to
intensity (a scalar) include the Sokolov tubes (1), liquid
surface holography (2,3), and those methods based upon displacement of solid surfaces (4) or particles (5). Amplitude
(a vector) sensitive methods include those based upon piezoelectric materials (6,7) and the displacement of thin membranes (8,9) and solid surfaces (10). Recording schemes
used before or after image formation have made use of photographic film (11), electronic storage tubes (12) and computer
based methods (13,14,15).

Imaging schemes which have been used include those
based upon transmission (7), reflection or scattering (16),
holography (10), electronic scanning (6), and cross section
reconstruction (15). Some of these methods require acoustic
lenses (7) while others do not (6,10).

This paper represents an attempt to develop an imaging
scheme for in-vivo human cardiac imaging which not only
incorporates the best features of previously known ultrasound
techniques, but also benefits from the use of several novel
imaging techniques. The imaging philosophy of recording
both phase and amplitude information throughout an aperture
is of advantage in a large class of general ultrasound and
acoustic imaging applications, including dynamic cardiac
visualization. The method to be described might be considered
holographic (10,12,13,14) since it records both phase and
amplitude. Yet, it is not unlike electronic scanning (6)
since it derives range information from the temporal recording
and mathematical processing of signals from multiple transducer elements.

The existence of such a data bank allows for several
selective image-forming algorithms to draw from common data
and produce images corresponding to unique properties of the
object under study. These algorithms may be used either
independently or interactively in a synergistic manner.
Furthermore, the use of a priori information concerning the
object and its environment is not precluded by use of
programmable algorithms. Unlike certain other schemes, no

approximate model of the propagation and interaction of
acoustic energy need be used and thus the method does not
suffer from the aberrations associated with the use of lenses
or those errors common to holographic techniques which depend
upon making the Fresnel approximation (13). This more general
imaging philosophy might be called "image synthesis." The
subset method applied to cardiac imaging is therefore termed
"synthetic focus" imaging.

MATHEMATICAL BACKGROUND

The mathematics for synthetic focus imaging may be
derived by reference to Figure 1. Let P_k define a measure
of the total energy scattered or reflected toward the detector
aperture from picture element k when illumination from each
transmitter element in the array aperture, measured at picture
element k, is normalized. For the case of a very narrow pulse
of ultrasonic energy, P_k is given by

$$P_k = \left[\sum_{m=1}^{N} \sum_{j=1}^{N} \sum_{i=s}^{d} V_{i,j,m} R_{j,k} T_{k,m} \right]^2 , \tag{1}$$

where $V_{i,j,m}$ is the ith voltage sample from the jth array
element (at sample time i) when the mth array element was
used as the transmitter, $R_{j,k}$ is the receiver-range azimuth
illumination normalization factor between the kth picture
element and the jth array element. Thus, $R_{j,k}$ is independent
of time, but is an approximate function of the square of the
magnitude of $r_{j,k}$ (where $r_{j,k}$ is the vector from array
element j to picture element k) and of cosine of the angle
between $r_{j,k}$ and $r_{1,N}$ (the vector from array element 1 (one)
to array element N). $T_{k,m}$ is the corresponding transmitter
function. Thus, $R_{j,k}$ and $T_{k,m}$ affect the energy normaliza-
tions referred to in the definition of P_k. N is the number
of array elements. The limits s and d are related to the
transmitter pulse width t_w. Let t_w be less than some arbitrary
bracketing interval t_B which contains the pulse of width t_w.
Then

$$s = \left[(r_{j,k} + r_{m,k}) / u - t_B / 2 \right] / \Delta t \tag{2}$$

and $\quad d = s + t_B / \Delta t \tag{3}$

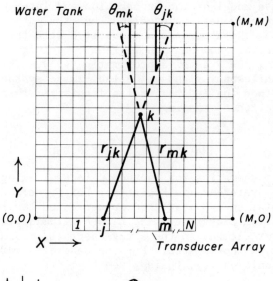

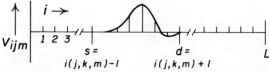

Figure 1   Geometry for derivation of equations for ultrasonic imaging by generalized synthetic aperture
methods. A transducer array with N elements is
in acoustic contact with a water tank. Positions
in the tank are described by coordinate system XY.
A typical ray path is shown for energy transmitted
by the mth transducer and scattered by picture
element k into the jth transducer. Here r_{jk} and
r_{mk} are the distances from j to k and from m to
k, respectively. Angle θ_{jk} is measured from r_{jk}
to the Y axis counter-clockwise. V_{ijm} is the
voltage sample from the jth transducer at time i
after transmission of a signal by the mth transducer. Here s and d are start and end times of a
time interval of length 2·l (2 times "el") which
contains i as a midpoint. Time L is greater than
the maximum time associated with all paths defined
by j, m, and k.

where t is the sampling interval time between successive
samples in the digitized signal, $r_{j,k}$ is the distance from
array element j to picture element k, and u is the effective
velocity of sound in the insonified medium.

One modification to the basic synthetic-focusing scheme
presented above would be to replace $V_{i,j,m}$ with the output
of a hardware or software correlator, allowing transmission
not only of pulses of energy, but arbitrary waveforms as well.
Thus, correlation of the detected form of $V_{i,j,m}$ with, for
example, a system transfer function corresponding to the
transmitted waveform would allow use of waveforms with less
peak power, but greater energy content, than that of a narrow
pulse. Matrix multiplication or correlation with other
functions will allow more optimum resolution and noise rejection
in the Wiener sense. An equivalent effect could be obtained
by performing the correlation operation as part of the sum
over index i in Equation (1). In this event, $V_{i,j,m}$ would
be replaced by the product $V_{i,j,m} C_{i,j,m}$ where $C_{i,j,m}$ acts
like a correlation kernel. Thus, we may write the more
general equation

$$V_{i,j,m} \leftarrow \sum_{i'} \sum_{j'} \sum_{m'} V_{i',j',m'} C_{i',j',m',i,j,m} \tag{4}$$

Another modification would allow Doppler processing by
spectral analysis of appropriate array signals from a time
sequence of raw array data, thereby gaining information as
to the velocity of reflecting regions in the image. It is
also possible to design a transmitted waveform and processing
kernel which provides for either Doppler sensitive or Doppler
invariant imaging (17).

Examples of the resolution and fidelity of images
produced with real data collected with an actual transducer
array are shown in the next section.

At a later date, bulk properties of biomaterials such
as refractive index, scattering parameters and impedance may
be incorporated in image analysis as input or as derived
information. An example will be given. A FORTRAN ray

tracing program for continuous media has been written and
extended to include possibilities of discontinuities in
refractive index. The new FORTRAN program will be tested
using the scanning device or an array to verify its predictions of refraction behavior using objects of known geometry
and refractive properties in the controlled environment of
a water tank. Thereafter, this program may contribute to
the capability of measuring impedance and sound velocity in
test objects of known geometry.

EXPERIMENTAL RESULTS

The simulation of the operation of a real-time cardiac
camera was undertaken using a moderately high-speed digital
computer, the CDC 3500, at Mayo. Both simulated and real
data were used. Image synthesis success using equation (1)
with simulated data prompted the early use of real data to
facilitate development of a feeling for characteristics,
such as signal to noise ratio, transducer bandwidth, etc.,
which are associated with real data. Through the use of
both inanimate and biological test objects, the characteristics of a practical real-time cardiac camera can be
studied and its operation simulated. In this paper only
results using inanimate test objects are reported.

A 3 MHz lead zirconium titanate piezoelectric ceramic,
linear transducer array of 32 elements with dimensions 2.4 cm
by 1.0 cm. was used for collection of data. The array was
coupled to a water tank through a thin latex window. Although
equation (1) allows for all N elements to be used as both
transmitters and receivers independently, it was found
convenient to only use 6 transducers (about every 6th element
as transmitters and all elements as receivers. The transmit
and receive circuits associated with the 6 elements used as
transmitters were isolated with SPDT reed relays. These
relays also served to multiplex the transmit elements. A
MOS analog multiplexer integrated circuit connected all
receive elements, including each reed relay and its receive
element, to a preamplifier (Panametrics PR/5050). The
functions of the transmit and receive multiplexers were under
computer control. The output of the preamplifer was amplifie
further by a swept gain amplifier and then digitized to 8
bits accuracy by an analog-to-digital converter (Biomation
8100) operating at a rate of either 10 or 20 samples per

microsecond.   Collection of digital output and other operating functions of the analog-to-digital converter was also
under computer control.   A data record of 1024 sample points
was taken from each receiver element.

   An example of the output of 32 elements acting as receivers is shown in figure 2.   Also shown in figure 2 is a
method of eliminating spurious reflection within the transducer array by subtraction of a clear field.   Thus, figure
2 is very similar to the standard format for displaying
data obtained from an array of geophones.

Figure 2   Computer-generated grayscale plots of the output
           of 32 elements of a transducer array.   Both the
           top and the bottom horizontal strip are composed
           of 32 horizontal lines.   The top strip represents
           256 sampled values from the raw unprocessed data
           with an object in the field of view of the array.
           Multiple reflections from end-to-end within the
           array may be seen.   The bottom strip shows the
           multiple reflections removed.   This was obtained
           by subtracting a clear field strip of data (not
           shown).   The clear field was digitized with the
           test object removed from the tank.

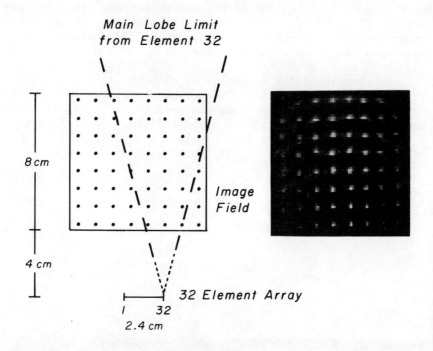

Figure 3   The effect of angular gain variation of each transducer element upon the sensitivity of the whole
array of 32 elements. The figure to the left
illustrates the relative position of the array,
a square grid of threads with 1.0 cm. spacing and
the main lobe limits of each element (in this
instance the 32nd element). The figure to the
right is a computer-generated gray scale plot of
the actual synthesized image. The threads are
brighter at the bottom because an inexact swept
gain function was used. The two top corner
threads are outside the field of view and therefore absent from the image. The rectangular grid
was composed of parallel 0.2 mm. diameter nylon
threads. This figure also illustrates the validity
of the assumption of linear superposition implicit
in equation (1). The functions $R_{j,k}$ and $T_{k,m}$ in
equation (1) may be considered to be unity except
for the effect of swept gain prior to digitization.

The effect of the main lobe pattern (angular sensitivity) of each element in the array upon the output in each region in a large field of view is seen in figure 3. The test object consisted of a square matrix with 1.0 cm. spacing between parallel nylon threads. The results are consistent with the measured field of view of 24° for each element. Each array element has a width of approximately one wavelength in water and a height of 20 wavelengths.

It can be seen from equations (1), (2) and (3) that P_k may be non-zero for some points in the synthesized image where no true scatterer or reflector exists in the experiment tank. This effect corresponds roughly to the existence of side lobes in the response of a real antenna of finite aperture. This property of equation (1) is due to the transmit-scatter-receive equivalent time delay for all points in the experimental tank which lie upon an ellipse with the transmit and receive elements at the two respective foci. Notwithstanding, the addition of amplitudes in each picture element (pixel) from different receiver-transmitter combinations tends to cancel out non-real contributions in each pixel. This is illustrated in figures 3 and 4.

The effect of increasing the number of receiver-transmitter combinations upon the amplitude and intensity of a synthesized image of a single thread from figure 3 is shown in figure 4. Also shown is a phaseless synthesis of the same object which indicates the importance of using un-rectified data.

Thus, it may be noted that a test object which has components in many regions of the test tank tends to produce images with a smooth background. However, test objects may be made with scatterers arranged to minimize this type of interference. Such a test object is synthesized in figure 5. It may be seen that a bow tie-shaped pattern results from the incomplete interference along ellipses of equal round trip time. All ellipses cross at the true thread position resulting in maximum constructive interference.

The bow tie-shaped ghosts may also be minimized by using more transmitting elements and by employing apodizing methods in synthesis equation (1). Apodizing is accomplished by both weighting the product terms along trajectories in the data space illustrated in figure 2 and by employing

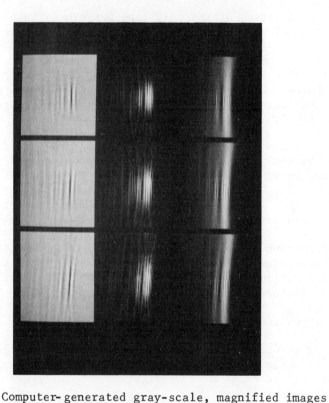

Figure 4 Computer-generated gray-scale, magnified images
of one thread in the rectangular array of threads
shown in figure 3. The top row represents amplitude. The middle row represents intensity. The
bottom row illustrates the intensity of a phaseless synthesis. The left-most column represents
data obtained from 2 transmitters and 24 receivers
per transmitter. The middle column corresponds
to 4 transmitters and 24 receivers per transmitter
The right-most column represents 6 transmitters
and 24 receivers per transmitter. Lateral
resolution is improved as more transmitter and
receiver elements are used. The phaseless synthes
was obtained by rectifying (taking the absolute
value of) the same data which were used to synthesize the images in the top and middle rows. The
field size is 5 mm. by 5 mm. The array is located
5 cm. below the bottom of the field.

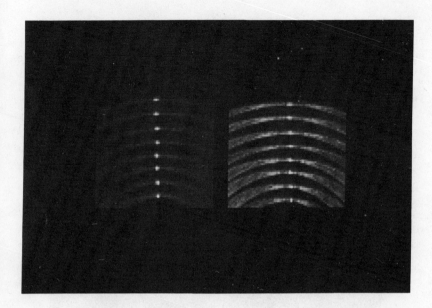

Figure 5   Computer-generated images of 8 nylon threads each
            normal to the transducer array imaging plane.
            The array is located several centimeters directly
            below the threads with its face normal to the line
            formed by the images of the threads. The spacing
            between threads is 1.0 centimeters. The bow tie-
            shaped pattern is a result of incomplete inter-
            ference along ellipse of equal round trip time.
            The image on the left was obtained through the use
            of 6 transmit and 24 receive elements per trans-
            mitter. Image on the right was the result of using
            6 transmit and only 2 receive elements per trans-
            mitter.

weighting in time along horizontal lines of figure 2. Both
weights may be incorporated by the proper choice of
$c_{i,j,m,i,j,m}^{'\,'\,'}$ in equation (4). In this sense apodizing may
be considered to be equivalent in effect to a type of deconvolution.

The resolution of the synthesized images is thus a
function of the amount of deconvolution which the signal to
noise ratio of the raw data will allow. Experimental resolution has been measured without deconvolution to be about
3 wave lengths in depth and about 90% of the Rayleigh limit
in the lateral direction beyond 3 cm. from the array.

The ability to synthesize a complex image with closely
spaced threads is shown in figure 6. Here the pattern of
nylon threads spells the word "MAYO".

The raw data which were used to synthesize the "M"
portion are shown in figure 7. Each wire in the "M" is
responsible for a hyperbolic-shaped pattern in each strip
of data. The six strips shown correspond to the 6 transmitters.

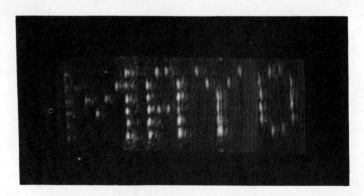

Figure 6   Computer-generated image of an arrangement of
            parallel nylon threads (diameter 0.2 mm.) in a
            water tank. The threads were arranged in a
            pattern to spell the word "MAYO", in a region
            1.25 cm. x 5.50 cm. The minimum center-to-center
            spacing of the threads was 2.5 mm. The array was
            situated 3.0 cm. from the left side of the image.

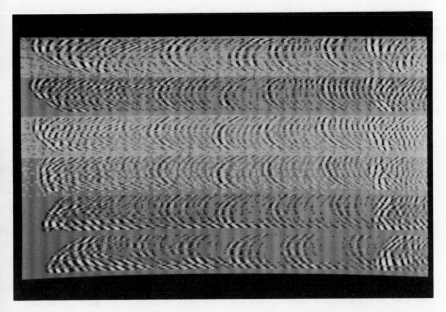

Figure 7   Computer-generated gray scale image of the raw
           data used to synthesize only the "M" in figure 6.
           Each horizontal strip is composed of 32 horizon-
           tal lines. Each horizontal line corresponds to a
           25.6 microsecond section of the received signal
           from the corresponding receiver. Each set of
           hyperbolic-shaped patterns in each horizontal
           strip corresponds to one thread in the "M".

Imaging biological tissue is more difficult than the imaging of threads in a tank because of the large variation in dynamic range between acoustic energy which is scattered or specularly reflected from tissue structure (16). This effect calls for preamplifiers and recording equipment capable of handling signals with at least three decades of dynamic range.

As has been seen, synthetic focus imaging provides great flexibility in image formation. The flexibility even allows in principle corrections for aberrations to the acoustic path of rays due to an inhomogeneous distribution of index of refraction in the object. Thus, if the index of refraction in the object space is known, the correct ray path may be traced and the corresponding acoustic path length determined. These two data are sufficient to correctly calculate for each pixel in image space the appropriately delayed amplitude from the receiver element to use in the image synthesis. There are at least two ways to obtain the necessary index of refraction data, namely by assuming a model which is corrected by iteration, or by reconstructing the index of refraction field from multiangular time of arrival data (18).

The past computational time requirements common to digital imaging methods imposed a practical and financial limitation upon their implementation in a real-time imaging system. However, powerful new minicomputer systems, special purpose arithmetic boxes, and fast microprocessors are already finding many new areas of application for digital image processing. The present rate of image computation with the Mayo CDC 3500 computer is about 4 picture elements per second. This rate can be increased to some extent by making use of the symmetry of the geometry of image formation (use of "lookup tables," FFT algorithms, etc). Additional improvements in speed can be attained with the implementation at Mayo of a new high-speed data interface (19). The speed per execution and cost per execution of a typical algebraic operation for various standard computers and new special purpose numerical processors are plotted in figure 8. Improved medical ultra-sound imaging systems should be forthcoming because of the trend toward faster and less expensive digital processors.

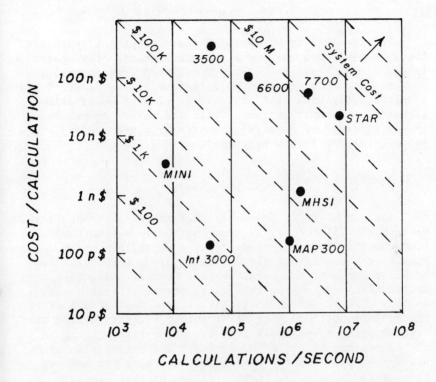

CALCULATIONS / SECOND

Figure 8 Plot of relative cost and speed of some existing
digital computers and processors. This plot
compares component characteristics by two standards,
but cannot be used to represent a component's
ideal performance under all conditions. The word
calculation represents a sequence of typical
operations used to compute a scalar produce between
two vectors. The scientific computers or processors
plotted are the CDC 3500, CDC 6600, CDC 7700,
CDC Star, a typical 1974 minicomputer, the Mayo
high-speed interface modification to a CDC 3500
(MHSI), a CSPI MAP 300 FFT processor, and the Intel
3000 series microprocessor. The cost of large
peripheral memory units is included in the cost
of the computers but not for the processors. The
system costs plotted are the original new purchase
price and are dependent upon the abscissa and
ordinate through a typical amortization schedule.

SUMMARY

The feasibility of applying the concepts of synthetic focus imaging to obtaining a dynamic sequence of stop-action images of the heart has been demonstrated with the multiplex system. Digitization of data from all receiver elements simultaneously with each transmitter element would allow stop-action dynamic imaging. With the higher computational speeds afforded by parallel computer architecture, real-time dynamic imaging is now practical.

ACKNOWLEDGEMENTS

The aid of P. S. Green of Stanford Research Institute, Menlo Park, California, in contributing to our work with transducer arrays is appreciated. A logarithim video amplifier and many stimulating discussions were furnished by R. S. Hughs of the Naval Weapons Center, China Lake, California. The analog-to-digital converter system used in this work is due in large part to the past effort of Dr. R. A. Robb (20). The first author is grateful for the several weeks spent in 1973 at the University of Utah in the laboratory of Dr. E. M. Eyring. The association with Dr. Eyring provided the conceptual beginning for much of this work. Dr. J. J. Tiemann of the General Electric Co. has provided valuable motivation and insight. Dr. T. C. Evans of the Division of Cardiovascular Diseases at the Mayo Clinic has provided much valuable advice and motivation for this work. We thank Mrs. Jean Frank and co-workers for their secretarial and graphic assistance. This work was made possible by the encouragement of Dr. Earl Wood, the support of his associated staff and the excellent laboratory facilities made available in the Biophysical Sciences Unit of the Mayo Clinic, of which he is chairman.

REFERENCES

1. Jacobs, J. E. and D. A. Peterson:
    Advances in the Sokoloff tube.
    Acoustical Holography, Vol. 5, Plenum Press, 1973.

2. Pille, P. and B. P. Hildebrand:
    Rigorous analysis of the liquid-surface acoustical holography system.
    Acoustical Holography, Vol. 5, Plenum Press, 1973.

3. Holbrooke, D. R., E. E. McCurry, and V. Richards:
    Medical uses of acoustical holography.
    Acoustical Holography, Vol. 5, Plenum Press, 1973.

4. Nigam, A. K. and J. C. French:
    Modified Sokolov camera utilizing condenser-microphone
    arrays of the foil-electret type.
    Acoustical Holography, Vol. 5, Plenum Press, 1973.

5. Cunningham, J. A. and C. F. Quate:
    Acoustic interference in solids and holographic
    imaging.
    Acoustical Holography, Vol. 4, Plenum Press, 1973.

6. Thurstone, F. L. and O. T. Von Ramm:
    A new imaging technique employing two-dimensional
    beam steering.
    Acoustical Holography, Vol. 5, Plenum Press, 1973.

7. Green, P. S., L. F. Schaefer, E. D. Jones and J. R.
    Suarez:
    A new, high-performance ultrasound camera system.
    Acoustical Holography, Vol 5, Plenum Press, 1973.

8. Mezrich, R. S., K. F. Etzold and D. H. R. Vilkomerson:
    Ultrasonovision.
    IEEE 1974 Ultrasonics Symposium Proceedings.

9. Altis, P.:
    Real-time acoustical imaging with a 256-256 matrix of
    electrostatic transducers.
    Acoustical Holography, Vol. 5, Plenum Press, 1973.

10. Kessler, L. W., P. R. Palermo and A. Korpel:
    Recent developments with the scanning laser acoustic
    microscope.
    Acoustical Holography, Vol. 5, Plenum Press, 1973.

11. Green, P. S., L. F. Schaefer and A. Macavski:
    Considerations for diagnostic ultrasound imaging.
    Acoustical Holography, Vol. 5, Plenum Press, 1973.

12. Weaver, J. L. and G. C. Knollman:
    Real-time reconstruction of images from hydroacoustic
    holograms.
    Acoustical Holography, Vol. 4, Plenum Press, 1973.

13. Fenner, W. R. and G. E. Stewart:
    An ultrasonic holographic imaging system for medical
    applications.
    Acoustical Holography, Vol. 5, Plenum Press, 1973.

14. Clark, C. S. and A. F. Metherel:
    Digital processing of acoustical holograms.
    Acoustical Holography, Vol. 5, Plenum Press, 1973.

15. Greenleaf, J. F., S. A. Johnson, S. L. Lee, G. T. Herman
    and E. H. Wood:
    Algebraic reconstruction of spatial distributions of
    acoustic absorption within tissue from their two-
    dimensional acoustic projections.
    Acoustical Holography, Vol. 5, Plenum Press, 1973.

16. Kossoff, G.:
    Ultrasonic visualization of the uterus, breast and
    eye by gray scale echography.
    Proc. Roy. Soc. Med., Vol. 67, p. 135, Feb. 1974.

17. Johnson, R. A. and E. L. Titlebaum:
    Range-Doppler uncoupling in the Doppler tolerant bat
    signal.
    IEEE 1972 Ultrasonics Symposium Proceedings.

18. Greenleaf, J. F., S. A. Johnson, W. A. Samayoa, F. A.
    Duck and E. H. Wood:
    Algebraic reconstruction of spatial distribution of
    acoustic velocities in tissues from time-of-flight
    profiles.
    Acoustical Holography, Vol. 6, Plenum Press, 1975.

19. Gilbert, B., M. Storma, C. James, L. Holbrock, E. Yang,
    K. Ballard and E. H. Wood:
    A real-time hardware system for digital picture pro-
    cessing of wide band video images.
    Sixth Annual Meeting of Biomedical Engineering
    Society, New Orleans, April 11-12, 1975 (Paper in
    preparation).

20. Robb, R. A., S. A. Johnson, J. F. Greenleaf, M. A.
     Wondrow and E. H. Wood:
     An operator-interactive, computer-controlled system
     for high-fidelity digitization and analysis of bio-
     medical images.
     Proceedings of the Society of Photo-Optical Instru-
     mentation Engineers, August 27-29, 1973. Volume 40,
     pp 11-26.

A SCANNING FOCUSED-BEAM SYSTEM FOR REAL-TIME DIAGNOSTIC IMAGING

K. Y. Wang   G. Wade

University of Houston   Univ. of California

Houston, Texas   Santa Barbara, Calif.

ABSTRACT

Several promising approaches to real-time acoustic imaging employ laser beams to read out the image information. In other approaches the readout is by piezoelectric transducers. Previous analyses have shown that, in terms of sensitivity, this latter category of systems is inherently superior to the former.

Among the piezoelectric systems, the positively-scanning-transmitter (PST) approach provides a system with the highest sensitivity and the lowest acoustic threshold contrast. Because of beam focusing and scanning, the PST system is especially well suited for a number of practical diagnostic problems, particularly those involving deep-lying targets within the human body. Consider, for example, transmission imaging of the kidney through the abdomen. For such a task, the PST approach is capable of producing effective real-time images with minimum tissue exposure to ultrasound.

This paper will present and discuss two schemes for implementing the PST approach. A key element in each of the schemes is an acousto-optic transducer which is activated by a light pattern. Scanning can be achieved either electronically or mechanically by putting the light pattern into motion. Focusing can be accomplished either with or without an acoustic lens. The system as visualized is sufficiently flexible to permit modification and tailoring

for enhanced capabilities corresponding to a variety of
specific applications.

INTRODUCTION

In previous analytical work,[1] a procedure was established to characterize the ideal performance of an imaging
system in terms of threshold contrast defined as the ratio
of the smallest difference in the acoustic transmittance
of two adjacent resolution element of the object when the
corresponding two adjacent resolution cells in the image
can just barely be distinguished from each other (i.e., as
having different shades of intensity) to the large transmittance of the two object elements.

The analytical procedure was applied to various acoustic imaging systems[1-3], including existing as well as hypothetical schemes that hold promise in real-time orthographic diagnostic imaging.

The schemes examined can be divided into two categories. The first category comprises the basic scanning
modes with piezoelectric detection: the Positively Scanning Transmitter (PST) System, the Positively Scanned Receiver (PSR) System, the Negatively Scanned Receiver (NSR)
System, and the Negatively Scanning Transmitter (NST) System. The basic difference in operation among these systems
in the first category is as follows. In the PST System,
the acoustic transmitter provides a focused, scanning beam,
focused at and scanned over the object plane. The dimension of the focus at the object plane corresponds to the
size of a resolution element. A transducer, or array of
transducers occupies the receiver plane to provide an electrical output to the detector. The time variations in the
output signal stem from the spatial variations in the object. These time variations are converted back into spatial variations in the image plane by means of synchronous
scanning. The principle of operation is analogous to that
of a scanning electron beam microscope.

In the PSR System, a broad collimated sound beam in-
sonifies the entire object plane. An array of receptors
constitutes the receiver, each receptor covering an area
corresponding to the size of a resolution cell. The re-

ceptors are activated one by one in sequence. The signal
output is processed as in the PST case, the image scanning
being synchronous with the receptor scanning.

In the NSR System, everything is the same as in the
PSR System except for the receptor scanning. Here all the
receptors are always kept "on" except for one. The position of the "turned off" receptor is scanned. The operation is a negative version of the PSR System at the receiver
where the scanning takes place.

The NST System was included for theoretical interest.
Its operation can be looked upon as the negative version
of the PST System at the transmitter where the scanning
takes place.

The basic scanning modes are conceptually simple and
involve a minimum number of conversion processes before detection. Most of the systems presently being built with
piezoelectric detection fall into the above broad category,
frequently hybrid arrangements of the basic scanning modes
described.

The second category includes systems that utilize
laser beam readout of the acoustic information. Three of
these systems have produced good real-time images. They
employ respectively static-ripple diffraction[4], dynamic-
ripple diffraction[5] and Bragg diffraction[6].

The threshold acoustic contract has been derived for
systems in both these categories[1-3]. In the first category,
the two positive systems were found to be the better theoretical candidates. For identical diffraction limited
resolution and integration time, both the PST System and
PSR System require the same amount of instantaneous insonification to visualize a particular acoustic contrast
in the object, but for reasons to be accented, the PST
System is the best candidate. The PST System was used as
the standard of excellence against which to measure the
performance of systems in the second category. It was
found that, with the compatible conditions assumed and
perfect components hypothesized, the inherent capability of
the systems in the second category are of the same order,
while the PST System is considerably better.

SAMPLE CALCULATIONS OF INHERENT SENSITIVITY

We will now compare the sensitivity of a PST System
with that of an existing system with impressive performance,
the ultrasonic camera of the Stanford Research Institute[7].
This SRI System is equivalent in sensitivity to a PSR System with improved integration time. Its inherent sensitivity can be calculated from information in the previous analysis[1].

To make this performance comparison, we will consider
a specific application. Assume both systems are used to
obtain transmission images of the human kidney through the
mid-section of the body. The total object thickness (i.e.
the thickness of the mid-section) will be assumed to be
30cm. The target depth (i.e. the distance between the
kidney and the mid-section surface nearest the transmitter)
will be assumed to be 10cm. These are reasonable figures
if we insonify from the back. We will assume that the
average attenuation of the sound in the bodily tissue is
3.3 db/cm for operation at 3 MHz. To simplify the calculation, the object will be assumed to be uniform in front of
and behind the target plane.

First, consider imaging target elements with the best
possible acoustic contrast, a contrast of unity. To make
these target elements distinguishable in the final image,
the sound intensity transmitted through the mid-section
must be at least the level of the minimum detectable intensity. This quantity is the same for both the PST System
and the PSR System. Previous results[1] gives 2×10^{-11} watt/
cm^2 at room temperature in water at 3 MHz.

As stated earlier, the SRI camera is equivalent to a
PSR System with improved integration time. This improvement is due to the use of storage in a one-dimensional receiving array, resulting in a reduction in the minimum detectable intensity by a factor equal to the number of elements in the array. Assuming 100 elements in the array,
the minimum detectable intensity needed to be transmitted
to the receiver becomes 2×10^{-13} watt/cm^2. The corresponding
intensity at the target plane that lead to such a transmitted level can be calculated via attenuation through 20cm

of tissue.   It is 8×10^{-5} watt/cm^2 for the PST System, 8×10^{-7} watt/cm^2 for the SRI System.

The relation between the front surface intensity and the target plane intensity is different for these two systems. This is due to the different approaches to insonification-the collimated-beam approach (Approach A) of the SRI camera, and the focused-beam approach (Approach B) of the PST System. This is illustrated in Fig. 1. In the collimated approach, the front surface intensity and the target plane intensities are related through attenuation. In the focused-beam approach, these two intensities are related through attenuation and concentration. The relations between these intensities are:

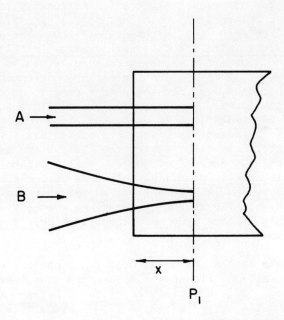

Figure 1   Collimated Beam Approach (A) and Focused Beam Approach (B) for Target Insonification

$$I_0 = 10^{-3.3} I_s \qquad \text{for the ultrasonic camera} \qquad (1)$$

$$I_0 = 10^{-3.3} (100/(6N^4\Lambda^2))I_s \qquad \text{for the PST System} \quad (2)$$

where I_0 is the target plane intensity, I_s the near surface intensity, N the focal number corresponding to the converging beam, and Λ the acoustic wavelength. Using $I_0 = 8\times10^{-7}$ watt/cm^2 for the ultrasonic camera, we get $I_s = 1.6$ mw/cm^2. Using $I_0 = 8\times10^{-5}$ watt/cm^2, N=2, $\Lambda = 500$ μm for the PST System, we get $I_s = 0.4$ mw/cm^2.

Therefore, in the case of this particular application, to observe a target with the maximum possible acoustic contrast, the SRI system needs an intensity of 1.6 mw/cm^2 at the surface nearest the transmitter, while the PST System needs only 0.4 mw/cm^2. When it comes to the risk for biological damage, some workers have proposed a maximum operating level for ultrasonic diagnostic instrumentation as low as 0.5 mw/cm^2 (average)[8]. If that safety specification is to be complied with, the SRI camera, although the existing complete system with the best performance, in this example, could not image the kidney successfully, while the PST System could. It is still not clear how ultrasound interacts with tissue, and how the damage depends on continued or pulsed sound. However, all investigators agree that whatever peak intensity is used, the risk of damage is lessened if the dosage is given on a pulsed basis with a low duty factor. In this regard, the PST System has a distinct advantage. Because of the transmitter scanning, each target element is insonified only on a periodic basis. Hence, the PST System is inherently a pulsed system with low duty factor, as far as the target-plane elements are concerned. This is true even when the system operates on a continuous-wave basis.

Let us take another example involving a patient with a reduced mid-section thickness. Assume the same type of object as before, but this time the distance between the far surface and the target plane is 15cm instead of 20cm. Assume that both systems operate under the condition that the object nowhere gets insonified with an average intensity greater than 0.5 mw/cm^2. Under these conditions the minimum acoustic contrast that could be imaged by the SRI ultrasonic camera would be approximately 24% while that for the PST System would be 12%.

GENERAL DISCUSSION

To summarize the calculations, the PST System permits
more efficient use of insonification for imaging. Examples
demonstrate that when it comes to imaging deep-lying targets including targets important in medical applications,
the PST System may be the only real-time inherently capable
of operating below a recommended maximum average level of
0.5 mw/cm^2.

In the above examples we used a strictly fundamental
PST System. A PST System in its most fundamental mode compares favorably with other systems. However, a PST System
can also be modified for enhanced capability. For example,
a one-dimensional focused beam plus a one-dimensional receiver array can improve the integration time (just as the
use of a one-dimensional receiver array improves integration
time in the SRI camera). The transmitter used reciprocally
as the receiver can constitute a focused transmitter focused receiver system. These variations on the basic system
could increase the system capability, if greater capabilities were needed.

EMBODIMENT SCHEMES

We will now describe, two possible front-end embodi-
ments of a basic PST System. The first embodiment is all-
electronic, does not use acoustic lenses, high-density
transducer arrays or sophisticated processing techniques.
Its key element is an opto-acoustic-transducer (OAT). An
example of such an element is a piezoelectric layer in contact with a photoconductive (PC) layer, both sandwiched
between electrodes, the one adjacent to the photoconductor
(PC) side being transparent. This is illustrated in Fig. 2.

A simplified explanation of the function of this
structure is as follows: Let the parameters be such that
when a voltage is applied across the electrodes under "dark"
conditions, most of the potential drop will appear across
the photoconductive layer. The piezoelectric layer will
send out only weak radiation. With a light beam addressing
the OAT from the transparent side, the resitivity of the
photoconductive layer in the light path can be greatly reduced, such that most of the applied voltage now will appear across the piezoelectric layer in the region addressed

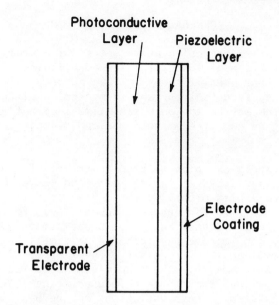

Figure 2   An Example of an Opto-Acoustic Transducer

by the light beam. The light beam, therefore, will acti-
vate the transducer via photoconductive switching. The
structure generates a near-field acoustic pattern in accordance with the optical pattern used to selectively acti-
vate regions of the transducer.

A second example of this key element, the OAT, is a
structure where a single layer of material that is both
photoconductive and piezoelectric takes the function of the
double layer just described. A transparent dielectric
separates this layer and the transparent electrode. The
transducer activation is a negative version of that in the
first example.

With an OAT, focusing can be effected by optical projection of a zone plate pattern* or other apodization patterns onto it. Scanning can be accomplished by setting the
projected light pattern into motion. For rapid scanning, a
laser scanner and a Fourier transform hologram can be used.

_____

*For focusing with a fixed <u>acoustic</u> zone-plate, see Ref. 9.

First, a Fourier transform hologram of the desired transparency (with an acoustic Gabor zone-plate pattern or other apodization patterns) is made with a point source reference. In the recording step, the carefully fabricated transparency (e.g. with an acoustic zone-plate pattern) is placed with a pinhole aperture. With plane wave illumination, a hologram is recorded at the Fourier-transform plane.

Next, an array of pinholes is made in an opaque mask. Each pinhole should be close in optical properties to the one used for recording the hologram. The position of the pinholes should properly match those of the laser scanner output.

During the reconstruction step, the optical setup is such that the scanning laser beam perpendicularly addresses the pinholes in a raster sequence. This is illustrated in Fig. 3. The laser beam passing through one of the pinholes constitutes the reconstruction beam, and projects the zone-plate pattern onto the OAT, generating a focused sound beam. As the laser beam moves from one pinhole to the next, the reconstructed zone-plate pattern at the OAT is translated spatially by an amount determined by the pinhole spacing. Thus the acoustic focus at the object plane is translated from one resolution element toward the next in synchronism with the laser scanning.

There are other ways of utilizing the OAT to produce a scanning focused acoustic beam.* One such way is illustrated in Fig. 4. A laser beam coming from the left, is split in two. In one arm, the beam is stationary, and passes through a pinhole in a fixed array. In the other arm, the laser frequency is shifted by an amount F equal to the desired acoustic frequency. This beam with shifted frequency goes through a laser scanner to sequentially

---

*The basic combination of an optical plate and an OAT leads to system flexibility. Different operations of the same system, when desirable, can be chosen by changing the optical plate. For each selected operation, the pattern for the transparency can be easily calculated and modified to optimize the performance of the combination. Although this paper discusses orthographic imaging, the application of such a combination can be extended to pulse-echo systems as well.

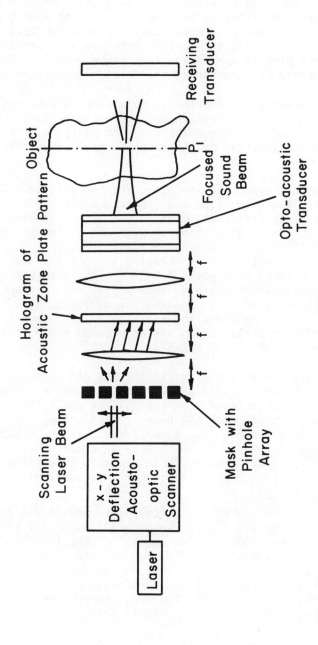

Figure 3   Schematic of the Zone-Plate Approach to Generate a Scanning Focused Beam

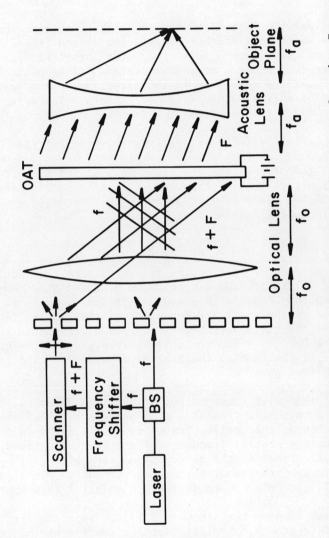

Figure 4 Schematic of the Sinusoidal-Fringe Approach to Generate a Scanning Focused Beam

address a pinhole array. The pinholes are in the front
focal plane of an optical Fourier-transform lens. At the
back focal plane of this lens is placed an OAT at resonance.

Consider the instant when we have the situation shown.
Two plane waves at different frequencies (f and f+F) illuminate the OAT structure. These two waves interfere with
each other to produce a moving sinusoidal fringe pattern on
the OAT. Let the x-axis be in the direction of the maximum
variation of this sinusoidal fringe pattern. The wave field
U at the OAT is then a sum of two terms:

$$U \propto Exp\{-j[2\pi(f+F)t-2\pi f_{xs}x]\}+Exp\{-j[2\pi ft-2\pi f_x x]\} \quad (3)$$

The first term represents the plane wave at the shifted
frequency with a spatial frequency f_{xs}. The second term
represents the plane wave at the original frequency with a
spatial frequency f_x. We have

$$UU^* \propto 1+cos[2\pi Ft-2\pi(f_{xs}-f_x)x] \quad (4)$$

The intensity (UU*) is a moving fringe pattern. It activates the OAT to produce a plane wave of sound at a temporal frequency F and with a spatial frequency equal to the
difference in the spatial frequencies of the two incident
waves. If the stationary beam is directed perpendicularly
toward the OAT ($f_x = 0$), the spatial frequency of the output sound is the same as the spatial frequency of the fre-
quency-shifted wave.

The plane wave sound output of the transducer is focused by an acoustic lens at the object plane. The laser
scanner changes the spatial frequency of the light in one
arm, hence the spatial frequency of the generated sound.
The acoustic focus at the object plane, therefore, is
scanned in synchronism with the laser scanning. This approach will be referred to as the sinusoidal fringe approach.

Let us compare the sinusoidal fringe approach with the
zone-plate approach. Similarities include the use of the
key element OAT, the use of a laser scanner and a pinhole
array. Furthermore, both are completely electronic, and
neither needs high density transducer arrays.

The major difference can be itemized as follows:

Pattern Used to Activate OAT

For the sinusoidal fringe approach, the pattern for each pixel is a moving-fringe pattern. The pattern is synthesized in real-time.

For the zone-plate approach, the pattern for each pixel is a stable pattern. Although the pattern is reconstructed in real-time, the zone-plate requires an extra step of fabrication or a recording. This does not imply a disadvantage.

Bias and Acoustic Frequency F

In the sinusoidal fringe approach, d.c. bias is used. The acoustic frequency F is derived from the motion of the moving fringes. It can be traced back to the frequency shifter. In the zone-plate approach, the voltage across the OAT is a.c.. The acoustic frequency f is derived from the a.c. source. Here the possibility of better impedance matching should permit a better electrical-to-acoustic power conversion.

Focusing and Optimum Resolution

In the sinusoidal fringe approach, acoustic lens focusing is used. In the zone-plate approach, we have lensless acoustic focusing. Assume ideal diffraction-limited resolution. In the sinusoidal fringe approach, the resolution will be limited by the numerical aperture of the acoustic lens. In the zone-plate approach, the resolution will be limited by the numerical aperture of the recorded zone-plate pattern. Because of lens aberrations, the zone-plate approach can achieve sharper focusing, hence better resolution.

Scanning

In the sinusoidal fringe approach, scanning is done by tilting and wobbling the acoustic plane wave. In the zone-plate approach, it is done by translating the projected zone-plate pattern (during reconstruction) at the OAT.

These differences set different requirements on the various parts of the system. For example, in the sinusoidal fringe approach, the optical spatial frequency determines the acoustic spatial frequency. We therefore have a wavelength ratio problem manifested in the pinhole spacing. The maximum pinhole spacing x_p is related to the acoustic lens focal length f_a, the optical lens focal length f_0, the resolution x_s, the optical wavelength λ, and the acoustic wavelength Λ through

$$\frac{x_p}{\lambda f_0} = \frac{x_s}{\Lambda f_a} . \tag{5}$$

For 3 MHz operation and a resolution of Λ, $f_0/f_a = 10$, x_p is of the order of 10λ, a figure which is too small to be practical. To obtain a practical pinhole spacing, a larger f_0/f_a ratio or a longer optical path length is necessary.

The requirement on the OAT also differs for these two approaches. Different patterns used for activating the OAT entail different minimum fringe spacing. This may lead to different demands on the OAT spatial response.

In the zone-plate approach, the minimum fringe spacing is the same for each pixel. In the sinusoidal fringe approach, the fringe spacing is minimum for pixels at the edge of the field.

Consider a numerical example. Assume a sound frequency of 3 MHz, a resolution of Λ, a focal distance of the sound of 10cm, and a capability of 200×200 resolvable spots. This would require a zone-plate of 72 zones with minimum zone spacing of 0.8mm. The corresponding minimum spacing of fringes for the sinusoidal fringe approach is 1mm. Although the zone-plate approach has slightly more stringent fringe-spacing requirements, the difference is not significant.

CONCLUSION

A more important difference is associated with the way scanning is achieved. This difference manifests itself in terms of differing requirements for the two approaches. In the zone-plate approach, a larger OAT dimension is required

and hence the cost of the OAT will be higher. To achieve
the same focal distance and the same central resolution,
the required OAT radius for the zone-plate approach exceeds
that for the sinusoidal fringe approach by the dimension of
the half-field. On the other hand, in the sinusoidal fringe
approach, a large acoustic-lens aperture will be required.
This letter requirement (on the acoustic lens) is particularly stringent; this requirement corresponds to an un-
reasonable focal number because of spherical aberration and
coma, unless edge resolution is sacrificed. With all factors considered, the zone-plate approach appears to be the
better of the two approaches.

Our next step will be to build a high-sensitivity
imaging system based on the zone-plate approach. Several
simplifications and modifications, of this approach, however, are currently under consideration.

REFERENCES

1. K. Wang and G. Wade, "Threshold Contrast for Various
   Acoustic Imaging Systems," pp. 431-462, Acoustical
   Holography, V. 4, Wade Ed., Plenum Press, New York, 1972.

2. K. Wang and G. Wade, "Threshold Contrast for Three
   Real-Time Acoustic Imaging Systems." p. 239, Acoustical
   Holography, V. 5, Green Ed., Plenum Press, New York,1974.

3. K. Wang and G. Wade, "Comparison of Ideal Performance
   of Real-Time Acoustic-Imaging Systems," Journal of the
   Acoustical Society of America, p. 922, V. 56, No. 3,
   September 1974.

4. B. B. Brendon, "Real Time Acoustical Imaging by Means
   of Liquid Surface Holography," pp. 1-9, Acoustical
   Holography, V. 4, Wade Ed., Plenum Press, New York,
   1972.

5. R. L. Whitman, M. Ahmed, and A. Korpel, "A Progress
   Report on the Laser Scanned Acoustic Camera," pp. 11-
   32, Acoustical Holography, V. 4, Wade Ed., Plenum Press,
   New York, 1972.

6. J. Landry, H. Keyani, and G. Wade, "Bragg-Diffraction Imaging: A Potential Technique for Medical Diagnosis and Material Inspection," pp. 127-146, Acoustical Holography, V. 4, Wade Ed., Plenum Press, New York, 1972.

7. P. Green, L. Schaefer, E. Jones, and J. Suarez, "A New High-Performance Ultrasonic Camera System," pp. 493-504. Acoustical Holography, V. 5, Plenum Press, New York, 1974.

8. See for example, Target Specification, Experiment No. 5, National Science Foundation.

9. B. A. Auld, S. A. Farrow, "An Acoustic Phase Plate Imaging Device," in this volume of Acoustical Holography.

OPTICAL VISUALIZATION OF ACOUSTICAL HOLOGRAMS

BY AREA HETERODYNING

Martin D. Fox

University of Connecticut

Storrs, Connecticut

ABSTRACT

A new acoustic holography detection mode is described which uses an expanded, premodulated laser beam to interrogate a mirrored plexiglas faceplate. Fresnel diffraction detects the acoustic field pattern, and a scanned array of discrete photodiodes using line subtraction processing picks up the resulting image. Expressions are derived to describe the principles of operation of the system. Experimental results are presented which show the real time visualization of the acoustic hologram of an off-axis point source. Sensitivity improvement through the use of line subtraction processing is demonstrated.

INTRODUCTION

Detection of ultrasonic fields by optoacoustic means has been accomplished in a number of different modes. In one approach sonically induced surface deformations at a liquid-air interface are picked up by an incident light field.[1] The main problems with liquid-air interfaces are instability due to their readily disturbable nature, acoustic streaming, lack of sensitivity and limited spatial frequency response.

In order to avoid the problems arising from liquid-air interfaces investigators [2,3] have explored the posibility of using liquid-solid or solid-air interfaces for acousto-optic image conversion. While in liquid-air inter-

229

faces the ripples represent a static image of the ultrasonic field pattern, in the liquid-solid or solid-air case,
the ripple pattern vibrates at the ultrasonic frequency.
Thus heterodyning is required to reduce the ripple frequency to a detectable value.

In the past, scanned laser beams were used to detect
the acoustic field at the solid-liquid or solid-air interface,[4] with subsequent injection of an electronic signal
to achieve the required heterodyning. Green et. al.[5] used
an expanded laser beam to interrogate a liquid-solid interface, but only shadow imaging was demonstrated. Problems
with the scanning approach include the loss of the capability for on-line optical Fourier processing or holographic
reconstruction, and lowered theoretically obtainable sensitivity due to the limited amount of time each picture
element is observed.

A new approach to opto-acoustic detection on solids
is presented here which uses area heterodyning. The probing light beam is expanded and premodulated at the acoustic
frequency so that it beats with the dynamic ripple pattern
resulting in a static image. Another novel feature is the
use of Fresnel diffraction behind a lens to convert the
phase modulation produced by the acoustic field at the surface of the faceplate into an amplitude modulation of the
light field in the image plane. In addition, a frame subtraction processing scheme is presented, which electronically separates signal information from statically scattered
light. This paper discusses the theory of the approach,
describes validating experiments, and presents the results.

It was found that the scheme could be used to visualize acoustic holograms in real time and the frame subtraction processing appeared to significantly improve system
sensitivity. The advantages include no laser scanning,
higher theoretically attainable sensitivity, capability
for on-line optical processing, and lessened mechanical
stability problems.

PRINCIPLES OF OPERATION

Consider the experimental arrangement illustrated in
Fig. 1. A 15 mw helium-neon laser produces a quasimono-
chromatic beam of coherent light. Assuming that an acoustic absorber is placed in the far end of the Raman-Nath cell
the acoustic field pattern can be represented in traveling

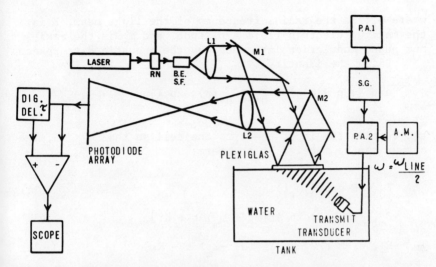

Fig. 1. Experimental arrangement for real time area heterodyne optical detection of acoustic holograms. RN Raman Nath
cell; B.E.S.F., beam expander and spatial filter; L1, L2,
lenses; M1, M2, mirrors; P.A.1, P.A.2, R.F. power amplifiers; S.G., R.F. signal generator; A.M., amplitude modulation.

wave form

$$P(x,t) = P_o \sin(\omega_m t + Kx) \tag{1}$$

where P_0 is the maximum pressure amplitude, ω_m is the radial
frequency of the acoustic wave, and K is the acoustic wavenumber.

For low acoustic intensities, the effect of the acoustic field on a beam of light propagating at right angles
is to impose a phase ripple on the wavefront of the beam
at the output of the cell. Thus the exiting light beam
can be represented

$$E_a + E \exp\{j[\omega_\ell t + m_m \sin(\omega_m t + Kx)]\} \tag{2}$$

where $\omega_\ell t$ is the radial frequency of the light beam, E is the maximum electric field amplitude, and m_m is the maximum phase modulation introduced by the acoustic disturbance. Using the identity

$$\exp (j\ p\ \sin\ q) = \sum_{n=-\infty}^{\infty} J_n (p)\ \exp (jnq)$$

and setting the input power to the cell so that $m_m \leq .3$, the output can be modeled by the terms

$$E_0 = E \exp (j\omega_\ell t),\qquad\qquad\qquad\qquad\qquad (3a)$$

$$E_{+1} = (m_m E/2)\exp j[(\omega_\ell + \omega_m)\ t + Kx],\qquad (3b)$$

$$E_{-1} = (-m_m E/2)\exp j[\ (\omega_\ell - \omega_m)\ t - Kx],\qquad (3c)$$

Obtaining Single Sideband Modulation

If the input monochromatic illumination to the cell is of unity amplitude within a circle of radius R, the electric field amplitude can be represented by

$$E(r) = \{ \begin{matrix} 1 & r \leq R \\ 0 & o.w. \end{matrix}\qquad\qquad\qquad\qquad (4)$$

Then when the output of the Raman-Nath cell passes through the transforming lens of the spatial filter, the theory of Fourier optics can be used to derive the distribution in the focal plane:

$$E_0(f_x,f_y) = ER\ J_1[2\pi R\ (f_x^2 + f_y^2)^{\frac{1}{2}}]\ (f_x^2 + f_y^2)^{-\frac{1}{2}} \quad (5a)$$

$$E_{+1}(f_x,f_y) = (m_m ER/2)J_1\{2\pi R\ [(f_x-K/2\pi)^2 + f_y^2]^{\frac{1}{2}}\}$$
$$[(fx - K/2\pi)^2 + f_y^2]^{-\frac{1}{2}} \qquad\qquad (5b)$$

$$E_{-1}(f_x,f_y) = (-m_m ER/2)J_1\{2\pi R[(f_x + K/2\pi)^2 + f_y^2]^{\frac{1}{2}}$$
$$[(f_x + K/2\pi\ ^2 + f_y^2]^{-\frac{1}{2}} \qquad\qquad (5c)$$

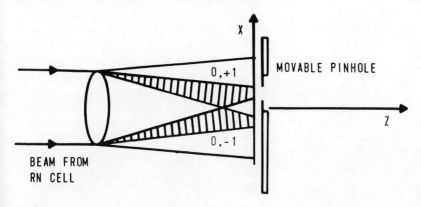

Fig. 2. Exaggerated illustration of production of collinear zero and first order light rays in the focal plane
of the spatial filter.

where $f_x = (x_f/f)$, x_f being the coordinate in the focal
plane, and f the focal length of the transforming lens.
As illustrated in Fig. 2, as long as the Bessel functions
do not separate in the transform plane, there will be a
point on the f_x axis where collinear rays of zero and first
order light of equal amplitude will exist. If the pinhole
is placed at this point, the output from the spatial filter
can be expressed

$$E_b = K[1 + \exp{(j\omega_m t)}] \exp{(j\omega_\ell t)} \qquad (6)$$

where K represents the maximum amplitude of the radiation.

                    Interaction with Plexiglas

    When (6) illuminates the mirrored plexiglas, a phase
ripple will be imposed upon it by the acoustic field in
the plate. The resultant waveform will be

$$E_c = K [1 + \exp (j\omega_m t)] \exp \{j[\omega_\ell t +$$
$$m_s \sin (\omega_s t + \phi(x,y)]\} \qquad (7)$$

where $\phi(x,y)$ is the spatial phase modulation imposed by acoustic irradiation of the mirrored surface of the plexiglas, ω_s is the temporal modulation and m_s is the modulation index of the acoustic field. For a single acoustic source, at infinity, radiating at an angle θ to the x-axis, and normally to the y-axis,

$$\phi(x,y) = 2\pi f_o x \qquad (8)$$

where f_o, the spatial frequency of the acoustic field, will be $\sin \theta/\Lambda$, Λ is the acoustic wavelength in water, and constant phase terms have been dropped since they do not contribute to the imaging process.

Modulation Conversion

In the form of Eq. 8, the acoustic field cannot be detected by a slow area sensor such as the eye or photographic film because it is time varying at an ultrasonic frequency, and because it is a pure phase modulation of the light beam. The imaging system presented here overcomes both of these difficulties. First we will discuss how phase modulation can be converted into amplitude modulation.

A pure phase modulation of a coherent light beam will be invisible since optical detectors are intensity sensitive. Techniques to convert optical phase modulation to amplitude modulation include Schlieren, Zernike phase plate and Fresnel diffraction.[6] Here it was necessary to employ Fresnel diffraction, since the spatial frequency of the acoustic signal was insufficient to separate the acoustically induced sidebands from the zero order in the transform plane, the focus of lens L2. These sidebands represent the reconstructed true and conjugate acoustic holographic images thus the same condition which requires Fresnel diffraction detection rules against the possibility of real time reconstruction. Although this limitation was found to hold in the present research, with the use of higher acoustic frequencies and better optical components separation should be achieved.

A modification of standard Fresnel diffraction detection was used in the present study; instead of simply allowing the beam to propagate in a linear fashion, a quadratic phase shift was introduced by the lens L2. This had the two primary effects, introducing a magnification and accelerating the Fresnel modulation conversion process, thus bringing the image closer to the dectection plane.

To show how the phase modulation of Eq. 8 can be converted into amplitude modulation in the Fresnel region behind a converging lens, consider the Fresnel diffraction formula[7].

$$U(x_2,y_2) = \int_{-\infty}^{\infty}\int U(x_1,y_1) \exp\left[-\frac{ik}{2f}(x_1^2, + y_1^2)\right]$$

$$\{\exp\left[i \frac{k}{2z}((x_2 - x_1)^2 + (y_2 - y_1)^2)\right]\}dx_1 dy_1 \qquad (9)$$

where $U(x_2,y_2)$ is the field amplitude in an arbitrary plane behind the lens, $U(x_1,y_1)$ is the field amplitude immediately in front of the lens, k is the optical wavenumber, f is the focal length of the lens and constant phase factors have been dropped. Defining the magnification of the system by

$$M = 1 - \frac{z}{f} , \qquad (10)$$

where z represents distance along the optical axis, make the change of variables

$$\tilde{x}_2 = \frac{x_2}{M} , \quad \tilde{y}_2 = \frac{y_2}{M} , \qquad (11)$$

in which case Eq. (9) can be expressed

$$U(\tilde{x}_2,\tilde{y}_2) = A(\tilde{x}_2,\tilde{y}_2) \int_{-\infty}^{\infty}\int U(x_1,y_1) \exp\{i \frac{k}{2}(\frac{1}{z} - \frac{1}{f})$$

$$[(\tilde{x}_2 - x_1)^2 + (\tilde{y}_2 - y_1)^2]\} dx_1 dy_1 \qquad (12)$$

where $A(\tilde{x}_2,\tilde{y}_2)$ is complex function of $\tilde{x}_2,\tilde{y}_2$. Then Eq. (7) can be seen to represent a convolution between the input

amplitude distribution $U(x_1, y_1)$ and the impulse response h
of the processor. Thus

$$U(\tilde{x}_2, \tilde{y}_2) = \Lambda(\tilde{x}_2, \tilde{y}_2) \; [U(x_1, y_1) * h(\tilde{x}_2, \tilde{y}_2)] \qquad (13)$$

where

$$h(\tilde{x}_2, \tilde{y}_2) = \exp \left[i \; \frac{k}{2} \; (\frac{1}{z} - \frac{1}{f}) \; (\tilde{x}_2^{\,2} + \tilde{y}_2^{\,2}) \right]. \qquad (14)$$

The transfer function of the Fresnel diffraction behind a
converging lens can then be calculated by Fourier transforming h to obtain

$$H(f_x, f_y) = \exp \left[-i\pi\lambda \; \frac{zf}{f-z} \; (f_x^{\,2} + f_y^{\,2}) \right]. \qquad (15)$$

Physically the effect of the lens is twofold; it causes
a linear magnification of the image, and more importantly, it
imparts a quadratic phase dispersion.
Thus from (15) Fresnel diffraction behind L2 results in
a phase delay of

$$\Delta\Theta = \frac{-\pi\lambda zf}{f-z} \; f_o^{\,2}. \qquad (16)$$

The effect of such Fresnel diffraction modulation conversion for a typical geometry are shown in Figs. 3 and 4.
Figure 4 illustrates the percent modulation conversion along
the optical axis of a typical system for an acoustic signal
of spatial frequency f_o = 6 cycles/cm. Note that in the
region from f = .9 - 1.2, no MTF is shown, since this zone
represents the Fraunhofer diffraction region.
In general, acoustic holograms will contain an ensemble
of spatial frequencies corresponding to the pass band of the
object under investigation. The fidelity of imaging such
an acoustic hologram will be dependent on the MTF of the
detection mode. Figure 3 shows a typical MTF for a lens
of 87 cm focal length, and z = 1.7f.

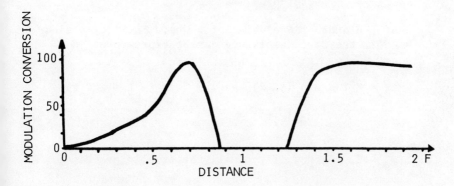

Fig. 3. Plot of percent modulation conversion versus distance along optical axis in unis of f, the focal length of lens L2, for f = 87cm. f_o, the spatial frequency detected, is 6 cycles/cm.

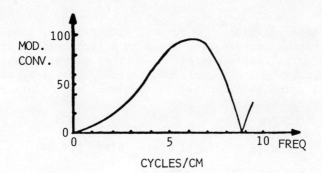

Fig. 4. Plot of predicted MTF for distance along optical axis, z = 1.7f, f = 87cm. Note the usable passband between 3.5 - 8 cycles/cm.

Heterodyne Image Detection

At the appropriate value of z where Eq. (16) yields $\Delta\Theta = -\pi/2$, the field amplitude can be represented

$$E_d = K[1 + \exp (j\omega_m t)] \exp [m_s \sin (\omega_s t + 2\pi f_o x)]$$

$$\exp (i\omega_\ell t). \tag{17}$$

With $m_s \leq .3$, which will typically be true in practice,

$$E_d \simeq K [1 + \exp (j\omega_m t)] [1 + m_s \sin (\omega_s t + 2\pi f_o x)]$$

$$\exp (i\omega_\ell t) \tag{18}$$

and the intensity will be,

$$I_d \simeq K^2 [2 + 2 \cos(\omega_m t)] [1 + 2 m_s \sin(\omega_s t + 2\pi f_o x)] \tag{19}$$

As mentioned earlier, the dynamic nature of the ripples at the plexiglas surface also poses an obstacle to their detection. The ripple pattern however is slowed or stopped by the beating with the premodulation. Thus if we expand Eq. (19) using the identity $\sin(a)\cos(b) = (\frac{1}{2})\sin(a + b) + (\frac{1}{2})\sin(a-b)$ we obtain

$$I_d \simeq K^2 \{2+2 \cos (\omega_m t) + 4 m_s \sin(\omega_s t + 2\pi f_o x)+$$

$$2m_s\sin[(\omega_s+\omega_m)t + 2\pi f_o x]+ 2m_s \sin[(\omega_s-\omega_m)t + 2\pi f_o x]\} \tag{20}$$

Utilizing a low pass sensor such as photographic film or the eye results in an observed intensity

$$I_d \simeq K^2 \{2 + 2 m_s \sin [(\omega_s-\omega_m)t + 2\pi f_o x]\} \tag{21}$$

which for $\omega_m = \omega_s$ is a static representation of the hologram on the surface of the plexiglas.

Line and Frame Subtraction Processing

Considering the dynamic nature of the acoustic field as compared to the static character of the noise terms in the system of Fig. 1 suggests a processing mode which can greatly enhance sensitivity of the imaging technique. If the acoustic hologram is tagged with an appropriate temporal modulation, and detected with a photodetector array, subtraction of a delayed signal from the present output will result in a cancellation of static noise and undesired background light and enhancement of the desired signal.

To demonstrate the application of frame subtraction processing to improve the contrast and signal to noise ratio of the image, assume that due to dust, dirt, lens aberrations, and nonumiformities in the mirrored plexiglas, a phase noise, $N(x,y)$ is produced, which, when included in the field of Eq. 7 yields

$$E_c = K[1 + \exp(j\,\omega_m t)]\,\exp\,\{j[\omega_\ell t +$$

$$m_s \sin(\omega_s t + \phi(x,y)] + N(x,y)\} \tag{20}$$

If the previous analysis of Fresnel diffraction detection which lead to Eq. (19) is now carried out, assuming $N(x,y) \leq .3$ and the use of a low pass sensor, the detected intensity can be expressed

$$I_d \simeq K^2\{2+2\,m_s \sin\,[(\omega_s-\omega_m)t + \phi(x,y)] + N(x,y)\} \tag{21}$$

where the fidelity with which $\phi(x,y)$ is replicated is dependent on the MTF of the modulation conversion process.

Two approaches can now be used to create a time variation in Eq. (21) consistent with frame subtraction processing. We can either let $\omega_s - \omega_m = \omega_f/2$, where ω_f is the radial frame rate, or we can let $\omega_s = \omega_m$ and vary the modulation index so that

$$m_s = m_s \cos [(\omega_f/2)t]. \tag{22}$$

The second technique will be followed in the analysis, since this was the approach which proved to be easiest to implement. Setting $\omega_s = \omega_m$ and substituting (22) into (21) yields

$$I_d \simeq K^2\{2 + 2 m_s \cos[(\omega_f/2)t] \sin [\phi(x,y)]$$

$$+ N(x,y)\}. \tag{23}$$

Now calculating I_d at $t = 0$ and $t = 2\pi/\omega_f$, the frame period, we obtain

$$I_d(t = 0) = K^2 \{2+2 m_s \sin[\phi(x,y)] + N(x,y)\} \tag{24}$$

and

$$I_d(t = 2\pi/\omega_f) = K^2 \{2-2 m_s \sin [\phi(x,y)] + N(x,y)\}. \tag{25}$$

Subtracting (25) from (24) leaves

$$\Delta I = I_d(t=0) - I_d (t=2\pi/\omega_f) = 4 K^2 m_s \sin[\phi(x,y)] \tag{26}$$

which is free of both zero order background and low amplitude noise produced by static effects. To use line subtraction processing, substitute ω_{line} for ω_f and use the same argument. Line subtraction processing yields the same advantages as frame subtraction, but requires less memory. In fact, point subtraction processing could be used, subtracting spatially adjacent points which are close enough to be within the Rayleigh limit of the acoustic image.

SENSITIVITY

High sensitivity is a key requirement for the imaging of deep biologic tissue structures in the reflection mode, and as a result, has become an important point of comparison between competing ultrasonic imaging modes directed toward biomedical application. Conceptually, the system introduced here should be capable of considerably higher sensitivity, than systems employing scanned laser beams due to the greater time of observation of each picture element.

To obtain an estimate of threshold sensitivity, take the minimum noise level to be set by the quantum noise inherent in the particle nature of light,[7]

$$E_q = \frac{h\nu}{q} \tag{27}$$

where E_q is the quantum noise energy, h is Planck's constant, ν is the light frequency, and q is the quantum efficiency of the detector. Then the signal to noise energy ratio will be

$$s/n = \frac{qA\tau I_i}{h\nu} \tag{28}$$

where A is the detector area, τ is the observation time, and I_i is the information bearing component of the received signal. Equation (28) shows that the signal to noise ratio, and thus the system sensitivity is directly proportional to observation time, τ, and information bearing intensity, I_i. While it would seem that sensitivity could be increased to the limit set by thermal acoustic noise, by increasing the incident laser intensity, and thus I_i, such a strategy fails in a practical system due to detector saturation. Given a fixed I_i, however, it is clear that the observation time τ for an array detector with storage is N times that of a scanned laser beam system, where N is the number of picture elements. Thus the reporting of a threshold of 5×10^{-9} w/cm^2 for a scanned laser beam system suggests an attainable sensitivity on the order of 10^{-13} w/cm^2, assuming 10^4 picture elements, for an area heterodyne mode of detection for equal laser power applied to each pixel and equal image bandwidth.

EXPERIMENTAL RESULTS

Two experiments will be described. The first was carried out to demonstrate real time detection of acoustic holograms using area heterodyning combined with Fresnel diffraction. The second shows the improvement in sensitivity resulting from line subtraction processing.

Visualization of Acoustic Hologram

Recall the schema illustrated in Fig. 1. An ultrasonic transmitting transducer with expanding acoustic lens
was placed 8 cm, from a mirrored plexiglas faceplate, thus
simulating a point source at about 50° from the normal. Premodulation of the incident beam was provided by a water
filled plastic Raman-Nath cell, while the spatial filter
was a standard Gaertner beam expander and spatial filter
with a 20x microscope objective and a 25μ pinhole. The
ultrasonic frequency was 900 kHz. Lens L2 had an 87 cm
focal length, and the ground glass was positioned 278 cm
from L2. A ½" thick plate of plexiglas mirrored on the
front surface was used.

The resultant image is shown in Fig. 5. The fringe
pattern corresponds to the expected acoustic hologram of
an off-axis point source. The measured average spatial
frequency of about 5 cycles/cm, agreed with the theoretically predicted value.

Line Subtraction Processing

The experimental arrangement was as shown in Fig. 1,
except lens L2 was omitted to keep the acoustic pattern
small. The line repetition period was set at 1.25ms, and
the ultrasonic signal to the transmitting transducer was
amplitude modulated at 400 cycles/second. The output from
the photodiode array was converted to digital form, held in
a semiconductor memory, and converted back to analog form,
the total time for this process being 1.25 ms. The digital
signal was a 10 bit representation so that the entire 1024
level dynamic range of the photodiodes was retained. The
resultant output from the digital delay line went to an
operational amplifier and was subtracted from the subsequent line.

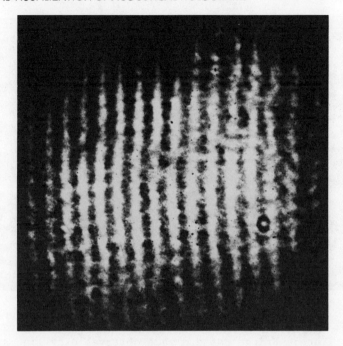

Fig. 5. Acoustic hologram of an off axis point source detected by an expanded, premodulated laser beam, at a mirrored plexiglas faceplate, using Fresnel diffraction detection. Acoustic frequency, 900 kHz; distance of point
source, 8 cm; average angle to normal, 50°.

The results of this operation are shown in Fig. 6. The
upper oscilloscope trace represents the unprocessed signal.
A 1/8" slab of plexiglas mirrored on the rear was used in
this experiment. Deterioration of the mirroring due to contact with the water in the tank resulted in a great deal
of static noise. Thus the upper trace did not reveal any
particular fringe pattern. The lower trace, shows the results of frame subtraction processing. Definite periodicity
is evident. Due to a lack of synchronization between the
amplitude modulator and the array in this preliminary experiment, the oscilloscope trace shown is a time averaged,
squared standing wave pattern. A marked improvement in signal to noise ratio is evident.

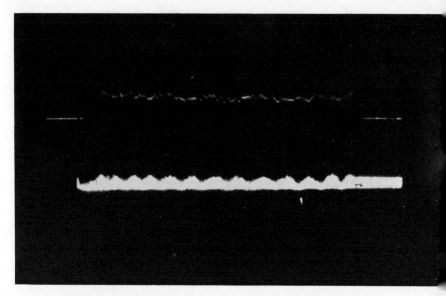

Fig. 6. Results of line subtraction processing. Upper
Trace: Unprocessed line output from photodiode array.
Lower Trace: Fringe pattern made observable by line sub-
traction.

DISCUSSION AND CONCLUSIONS

The new ultrasound imaging technique presented here
has considerable potential for real time imaging in biomedical or nondestructive testing applications. Instability problems such as those associated with liquid-air
interfaces are bypassed through the use of a solid faceplate. Conceptually, the approach suggested here should
provide N, where N is the number of photodetectors, times
more sensitive detection than scanned laser beam techniques, for equivalent laser power, due to the longer time
interval over which each element can sense the signal.
Frame subtraction processing provides a means to greatly
diminish the effects of static phase anomalies in the optical system electronically.

Since the present system only detects the acoustic
field pattern, it does not provide acoustic holographic

reconstruction. Theoretically, such a reconstruction does exist near the focus of L2, but we have thus far been unable to detect it.

In the absence of on-line reconstruction, the most immediate applications will be in the observation of focused acoustic images, shadow imaging, and acoustic holography in which delayed reconstruction is acceptable.

Ultimately, the utility of the technique will hinge on the attainable sensitivity and resolution. If the spatial frequency of the sonic pattern was high enough to throw the sideband off-axis appreciably, optical spatial filters would improve sensitivity considerably. Studies are currently underway to determine the practical limitations on sensitivity and to increase resolution through the use of higher acoustic frequencies.

ACKNOWLEDGMENTS

This work was supported in part by NSF grant GK42114 and University of Connecticut Research Foundation grant 35 - 146. The author is grateful to R. Northrop, and J. Walters for many helpful discussions, and L. Puffer and K. Taylor for their assistance in the construction of the experimental apparatus.

REFERENCES

(1) Brenden, B. B., Acoustical Holography Vol. 4, ed. by G. Wade, p. 1 Plenum Press, 1973.

(2) Whitman, R. L. and Korpel, A., Appl. Opt., 8, 1567, 1969.

(3) Massey, G. A., Proc. of the IEEE, 56, 2157, 1968.

(4) Korpel, A. and Desmares, P. J. of Acoust. Soc. of Amer., 45, 881, 1969.

(5) Green, P. S.,Macovski, A. and Ramsey, S. D., Appl. Phys. Lett., 16, 265, 1970.

(6) Damon, R. W.,Maloney, W. T. and McMahon, D. H., Physical Acoustics, Vol. VII, ed. by W. P. Mason and R. N. Thurston, p. 336, Academic Press, New York, 1970.

(7) Goodman, J. W., Introduction to Fourier Optics, p. 60,
     McGraw Hill, 1968.

(8) Mezrick, R., Etzold, K., and Vilkomerson, D., Acoustical
     Holography, Vol. 6, ed. by N. Booth, Plenum Press, in
     Press.

(9) Korpel, A. and Kessler, L., Acoustical Holography, Vol.
     3, p. 23, Plenum Press, 1972.

REAL-TIME DOPPLER IMAGING FOR UNAMBIGUOUS MEASUREMENT OF

BLOOD VOLUME FLOW

C.F. Hottinger, J.D. Meindl

Stanford University

Stanford, California 94305

ABSTRACT

A new ultrasonic method involving real-time Doppler imaging in the transcutaneous measurement of blood volume flow is described. The approach involves examination of an arbitrary sample plane cutting across the vessel lumen. Two general techniques utilizing this routine are described.

The first technique measures separately (1) the average particle velocity normal to the plane, $< \bar{v} \cdot \bar{n} >$, and (2) the projected area of the vessel lumen, A_{PROJ}; the product of these quantities is the volume flow. This routine enjoys several advantages over conventional methods; in particular, estimates of volume flow are not a function of vessel shape, orientation, or the velocity profile shape.

Two implementations of this technique are under development. The first involves a two-dimensional array of pulsed-Doppler transducer elements; each element is range-gated to examine an incremental element of the sample plane. The returning signals are first processed to generate a Doppler C-scan, and the area displayed is summed to estimate A_{PROJ}. The signals returning to each element are also summed, and the centroid of the Doppler power spectrum indicates $< \bar{v} \cdot \bar{n} >$. The second implementation directly measuring A_{PROJ} uses a linear array to generate a Doppler B-scan of the sample plane. A large single element transducer uniformly illuminates the sample plane, and the centroid of the back-scattered Doppler signal is measured to estimate $< \bar{v} \cdot \bar{n} >$.

The second technique involving examination of an arbitrary sample plane does not rely on separate measurements

of A_{PROJ} and $< \overline{v \cdot n} >$. Instead, a measurement is made of
the detected power backscattered from a known volume of
blood. When the sample plane is simultaneously isonified
with a uniform beam pattern, the unnormalized first-moment
of the Doppler power spectrum can be properly scaled to
yield an unambiguous estimate of volume flow.

1. INTRODUCTION

The clinical importance of measuring blood volume
flow through the different vessels in the body has long
been recognized. This capability would allow both study of
normal vascular conditions as well as the diagnosis of a
variety of pathological states.[1] Most of the modalities
proposed for this task suffer from one or more inherent
drawbacks. Techniques which use the electromagnetic cuff,
the ultrasonic transit-time detector, or radioactive dye
dilution measurements are inherently invasive and not
suited to transcutaneous modes of use. Such non-invasive
methods as plethysmography, nuclear magnetic resonance
detection, and electrical impecance techniques are insuffi-
ciently specific; distinguishing between flows through the
internal and carotid arteries, for instance, is not possible.

During the past fifteen years, the ultrasonic Doppler
shift has been exploited as a modality for blood flow
measurement.[2,3,4,5] While not suffering from the inherent
difficulties that plagued other methods, ultrasonic Doppler
shift techniques have until now required careful measurement or estimation of parameters such as vessel orientation, size, and shape, as well as the shape of the velocity
profile across the lumen. Without determination of these
parameters, an accurate estimation of volume flow was not
possible. This paper outlines a new method utilizing the
ultrasonic Doppler shift for both velocity measurement and
moving-target imaging[6] that overcomes this problem.

2. RAYLEIGH SCATTERING OF ULTRASOUND BY MOVING PARTICLES

The mechanism responsible for scattering ultrasound
below 16 MHz in blood has been shown by Reid[7] to be first-
order Rayleigh scattering by eurythrocytes. A particle's
speed and direction determine the detected Doppler shift,
while its size and location control the amplitude of the
scattered wave. Because of these relationships, the
detected Doppler power spectrum can be shown to contain the
required information for flow determination.

When a particle of velocity $\bar{v}$ scatters an incident beam with wave vector k_i and wavelength λ, the vector of the detected scattered wave $\bar{k}_s$ has been shifted in frequency[8]

$$\Delta f = (2\pi)^{-1} \bar{v} \cdot (\bar{k}_s - \bar{k}_i) \qquad \text{where} \qquad (1)$$

$$\left| \bar{k}_s \right| = \left| \bar{k}_i \right| (1 - \bar{v} \cdot \bar{k}_i) \qquad (2)$$

In the usual condition where $\left| \bar{v} \right| \ll c$ (the propagation velocity), (1) can be reduced to

$$\Delta f = \lambda^{-1} \left| \bar{v} \right| (\cos \theta_s - \cos \theta_i) \qquad (3)$$

In this situation $\left| \bar{v} \right| (\cos \theta_s - \cos \theta_i)/2$ is the component of the particle velocity in the direction defined by $(\bar{k}_s - \bar{k}_i)/\left| \bar{k}_s - \bar{k}_i \right|$. If the sound wave is simply back-scattered ($\bar{k}_s$ and $\bar{k}_i$ are anti-parallel), then

$$\Delta f = 2\lambda \left| \bar{v} \right| \cos \theta_s \qquad (4)$$

In this case, the detected Doppler shift indicates the velocity component along the beam axis.

The relative power scattered by a particle is determined not by its velocity, but by its size, location, and the scattering angle. For a particle of volume $\mathcal{U}$, at distance r from the detector, the scattered power per steradian is given by

$$\left(\frac{I_s}{I_o} \right) = \frac{\bar{k}_i^{\,4}}{16\pi} \frac{\mathcal{U}^2}{r^2} \left\{ \left(\gamma_K + \gamma_\rho \frac{\bar{k}_i \cdot \bar{k}_s}{\left| \bar{k}_i \right| \left| \bar{k}_s \right|} \right) \right\} \qquad (5)$$

Here γ_K and γ_ρ are the scattering co-efficients determined by the particle compressibility and density relative to the medium.

Since the particles within a sample volume are randomly located[9], the average scattered power received from uniformly isonified particles is proportional to the second moment of the particle volume distribution

$$\left\langle \sum_{\substack{N \\ \text{particles}}} \left(\frac{I_s}{I_o} \right) \right\rangle \sim \sum_{i=1}^{N} \mathcal{U}_i^{\,2} \qquad (6)$$

Normally, this second moment per unit volume is considered
uniform, on the average, through a fluid such as blood, so
that

$$\left(\frac{I_s}{I_o}\right) \sim \sum_{i=1}^{N} \upsilon_i^2 \sim N \tag{7}$$

In addition, particle motion is assumed to be due to
convection, so that volume flow of blood can be based on
measurements of particle movement. Under these conditions,
the Doppler power spectrum S(f) returning from a sample
volume has the property

S(f)df ~ number of particles causing Doppler shifts
        between f and f + df

If the particles within the sample volume are uniformly
isonified, then the back-scattered spectrum has as its
centroid

$$<f> = \frac{\int fS(f)df}{\int S(f)df} = \frac{2 \ |\bar{v}| \cos \theta}{\lambda} \tag{8}$$

Thus, the centroid of the back-scattered Doppler power
spectrum indicates the average velocity component parallel
to the beam axis within the sample volume.

But how should Doppler shift measurements be applied
to the measurement of volume flow? The optimum approach
would appear to be suggested by the general definition

$$\dot{Q} \ [ml/sec] \overset{\Delta}{=} \int \bar{v}\cdot\bar{n} \ da \tag{9}$$

where n is normal to incremental area da making up the
arbitrary surface S. Briefly stated, an unambiguous estimate of $\dot{Q}$ can be achieved by estimating the volume flow
normal to an arbitrary plane of examination.[10] This can
be achieved by choosing a sample volume as mentioned above
in the form of a sample "plane" cutting the flow. With
the beam axis normal to the plane, the Doppler power centroid then indicates the average fluid velocity normal to
the plane. The <u>quantity</u> of fluid having this average velo-

city can be determined in any of the three techniques that are outlined in the following sections.

3. DOPPLER IMAGING TO MEASURE PROJECTED LUMEN AREA

Equation (9) can be separated directly into the form

$$\dot{Q} = \int_{s} \overline{v} \cdot \overline{n} \, da = < \overline{v} \cdot \overline{n} > A_{PROJ} \tag{10}$$

where A_{PROJ} is the projected area of the vessel lumen onto the sample plane S, and $< \overline{v} \cdot \overline{n} >$ is the area-average fluid velocity normal to the plane. Stated in this form it is apparent that precise knowledge of the lumen size, shape, and orientation is not necessary for an accurate measurement of $\dot{Q}$. Instead, it is sufficient to know the area of the lumen projection A_{PROJ} onto an arbitrary sample plane. Likewise, precise determination of the two-dimensional velocity field across the lumen is not necessary. Rather, only the area-average velocity component $< \overline{v} \cdot \overline{n} >$ normal to the sample plane need be measured.

Figure 1 shows one implementation relying on Doppler imaging to measure the lumen projection A_{PROJ}. In this case, a two-dimensional array of near field transducer elements is range-gated to examine a single depth. When the return to each element is separately processed, a Doppler C-scan[12] can be generated to display the portion of the sample plane through which flow occurs.

A block diagram of such a system is shown in Figure 2. As indicated, a pulsed coherent Doppler detector is simultaneously multiplexed through an array of transducer elements as well as a bank of audio energy detectors. In this case, each audio energy detecting channel is range-gated to examine a single depth, and the output unblanks a point on the CRT display of the C-scan. The unblanked area of the display can be measured as shown to compute A_{PROJ}.

A less complex implementation of the same general principle is outlined in Figure 3. In this case a linear array of transducer elements is used to generate a Doppler B-scan[12] of the sample plane. The display is identical to the C-scan shown in Figure 1 since the same vessel section is to be visualized.

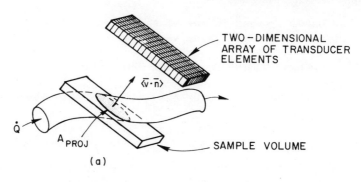

(a)

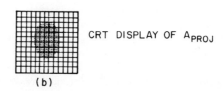

(b)

Fig. 1. (a) Pulsed-Doppler scanning of vessel lumen using
two-dimensional transducer array. (b) Doppler C-scan
display of vessel lumen projection on sample plane.

In this configuration, a large single element transducer is positioned to uniformly isonify the sample plane.
Again, the returning range-gated video is processed to
determine the average Doppler shift, and, hence, the area-
average particle velocity across the sample plane.

The systems outlined in Figures 1 and 3 both rely on
accurate Doppler imaging of the lumen interior to permit
calculation of the area of the lumen-projection. A chief
source of error in the estimation of volume flow by these
methods, however, arises from the measurement of the
projected lumen area.

The projected area can be estimated from the number of
unblanked points on the CRT; alternate schemes can be used
instead, but all are still imited in resolution by the
non-zero size of the Doppler-imaging sample volumes. As
shown by

$$\frac{\Delta A_{PROJ}}{A_{PROJ}} = \frac{\Delta a}{a} + \frac{\Delta b}{b} + \frac{\Delta a \Delta b}{ab} \tag{11}$$

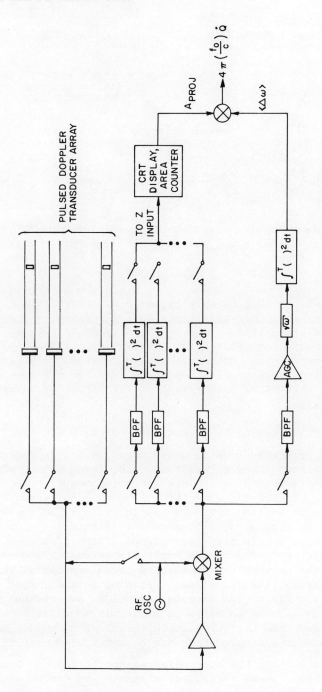

Fig. 2. Ultrasonic system for real-time Doppler-imaging and measurement of Doppler power-spectrum centroid.

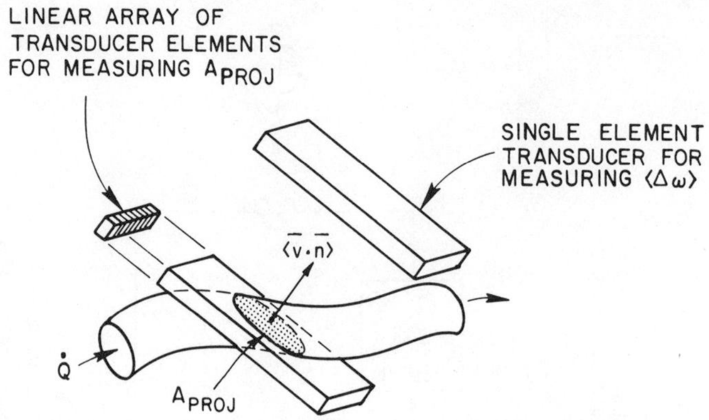

Fig. 3. Pulsed-Doppler scanning of vessel lumen using
         linear array and uniform-illumination transducer

in measuring the area of an ellipse, a 10% over-estimation
of the major and minor axis leads to a 21% error in the
volume flow estimation. Thus, while Doppler imaging is
still desirable for determining a vessel's location, as
well as general size, orientation, and relation to other
vessels, it is advantageous not to rely on lumen visualization for quantitative measurement of the vessel is internal
projected area.

4.   DIRECT ESTIMATION OF $\dot{Q}$ FROM THE DOPPLER POWER SPECTRUM

These problems prompted a new approach not dependent
on Doppler imaging for estimation of the lumen projected
area. In this method, a two-element transducer is used
(as shown in Figure 4), with both elements simultaneously
range-gated to examine the same sample plane. The smaller
central zone, element 2, is designed to locate its sample
volume totally within the vessel lumen. Back-scattered
Doppler power detected by the small transducer is proportional to the unit-volume scattering cross section of the
blood; however, like the signal returning to the larger
element, it suffers from attenuation due to absorption.
But if it is assumed possible to uniformly isonify the
sample plane, then the attenuation losses with respect to
all points on the sample plane can be considered uniform.

Figure 5 shows the block diagram specifying how this
system is used. Both transducer elements are driven with

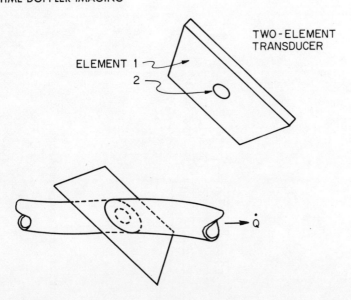

Fig. 4. Transducer configuration for measurement of $\dot{Q}$
with compensation for round-trip attenuation losses.

the same sinusoidal burst, but the returning back-scattered
signals are mixed separately. The video signals are likewise range-gated simultaneously after a delay corresponding
to the range of interest.

When the two signals are summed, as shown in the upper
signal path, uniform illumination of the sample plane can
be approximated. This composite audio signal, with power
spectrum $S_{1+2}(f)$, is passed through a $\sqrt{\omega}$ discriminator as
shown. The resulting signal power is then detected to
achieve an unnormalized average velocity. After the gain
in this channel is adjusted to compensate for the different
antenna beam gains of the two transducer elements, the
unnormalized average velocity estimate is divided by the
Doppler power level detected by the small XDCR element.

This operation is expressed as

$$4\pi \left(\frac{f_o}{c}\right) \dot{Q} = \frac{\int |G_2(r)|^2 dr}{|G_{1+2}(0)|^2} \frac{\int f\, S_{1+2}(f)\, df}{\int S_2(f)\, df} \tag{12}$$

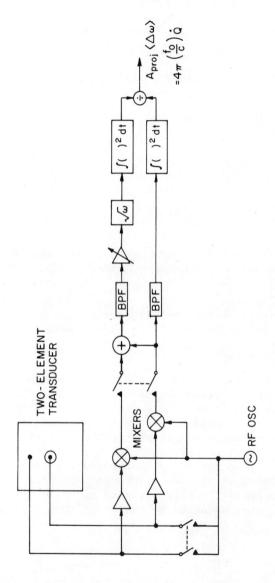

Fig. 5.   System for measurement of $\dot{Q}$ utilizing compensation for attenuation.

While f_o is the transmitted ultrasonic frequency, $\int \left| G_2(r) \right|^2 dr$ is the antenna beam factor of element 2 integrated over its entire sample volume, $\left| G_{1+2}(o) \right|^2$ is the beam factor per cm^2 of sample area when elements 1 and 2 are summed, and the integral $\int S_2(f) df$ indicates the Doppler power received by element 2 from within the lumen.

5. CONCLUSION

The roles that real-time Doppler imaging can potentially play in the measurement of blood flow can be summarized as follows. The first two configurations described required either a Doppler B-scan or C-scan to locate the vessel and determine accurately its projected area. Because of the judicious choice of sample volume and transducer alignment, precise knowledge of the lumen orientation, size, and shape is not necessary. Nor is precise determination of the fluid velocity profile across the lumen required, since only $< \bar{v} \cdot \bar{n} >$, the average velocity component across the sample plane, need be measured to estimate volume flow.

The third system described has freed the user from dependence on Doppler imaging for accurate estimation of the projected lumen area as well. Imaging is still required for proper positioning and alignment of the two-element transducer when used in the transcutaneous mode. Proper positioning of the transducer is critical since the sample volume of the smaller transducer must lie totally within the lumen of the vessel of interest, and the sample volume of the larger element cannot include any other vessel.

ACKNOWLEDGMENT

This investigation was supported by PHS Research Grant No. 1P01 GM17940-04 from the Department of Health, Education, and Welfare, and by NASA Research Grant No. NGR-05-020-615.

REFERENCES

1. V.C. Roberts, "A Review of Non-Invasive Measurement of Blood Flow," Biomedical Engineering, p. 332, August 1974.

2. D.L. Franklin, W. Schlegel, and R.F. Rushmer, "Blood
    Flow Measured by Doppler Frequency Shift of Back-
    Scattered Ultrasound," Science 134, pp. 564-568, August
    1961.

3. P.A. Peronneau and F. Leger, "Doppler Ultrasonic Pulse
    Blood Flowmeter," Proc. 8th ICMBE & 22nd ACEMB,
    Chicago, Ill., 1969.

4. G. Cross, L.H. Light, "Non-Invasive Intra-Thoracic
    Blood Velocity Measurement in the Assessment of Cardio-
    vascular Function, Biomedical Engineering, pp. 464-477,
    October 1974.

5. K. McCarty, J.P. Woodcock, "The Ultrasonic Doppler Shift
    Flowmeter - A New Development," Biomedical Engineering,
    pp. 336-341, August 1974.

6. C.F. Hottinger, J.D. Meindl, "An Ultrasonic Scanning
    System for Arterial Imaging," Proc. IEEE Ultrasonics
    Symposium, Monterey, Ca., 1973.

7. J. Reid, et.al., "Scattering of Ultrasound by Human
    Blood," Proc. 8th ICMBE and 22nd ACEMB, Chicago, Ill.,
    1969.

8. P. Morse, K. Ingard. Theoretical Acoustics. New York:
    McGraw-Hill, 1968.

9. W.R. Brody and J.D. Meindl, "Theoretical Analysis of
    the CW Doppler Ultrasonic Flowmeter," IEEE Trans.
    Biomed. Eng., vol. BME-23, pp. 183-192, May 1974.

10. C.F. Hottinger and J.D. Meindl, "Unambiguous Measure-
    ment of Volume Flow Using Ultrasound," Letter to Pro-
    ceedings of the IEEE (in press).

11. L. Gerzberg, "Development of the First Moment Detector,"
    private communication, Stanford Electronic Laboratories,
    Stanford, Ca., Sept. 1973.

12. P.N.T. Wells. Physical Principles of Ultrasonic Diagnosis
    London:Academic Press, 1969, p. 141-187.

AN ACOUSTIC PHASE PLATE IMAGING DEVICE

S. A. Farnow and B. A. Auld

Stanford University

Stanford, California 94305

INTRODUCTION

Fresnel zone plates have long been used in the field of classical optics and more recently have had applications in such varied fields as gamma ray imaging[1] and laser micromachining.[2] Acoustic applications have been investigated at frequencies of 1.5 GHz, where zone plates were used to focus volume acoustic waves in sapphire rods.[3] We have recently begun exploring the use of zone plates to focus sound in water at frequencies near 10 MHz. Our original approach to this problem was to use a zone plate pattern of appropriate design as one electrode on a disk transducer.[4] We have found, however, that the performance of these devices may be greatly enhanced by converting them to acoustic Rayleigh-Wood phase-reversal zone plates,[5] or acoustic phase plates. We shall describe how this conversion is easily effected through the use of a simple poling procedure. Our imaging system, the heart of which is this phase plate transducer, has resolution capabilities of 0.27 mm or approximately 1.8 wavelengths. Images have been obtained in both transmission and reflection modes.

PROPERTIES OF FRESNEL ZONE PLATES

The concept of a zone plate evolves directly from the diffraction theories put forward by Fresnel in 1815. He approached diffraction problems by regarding each point on a wavefront as being a source of secondary wavelets, Huygens'

259

wavelets. Treating the monochromatic wave at the aperture plane of Fig. 1 in this way, we see that radiation of all phases reaches the point P . The intensity at P will be increased if all but in-phase radiation is blocked at the aperture. From the diagram, if the regions between r_2 and r_1 , r_4 and r_3 , etc. (or the regions between r_1 and r_0 , r_3 and r_2 , etc.) are made non-transmitting, and if $R_n = f + n\lambda/2$, all radiation at P will constructively interfere. From the geometry in Fig. 1,

$$r_n^2 = R_n^2 - f^2 = n\lambda f + \frac{n^2\lambda^2}{4} . \tag{1}$$

For most cases of practical interest, $f \gg n\lambda/4$ and

$$r_n^2 \simeq nr_1^2 \simeq n\lambda f . \tag{2}$$

Apertures constructed in this way are called zone plates.[5]

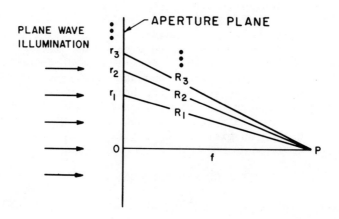

PLANE WAVE ILLUMINATION

APERTURE PLANE

FIG. 1--Fresnel zone plate construction.

Using Eq. (2), we may write the transmittance function for a zone plate of infinite extent as

$$t(r) = \frac{1}{2} \pm \frac{j}{\pi} \sum_{\substack{n=-\infty \\ (\text{odd})}}^{\infty} \frac{1}{n} \exp(-jn\pi r^2/r_1^2) \qquad (3)$$

where $+$ denotes a positive, or transmissive central zone, zone plate and $-$ denotes a negative, or non-transmissive central zone, zone plate. Comparison of Eq. (3) with the transmittance function for an ideal thin lens of focal length f',

$$g(r) = \exp(-j\pi r^2/\lambda f') , \qquad (4)$$

demonstrates that the zone plate transforms an incident plane wave into an undiffracted plane wave (the $1/2$ term) and a superposition of spherical waves with focal lengths

$$f_n = \frac{r_1^2}{\lambda n} = \frac{f}{n} \qquad n = -\infty; \ldots; \infty; \text{odd} \qquad (5)$$

where $f_1 = f$ is the primary focal length. Thus, in addition to the primary focus, we have an infinite number of real and virtual foci. The first few real foci are shown in Fig. 2, along with the plane wave, or so-called DC term.

In practice, of course, one uses a finite number of zones. Mathematically, this may be expressed by multiplying the transmittance function by an appropriate pupil function,[6] such as the Heaviside step function $H(D/2-r)$, where D is the diameter of the outermost ring. From Eqs. (3) and (4), then, we should expect that the properties of the primary focus should resemble those for an ideal lens of the same diameter and focal length. The normalized focal plane intensity distribution of such a lens is

$$I(r) = \left[2 \frac{J_1(\pi Dr/\lambda f)}{\pi Dr/\lambda f} \right]^2 . \qquad (6)$$

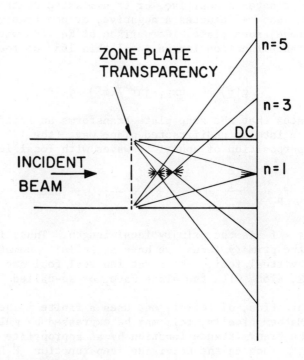

FIG. 2--The focusing properties of a Fresnel zone plate.
(Virtual foci not shown.)

If the number of active zones is kept above seven, Eq. (6) may be found to also be an adequate description of the zone plate focal plane intensity distribution.[2,7]

ACOUSTIC ZONE PLATE (AZP) TRANSDUCERS

To realize an acoustic zone plate, it is necessary to excite a transducer only in regions corresponding to the transmissive zones of the zone plate pattern. Ours were fabricated by using photolithographic techniques to evaporate a gold, 10 zone, negative zone plate as one electrode on a PZT-5A disk, with a full-face counterelectrode. The gold annuli which made up the zone plate electrode were electrically interconnected with 2-mil gold wire. The AZP transducers were designed to have focal lengths of 3 cm in water at a frequency of 10 MHz, giving us the equivalent of $f/1.5$ lenses. Since the focal length is $\sim 200\lambda$, the approximation in Eq. (2) is seen to be valid.

While we operated at 10 MHz, the PZT-5A disks actually had fundamentals of 7.35 MHz. Experimentally, we found that the focal spot diameter was reduced by a factor of about two when we operated at 1.4 times the fundamental as opposed to on fundamental. This behavior may be related to the spatial frequency response, the so-called transfer function H ,[6] of the AZP transducers. The zone plate response may be considered constant up to a cutoff, determined by its diameter and focal length,[1] Fig. 3a. The PZT-5A disk, however, has its own spatial frequency response which must be considered. Following methods of references 8 and 9, we found that the actual AZP transducer transfer functions for operation on fundamental (Fig. 3b) and for operation at 1.4 x fundamental (Fig. 3c) were very nonuniform. The latter curve demonstrates a better response to high spatial frequencies, hence the narrowed focal spot. (The actual intensity distribution is predicted by the Fourier transform of the transfer function, up to a scaling factor.[6]) As we shall see, however, a transfer function such as that shown in Fig. 3c can lead to spurious results in a single element imaging system.

The nonuniformity in the spatial frequency response of the AZP transducers may be somewhat alleviated by using 1/4 wave acoustic impedance matching layers.[9,10] A double layer structure could be bonded on the water side of the

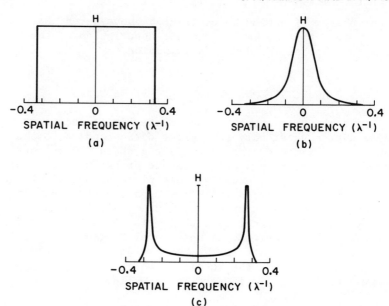

FIG. 3--Cross sections of radially-symmetric transfer functions for (a) the zone plate pattern; (b) the AZP
transducer operating on fundamental; and (c) the
AZP transducer operating at 1.4 × fundamental.

PZT-5A disk, resulting in a transfer function which closely
approximates Fig. 3a.

ACOUSTIC PHASE PLATE (APP) TRANSDUCERS

The major difficulty with the AZP transducers relates
to their generation of the DC term. If we were to place an
object in the primary focal plane of one of these transducers and try to monitor reflection from the primary focus,
we see from Fig. 2 that, in addition to that signal, we are
apt to receive a substantial reflected signal from the DC
term. The other foci do not present a difficulty, but the
DC term, which contains 25% of the power as opposed to ~10%

in the primary focus, severely restricts the device's imaging capabilities. The same problem will exist in a transmission arrangement. One possible approach to this problem
is to convert from AZP transducers to off-axis AZP transducers, which would result in the DC term being spatially
separated from the primary focus in the primary focal plane.
We have found more convenient, however, the elimination of
the DC term altogether, by replacing our AZP transducers
with acoustic phase plate (APP) transducers.

For a classical Rayleigh-Wood phase plate, the opaque
zones of the zone plate aperture are made transmissive but
phase-retarding. Referring back to Fig. 1, we see that if
the phase delay is made 180°, all radiation incident on the
aperture will constructively interfere at P . In addition to throwing more relative power into the primary focus,
this construction eliminates the troublesome DC term. We
may see this by noting that a phase plate is simply a superposition of a positive zone plate and a negative zone plate,
180° out of phase with each other. Thus from Eq. (3), the
phase plate transmittance function becomes

$$
\begin{aligned}
t(r) &= \frac{1}{2} + \frac{j}{\pi} \sum_{\substack{n=-\infty \\ (\text{odd})}}^{\infty} \frac{1}{n} \exp\left(-jn\pi r^2/r_1^2\right) \\
&+ \exp(j\pi)\left[\frac{1}{2} - \frac{j}{\pi} \sum_{\substack{n=-\infty \\ (\text{odd})}}^{\infty} \frac{1}{n} \exp\left(-jn\pi r^2/r_1^2\right)\right] \\
&= \frac{2j}{\pi} \sum_{\substack{n=-\infty \\ (\text{odd})}}^{\infty} \frac{1}{n} \exp\left(-jn\pi r^2/r_1^2\right) \quad .
\end{aligned}
\tag{7}
$$

The $\frac{1}{2}$ term, which led to the DC term, has vanished while
the properties of the primary focus remain the same, apart
from an increase in relative power.

Our APP transducers are realized by means of a novel
poling procedure. We use the fact that PZT-5A is a poled
ceramic and its deformation (i.e., either expansion or contraction) when a voltage V is applied across its electrodes, is determined by whether V has the same or the
opposite polarity as the original poling voltage. An AZP

transducer (uniformly poled) must first be constructed, as
seen in Fig. 4a. A high DC voltage, having the opposite
polarity as the original poling voltage is then applied
across the zone plate electrode and its counterelectrode,
reversing the polarization only in the regions between the
two electrodes, Fig. 4b. The zone plate electrode may then
be replaced with a simple disk electrode, evaporated over
the pattern now poled into the PZT-5A disk, Fig. 4c. When
rf is applied across the transducer, adjacent zones will
have opposite deformations, the equivalent of being 180°
out of phase, resulting in phase plate action.

We constructed our APP transducers from the 10 MHz,
f/1.5 AZP transducers described earlier. A small acoustic
point probe was used to map the acoustic field generated by

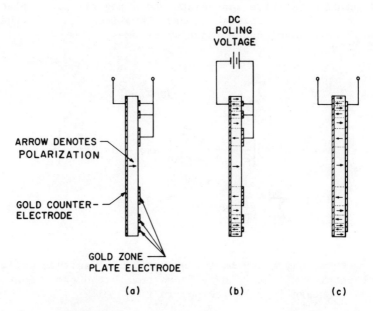

FIG. 4--Fabrication of an APP transducer. (a) The PZT
        polarization is determined. (b) A DC poling volt-
        age reverses the polarization in regions between
        the ZP electrode and its counterelectrode.
        (c) A simple disk electrode is placed over the
        phase plate pattern.

these devices. The probe consisted of a PZT-5A transducer
of the same fundamental as the APP transducer, but with
lateral dimensions of 0.2 mm (or close to one wavelength in
water at 10 MHz). The APP transducer was mounted into the
side of a water tank and the point probe translated about
the primary focus. By monitoring acoustic transmission as
a function of position, we obtained focal plane intensity
distributions such as the one shown in Fig. 5.

ACOUSTIC IMAGING RESULTS

Acoustic imaging has been performed in both transmission and reflection modes with our APP transducers.
Initially, as seen in Fig. 6, the system consisted of a
small water tank with the object to be imaged mechanically
raster scanned in the focal plane of the APP transducer.
The object was illuminated from one side by a plane wave
transducer with the APP transducer used as the receiver on
the opposite side. The image is formed by synchronizing
the spot location on a CRT display with the mechanical
scanner and modulating the spot intensity with the transmitted power. Thus an image is formed where brightness
denotes acoustic transmissibility.

An interesting image was obtained of a stainless-steel
wire mesh of 1.5 mm periodicity, Fig. 7. The unexpected
spots at the center of each cell and at the wire inter-
sections, in the acoustic image, are a direct result of the
transfer function of Fig. 3c. Only the first four terms in
the spatial frequency spectrum of the transmittance function for the grid are within the passband of the receiver.
Normally, this would allow for a fairly faithful image.
Those spatial frequencies within the passband, however, are
subject to the nonuniform response of the plate. If the
spatial frequency spectrum of the grid is multiplied by the
transfer function, Fig. 3c, the resulting image spectrum is
found to predict the spurious spots in our image.

This difficulty may be overcome by replacing the plane
wave transmitter with another APP transducer, positioned
such that its axis and focus are coincident with those of
the receiver. An analysis of this arrangement shows that
the system transfer function becomes the autocorrelation of
the function in Fig. 3c, a surprisingly level function with

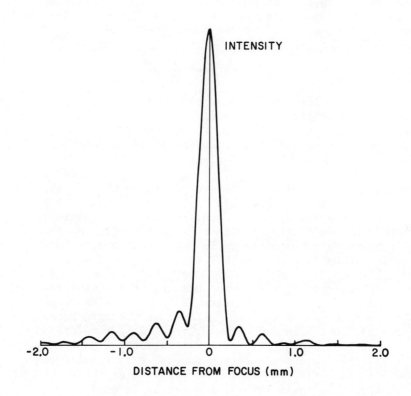

FIG. 5--Focal plane intensity distribution for the APP
    transducer, operating at 10 MHz with a 3 cm focal
    length. The left sidelobes are slightly enhanced
    due to a slight overlap of the disc electrode on
    that side of the APP transducer.

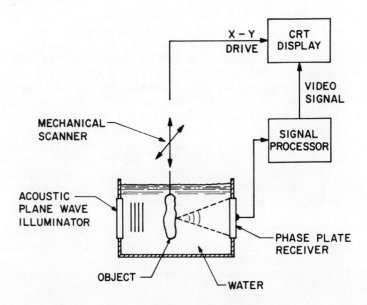

FIG. 6--Schematic of imaging system operating in transmission with a plane wave transmitter.

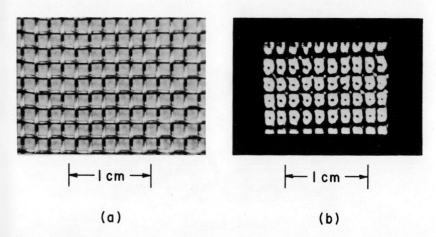

(a)   (b)

FIG. 7--Comparison of (a) optical and (b) acoustic transmission images of a stainless-steel wire mesh,
made with the plane wave transmitter.

a frequency cutoff of $0.68\lambda^{-1}$. The result is an image spectrum which faithfully reproduces the object transmittance
function, Fig. 8. Imaging may be performed in reflection
also. To accomplish this, we use only one APP transducer
and operate in a pulse-echo mode. The power in the first
echo is used to intensity modulate the CRT so that image
brightness denotes acoustic reflectivity. The frequency
analysis of this system is identical to that of the APP-
transmitter transmission system. A sawtooth pattern, Fig. 9,
punched into a 3-mil nickel-plated copper sheet is seen to
be faithfully reproduced. In Fig. 10, we see that contours
in a relief pattern etched into the nickel-copper sheeting
may also be discerned. Most of the detail is resolvable if
one is willing to sacrifice contrast.

The resolution of this imaging system with a focused
transmitter (either reflection or transmission) has been
determined through the use of resolution charts to be 0.27 mm
or 1.8λ. Improvements in the mechanical scanner should reduce this further.

DISCUSSION

This acoustic imaging system would be considerably
more practical if the object could remain stationary.
Since the APP transducer is small, but relatively rugged,
the system could easily be modified such that the trans-

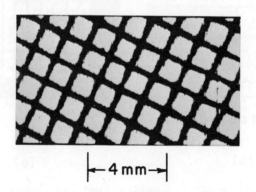

$\vdash$—4 mm—$\dashv$

FIG. 8--Acoustic transmission image of the wire mesh of
Fig. 7, made with the APP transmitter.

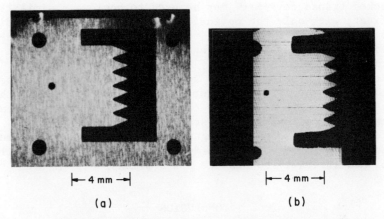

FIG. 9--Comparison of (a) optical and (b) acoustic reflection images of a sawtooth pattern punched into a
3-mil nickel-plated copper sheet.

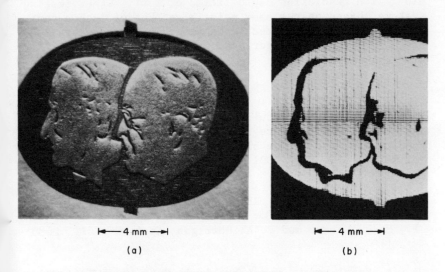

FIG. 10--Comparison of (a) optical and (b) acoustic reflection images of a relief pattern etched on a 3-mil
nickel-plated copper sheet.

ducers are scanned, and not the object. Photoconductively-
backed transducers[11,12] may be used to achieve a similar
effect, with only the optical image of a zone plate needing
to be scanned.[4] (This will result in a scanning AZP system, but with faster scanning rate capabilities. Electronic filtering methods may then be employed to handle
the DC term problem.) Fast scanning systems are also possible with the APP transducer. We recall from Eq. (2) that
the focal length of the APP transducer is frequency dependent. This allows for a system where the focus may be
electronically scanned along the axis, providing one dimension of an image. The transducer may then be mechanically
scanned laterally to produce a two-dimensional image.

CONCLUSION

We have described the theory and fabrication of an
acoustic phase plate transducer and demonstrated its usefulness in the field of acoustic imaging. The mechanically-
scanned imaging system described above offers simplicity
along with resolution capabilities of 1.8λ. Other imaging
systems, based on the same principles, but of such a construction that the object to be imaged may remain stationary, have also been described.

ACKNOWLEDGEMENTS

The authors gratefully acknowledge the fine technical
assistance of F. Futtere and L. Goddard. This work was
supported by the Office of Naval Research under Contract
No. N00014-67-A-0112-0001.

REFERENCES

1. H. H. Barret and F. A. Horrigan, Appl. Opt. 12, 2686
    (1973).

2. A. Engel and G. Herziger, Appl. Opt. 12, 471 (1973).

3. G. Chao, B. A. Auld, and D. K. Winslow, 1972 Ultra-
    sonics Symp. Proc. p. 140-141, IEEE Publication 72
    CHO 708-8 SU.

4. S. A. Farnow and B. A. Auld, Appl. Phys. Letters 25,
    681 (1974).

5. R. W. Wood, Physical Optics (Dover, New York, 1975),
    p. 37-39.

6. J. W. Goodman, Introduction to Fourier Optics (McGraw
    Hill, New York, 1968).

7. D. J. Stigliani, Jr., R. Mittra, and R. G. Semonin,
    J. Opt. Soc. Am. 57, 610 (1967).

8. M. Ahmed, R. L. Whitman, and A. Korpel, IEEE Trans.
    Sonics Ultrasonics SU-20, 323 (1973).

9. B. A. Auld, M. Drake, and C. G. Roberts, Appl. Phys.
    Letters 25, 478 (1974).

10. B. A. Auld, C. DeSilets, and G. S. Kino, 1974 Ultra-
    sonics Symp. Proc. p. 24-27, IEEE Publication 74
    CHO 896-1 SU.

11. C. G. Roberts, "Optically Scanned Acoustic Imaging,"
    Ph.D. Dissertation, Stanford University (1974).

12. K. Wang and G. Wade, Proc. IEEE 62, 650 (1974).

AN ELECTRONICALLY FOCUSED TWO-DIMENSIONAL ACOUSTIC IMAGING SYSTEM

J. Fraser, J. Havlice, G. Kino, W. Leung, H. Shaw,
K. Toda, T. Waugh, D. Winslow, and L. Zitelli

Stanford University, Stanford, California 94305

I. INTRODUCTION

We described a new type of electronically focused and scanned acoustic imaging device in last year's conference. This device made use of a surface acoustic wave delay line to provide the necessary phase references and time delay for the imaging system and for scanning; it employed mechanical scanning in one direction and electronic scanning in the other, and was only focused electronically in one direction. Since that time, we have developed a two-dimensional electronically focused C-scan device operating in a transmission mode using a separate electronically focused receiver and transmitter. The receiver is focused and scanned in the x direction and the transmitter focused and scanned in the y direction. By using this combination of transmitter and receiver, it is possible to scan out a raster and obtain M × N resolvable spots with only M + N elements in the transmitter and receiver arrays.

At the same time we have been constructing a 100 element receiver array, which is intended to be used either with the electronically focused transmitter or with a mechanically scanned transmitter focused with a lens. In both devices we have demonstrated that the principles of the system are valid; we have improved the sensitivity of the receiver system by several orders of magnitude over the earlier devices, by employing double balanced mixers and amplifiers on every element. However, we have encountered the major difficulty common to all phased array systems — the problem of sidelobes. So far, in both the transmitter

275

and the receiver array, the sidelobe level is typically
13 dB lower than the main lobe when the system is unapodized,
as it should be according to the simple theory of a uniformly
excited array.

We have investigated apodizing the system. In theory,
by using Hamming weighting, we should obtain a sidelobe
level of -43 dB. In practice, we have not succeeded in obtaining better than -20 dB sidelobe levels. The basic reason for this problem is associated with missing elements and
errors in the system. So in this paper we carry out a fairly
detailed analysis of the problem of missing elements, a problem which will be common to all electronically focused systems. We show that it is possible to predict very closely
the amplitudes of the sidelobes when there are elements missing, both by simple techniques and more sophisticated
numerical procedures. We also show how it is possible to
evaluate a particular system with missing elements and discuss possible procedures for eliminating errors.

II. PRINCIPLES OF THE ELECTRONICALLY FOCUSED SYSTEM

The electronically focused and scanned system on which
we have worked can be used either in a transmission mode or
a receiver mode, or for both transmission and reception.
In order to explain the principles of operation, we will
consider a receiver in which the object is illuminated with
an acoustic wave of frequency ω_s , as illustrated in Fig. 1.
The signal is received by a set of piezoelectric transducers. The electronic system employed scans the signals
arriving at these transducers and provides phase compensation for the different rays arriving from a point x, z .
We use, as a phase reference, a tapped acoustic surface
wave delay line, one tap for each PZT transducer. The reason for the use of an acoustic surface wave delay line is
its convenience, flexibility, and ready availability. It
is also possible to employ a CCD in the same manner, or to
employ a shift register in a very similar system. The output at a frequency ω_s from each transducer is mixed with
the output from a corresponding SAW tap in a mixer. Originally these mixers were simple diodes; now we employ balanced integrated circuit mixers, and the outputs from each
mixer are summed in an output circuit.

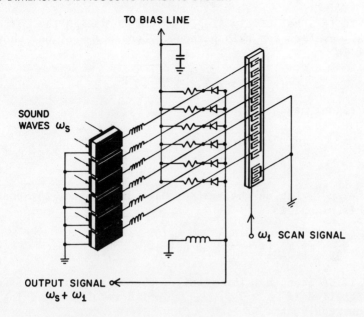

FIG. 1--Schematic-pictorial diagram of acoustic imaging
    system.

Suppose we wish to detect a signal from the point x ,
z , at the plane of the transducer array $z = 0$. The
phase of the signal at an element with coordinates x_n , 0
is

$$\phi_{sn} = \frac{\omega_s}{v_w} \left[\sqrt{(x - x_n)^2 + z^2} \right] , \qquad (1)$$

where v_w is the wave velocity in the medium of interest.
If we insert a signal on the acoustic delay line with a
phase ϕ_{An} such that $\phi_{An} + \phi_{sn} =$ constant , then in this
case, all the signals from the outputs of the mixer are in
phase and can be added. We have therefore constructed a
matched filter, matched to a source at the point x , z ,
i.e., we have constructed an electronic lens.

It is a relatively simple matter to design the correct signal waveform for this purpose. Consider a signal whose phase is

$$\phi = \omega_1 t + A \sqrt{t^2 + B^2} \quad . \tag{2}$$

This signal has a frequency

$$\omega = \frac{\partial \phi}{\partial t} = \omega_1 + \frac{At}{\sqrt{t^2 + B^2}} \quad . \tag{3}$$

At the nth tap on the surface wave delay line, the phase of this signal is

$$\phi_{An} = \omega_1 \left(t - \frac{x_n}{v} \right) + A \sqrt{\left(t - \frac{x_n}{v} \right)^2 + B^2} \quad . \tag{4}$$

It will be seen that the required phase matching condition is satisfied if $\omega_1 x_n / v = 2n\pi$, which requires choosing ω_1 correctly, and

$$t = x/v \tag{5}$$

with

$$A = \frac{\omega_s v}{v_w} \tag{6}$$

$$B = \frac{z \, \omega_s}{A \, v_w} = \frac{z}{v} \tag{7}$$

It will be seen from these results that by using this electronic signal processing technique, the system focuses on the point x , z at a time t = x/v and, therefore, scans along the plane z at a velocity v . At the same time, with the correct choice of the parameter A , adjustment of the parameter B of the signal waveform inserted into the acoustic surface wave delay line is equivalent to adjusting the focal length of the lens. In this case, by using the correct waveform, no paraxial approximation is required, and the lens should not suffer from spherical aberration.

In our devices so far, we have not used this waveform. Instead, we have simplified the approach and used a linear FM chirp with a frequency

$$\omega = \omega_1 + \mu t \tag{8}$$

and phase

$$\phi = \omega_1 t + \frac{\mu t^2}{2} . \tag{9}$$

In this case, which is equivalent to the paraxial approximation, i.e., assuming that $z^2 \gg (x - x_n)^2$, the chirp rate is varied to vary the focal length of the lens and we find to focus on the plane z,

$$\mu = \frac{v^2 \omega_s}{2v_w} . \tag{10}$$

It should be noted that a still better approximation to the ideal signal waveform, which is easier to realize and which is accurate to terms in $(x - x_n)^4$, is to use an input signal with frequency

$$\omega = \omega_1 + \frac{\omega_s v}{v_w \sqrt{3}} \sin\left(\frac{vt}{z} \sqrt{3}\right) . \tag{11}$$

This is normally accurate enough for most practical purposes.

In our first experiments, we employed a receiver system of this kind in a transmission mode with an object placed in front of a narrow strip transducer, the length of the strip being parallel to the length of the array. The object was moved up and down mechanically in the y direction and the receiver was electronically focused and scanned in the x direction. An alternative to this procedure is, of course, to use a focused transmitter, focused on a line within the object. Then either the transmitter can be moved up and down in the y direction, or the objects can be moved up and down in the y direction to obtain a two-dimensional scan. Of course, such a system can also be used in reflection mode with the transmitter placed in the same plane or near to the receiver's array.

A second alternative on which we have been working is
to use an electronically focused and scanned transmitter in
the arrangement shown on the left of Fig. 2. By inserting
a signal in what normally would be the output port of the
receiver and a chirp signal on the delay line, it is possible to obtain a focused and scanned transmitting beam.
A full two-dimensional image is obtained with the rest of
the system of Fig. 2. However, the disadvantage of such a
system is that the scan rate is comparable to that of the
acoustic velocity along the delay line. Ideally we would
prefer to work with a system in which the scan rate in the
vertical direction is relatively slow, and comparable to the
frame rate required, while the scan rate in the horizontal
direction is fast and corresponds to the line time.

An additional desirable feature would be to excite all
the transmitting transducers at the same frequency; this
would minimize aberrations and also place less severe requirements on the transducer bandwidth. Thus it would be
possible to scan out a normal type of TV raster.

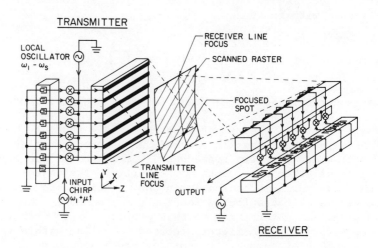

FIG. 2--Schematic-pictorial diagram of an electronically
            scanned, electronically focused imaging device.
            On the left side is the transmitter which provides
            a scan rate comparable to the acoustic velocity on
            the delay line.

In order to see how to carry out such an operation with chirp signals that travel along the delay line at the acoustic surface wave velocity, we can consider the transmitting chirp to behave like a moving lens, or as if it produces a moving focused beam from a parallel beam source with a focal length $f = \omega_s v^2/\mu v_w$, as shown in Fig. 3. Now suppose that a second chirp was inserted into the opposite end of the delay line; this would be like placing an additional lens in front of the focused beam, thus making up a compound lens with a focal length of $f/2$. If now one lens moves in the $+y$ direction, and the other lens in the opposite direction, it follows by symmetry that the focal point of the compound lens will not move, and the focal point will be fixed in position. Finally if we arrange that one lens moves at a slightly different velocity than the other (equivalent to changing one of the chirp rates slightly) the focal point can be made to move at an arbitrary velocity parallel to the transducer array.

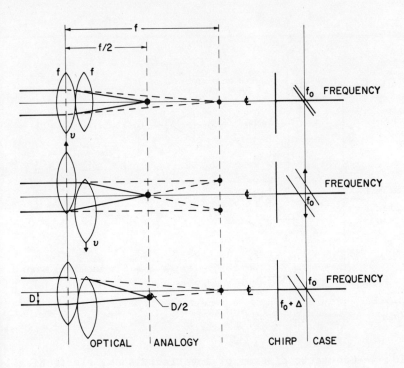

FIG. 3--Optical analogy for a variable scan rate transmitter.

A schematic of the circuit required to scan and focus
in this manner is shown in Fig. 4. The theory of the system
has been given in detail in an earlier paper.[1] In this circuit two sets of mixers are used at each tap, to mix three
different frequencies together. Initially, a chirp of frequency $\omega = \omega_1 + \mu t$ is mixed with two signals of frequencies
ω_2 and ω_3 . This gives rise to chirps with frequencies
$\omega_1 - \omega_2 + \mu t$ and $\omega_3 - \omega_1 - \mu t$, respectively, i.e., chirps of
opposite signs. These signals are amplified and inserted at
opposite ends of the delay line. When they are mixed in the
first mixer on each tap, they give rise to a signal at the
tap x_n with a frequency $2\omega_1 - \omega_3 - \omega_2 + 2\mu(t - L/v)$ where
L is the length of the delay line. If this signal were
further mixed in a second mixer with a similar signal from
a reference tap, it would give rise to a zero frequency output. By modulating the signal from the reference tap with

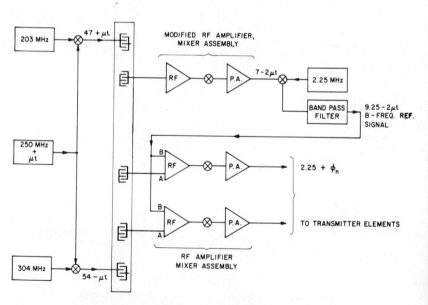

FIG. 4--Schematic diagram of a variable scan, electronically
focused transmitter system.

a frequency ω_s , filtering it so as to only keep the upper sideband, and then inserting it into the second mixer, we obtain outputs into all the transducers of frequency ω_s .

An analysis of the phase variation of the signals passing along the delay line shows that the phase of the signals at the nth tap are

$$\phi_2(x_n) = (\omega_1 - \omega_2)\left(t - \frac{y_n}{v}\right) + \frac{\mu}{2}\left(t - \frac{y_n}{v}\right)^2 \qquad (12)$$

$$\phi_3(x_n) = (\omega_3 - \omega_1)\left(t + \frac{y_n}{v} - \frac{L}{v}\right) - \frac{\mu}{2}\left(t - \frac{L}{v} + \frac{y_n}{v}\right)^2 \qquad (13)$$

After these signals are mixed together, we obtain a signal with phase $\phi_2(y_n) - \phi_3(y_n)$. This in turn is mixed with the similar signal from the tap at $y_n = 0$, after it has been modulated with a frequency ω_s . The phase of the resultant signal at the nth tap is

$$\phi(x_n) = (\omega_2 + \omega_3 - 2\omega_1)\frac{y_n}{v} + \frac{\mu}{v^2}\left(y_n - \frac{L}{2}\right)^2 . \qquad (14)$$

Thus, the output signal has a square law variation of phase along the system, which is equivalent to a lens producing a beam focused at the plane $z = \omega_s v^2/2\mu v_w$. In addition there is a linear phase term, which controls the y position of the focal point. If $(\omega_2 + \omega_3 - 2\omega_1)y_n/v = 2n\pi$, the lens focuses on the point $y_n = L/2, z$. Alternatively, if ω_3 or ω_2 is varied slowly the focal point moves along the y axis. Thus by changing one of these frequencies slowly the beam may be scanned along the y axis; this is equivalent to using a slow chirp for the frequency ω_2 and a fast chirp for ω_1 which can be changed at will to vary the focal length of the lens. In practice the chirp lengths are chosen to be of the order of 100 μsec long, sufficient to produce a beam focused on one line of the image, which itself is scanned by the receiver array within this time. Then with a delay of the same order, the chirps are re-inserted in the transmitter delay line and the whole process is repeated, now with the line moved one line along the y axis.

III.   EXPERIMENTAL SYSTEMS

We have been working with two types of experimental
arrays, an 80 - 100 element receiver system, and a 22 - 29
element transmitter system.   It is intended that systems
of these types will be used together in a C-scan transmission or reflection mode, as well as employing the receiver array with a mechanically scanned transmitter for two-
dimensional C-scan imaging.

A schematic of the basic components of the receiver
system is shown in Fig. 5.   Each transducer element consists of a cube of PZT 5A 20 mils square and 16 mils high,
glued with epoxy to a mycalex backing, and covered on its
front side with a mylar film.   The elements have a center-
to-center spacing of 47 mils; they are square in cross
section so that the acceptance angle (measured to be ± 30°)
is approximately the same in the horizontal and vertical
directions, thus giving approximately the same field of
view in both planes.   The array itself is mounted on the
side of a water tank.

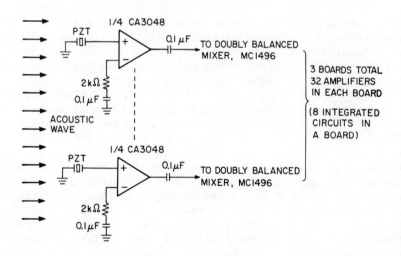

FIG. 5--Schematic diagram of PZT preamplifier system.

The impedances of the individual elements of the array are of the order of 5000 Ω, with a center frequency of 2.25 MHz. The system is, in fact, operated anywhere in the range from 1.6 MHz to 2.5 MHz, with most tests being carried out at 1.8 MHz. Amplifiers for the individual elements (CA3048), 4 to a package, are mounted on the back of the transducer array. Their outputs are taken through coaxial lines into doubly balanced mixers (MC1496), each mixer being driven via an MC1350 amplifier from a tap on a 50 MHz, 100 tap surface wave delay line approximately 5 inches long, made of BGO. The taps are arranged in two rows of 50, side by side, and the total delay through the line is 60 μsec. The outputs of 4 mixers at a time are summed with a 1350 summing amplifier, whose outputs are in turn connected in parallel. The output of this system is mixed down to 30 MHz, passed through a 1.5 MHz bandwidth filter with 60 dB out of band rejection, to eliminate any feed through from unwanted signals, and the output then passed into an IF amplifier. After detection, this output signal is used to modulate the intensity of a cathode ray display, whose horizontal sweep is synchronized from the FM chirp trigger, and whose vertical motion is controlled either from a mechanical sweep or from the transmitter synchronizing circuits.

The receiver array performed basically as designed. The measured sensitivity of the individual elements was 10^{-11} watt/ cm^2, and the 3 dB width of the main lobe of the focused beam, when illuminated from a 1.8 MHz narrow strip source pulsed for 100 μsec was about 1.5 mm at a distance of 25 cm. As will be discussed, the sidelobe level was higher than we had hoped thus severely limiting the dynamic range of the receiver system, and the effective uniformity of the illumination of an object placed in front of a uniformly illuminated large area transmitter. For this reason we have, so far, been employing the device to observe objects with relatively good contrast. This is done by placing the object in front of a thin strip transducer approximately 4 inches long, and moving the object up and down mechanically.

One picture of a piece of rubber in which are cut a triangular, circular, and square hole is shown on the top of Fig. 6. On the bottom is the acoustic photograph of a thin metal plate with a number of small holes. It will be observed that the definition of the system is of the order of 1.5 mm.

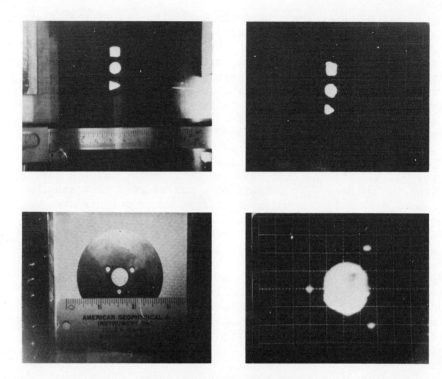

FIG. 6--Acoustic photographs using an 84 element receiver
        array. On the left are optical photos and on the
        right are acoustic photos.

The sidelobe level in this system was initially very
high and of the order of 13 dB, both near in to the main
lobe as well as further out. We attempted to apodize the
array with either Hamming weighting of the chirp amplitude
itself, or better still, by Hamming weighting of the array
taps by means of trimmer pots on each element. Our best
attempts to date have yielded a sidelobe level 20 dB down
from the main lobe. This is mainly due to the fact that
initially 14 elements out of 100 were inoperative, and later
approximately 6 elements out of 84 were inoperative. This
leads to a sidelobe level somewhat worse than we would
expect from the theory given in the next section. In part
this is due to the difficulty of apodizing accurately, to
phase errors as well as amplitude errors, and to some cross
coupling in the delay line. But the principle problem is

the missing elements themselves, caused by both PZT transducer failure and SAW tap failure. We are now in the process of improving the original system and we hope to obtain
a sidelobe level of the order of 30 dB down from the main
lobe.

An important application of these scanned focused array
systems is for nondestructive testing. At the present time
one technique employed for looking at disbonds in laminated
samples is to transmit a focused beam through the object,
receive it with a focused transducer, and carry out a raster scan of the object, line-by-line scanning in both directions mechanically. If definitions of the order of 1 mm
are required, it takes several hours to scan large objects.
If electronic scanning could be used, a line of the order
of 10 cm in length could be scanned in approximately 60 μsec,
at least 100 times faster than would be possible with mechanical scanning. Scanning in the other direction could
still be carried out mechanically.

In order to prove the feasibility of this idea, we
carried out tests on a sheet of a titanium-boron composite,
which consists of a sheet of titanium bonded to a boron
fiber epoxy composite. This sheet had faults in the bond-
ing, deliberately introduced. These disbonded sections had
already been mapped out acoustically in a mechanically scanned focused system by E. Caustin of the North American Rockwell B-1 project, who supplied us with the test piece. The
original experiments by Caustin were carried out at a higher
frequency than 2.25 MHz. This gives better contrast due to
the bigger change in attenuation on the partially disbonded
layers at higher frequencies. We therefore operated our
electronic scanning array at a higher frequency resonance
of the transducers (6.5 MHz) and were able to obtain a well
focused beam with a ± 10° acceptance angle but a relatively
small field of view. As will be seen from Fig. 7, we show
a result obtained on the NDT sample. It will be noted that
a square disbonded region shows up extremely well with this
technique.

A transmitter system based on the ideas for two-
dimensional electronic focusing given in Section II has been
constructed. In the initial system we employed a 29 element
transmitter array approximately 4 inches square, with each
element consisting of a long rectangular strip 20 mil square

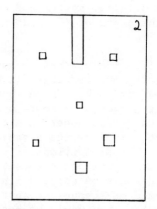

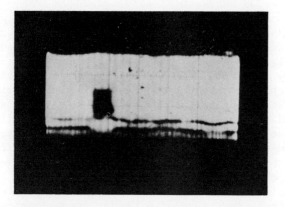

FIG. 7--Acoustic photographs of an NDT sample. At the top, a schematic of part of a sheet of a titanium-boron composite with various disbonds is shown. A photograph obtained acoustically at 6.5 MHz of one of the disbond regions (2nd hole down from the top at the right hand side) is shown. The field of view is limited in this photograph because of the relatively large spacing between the PZT array elements for the frequency used, and the small transmitter source available at this frequency.

epoxy bonded to a mycalex backing. The array is covered with a thin mylar film. The elements have a center-to-center spacing of 47 mils, but only every other element was excited in the initial system.

In the first system we used FET amplifiers driven from taps on the SAW line, followed by diode mixers for the three frequencies involved, and transistor amplifiers. The system was subject to some phase and amplitude errors and leakage of signals at unwanted frequencies, as well as a low output. But it served to prove the feasibility of the ideas involved. Initially, we used an early developmental model of the receiver array already described with only 22 elements.

By placing an object between the transmitter and receiver, we were able to obtain a focused image, with a definition of approximately 2 mm in the x and y directions, a 60 cycle frame rate and a line time of approximately 100 μsec. As predicted we could obtain a stationary focus in the y direction or move the line up and down at will by changing the frequency ω_2 manually, thus making it possible to examine an object slowly or fast. By using a slow chirp repeated at a 60 cycle rate we obtained a frame time of 60 cycles.

A picture of a letter S cut in a piece of rubber located 7 cm from the transmitter and 10 cm from the receiver is shown in Fig. 8. By changing the chirp rate by 25% in the horizontal direction, the letter S is seen to be defocused in the horizontal direction. Similarly by changing the chirp rate of the transmitter by 25%, the letter S is seen to be defocused in the vertical direction, as would be expected.

The transmitter system was rebuilt with different mixers and amplifiers, and a 24 element 1.8 MHz array with elements 47 mils apart. In this case, we used double balanced mixers (TI76514), which eliminated the problems with unwanted sideband frequencies. The mixers were followed by transistor amplifiers, (2N3906), and a signal with 20 V peak-to-peak output could be obtained at the transducers. We also made provision for trimming the output amplitude.

With this system we obtained, with Hamming weighting, a 20 dB sidelobe level. We took pictures of simple objects cut in rubber with the system, using a strip receiver mechanically scanned in the horizontal direction. Such a

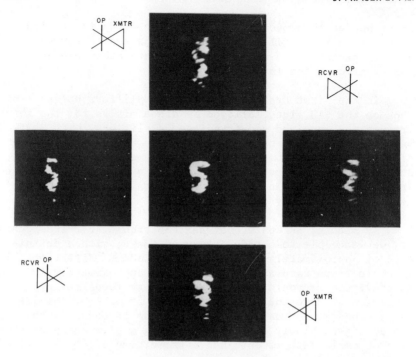

FIG. 8--Acoustic images obtained with an electronically
       scanned, electronically focused two-dimensional
       real time system. Corner photos show the effect
       of 25% defocus in the imaging system.

picture is shown in Fig. 9. We have not as yet tested the
transmitter with an electronically focused receiver array,
but hope to do so shortly.

IV. WEIGHTING OF THE ARRAY AND ERROR ANALYSIS

     One of the principle problems associated with any
phased array system is that associated with sidelobes. If
the sidelobe level is R dB down from the main lobe, and an
attempt is made to image two points A and B with point
B more than R dB lower in intensity, the image of the
point B may be obscured in the sidelobes of the stronger
point source A . Thus, the dynamic range of an imaging

OPTICAL IMAGE

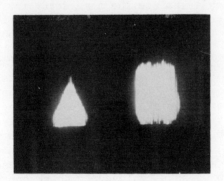

ACOUSTIC IMAGE

FIG. 9--Image obtained with Hamming weighted transmitter system.

system is limited by the sidelobe level.

There are several possible approaches to lowering the
effective sidelobe level. One is to apodize the array,
e.g., amplitude weight the response of the array. In an
acoustic lens system, this is done in part automatically,
because a convex lens tends to have more attenuation near
its edges. With any electronic or physical lens, the response also tends to fall off with angle; this provides
further apodization. A second approach, that used by
Thurston,[2] is to use logarithmic amplifiers on each receiver element, and add the resultant signals after the appropriate phase matching, as already described. As all rays
from a point source may be assumed to have equal amplitudes,
this gives an image of a point source with a sidelobe level
of just the same value as if the amplifiers were linear.
However, now the signal from another point source would give
an output R dB down in level only if it were far weaker
relative to the main image. This is because of the logarithmic response of the amplifiers; if a 50 dB change in
input signal corresponds to a 10:1 change in output power
from the amplifier and a linear system had a 10 dB sidelobe
level, this system would have a dynamic range of 50 dB.
The full implications of this nonlinear processing technique
are not entirely clear, but they do imply a lower effective
sidelobe level at the expense of some loss in grey scale.
A third approach is to use the same array for both transmission and reception. This tends to square the response
of the system, and lower the effective sidelobe level to
2 R dB .

Even with careful design, certain difficulties can
arise when elements are missing, there is coupling between
the elements, or there are phase errors in the system. By
far the worst type of error is that due to missing elements.
In this section we shall analyze the response of an array to
a point source, and a distributed source, and show how this
is affected by missing elements. We shall describe an analog
technique which we have employed to check operating arrays,
and a simple theory which gives easily calculated and accurate estimates of the effect of missing elements.

We consider a receiver array in which the amplitude
response of the array at a point x is $F(x)$. We suppose
that a signal of frequency ω_s from a source at the plane z

of amplitude $G(x')$ is incident on the array, and the array is focused by a chirp signal $\exp j(\omega_1 t + \mu t^2/2)$ which is mixed with the signals reaching it. The output of the array when focused on the plane z, $(\mu = \omega_s v^2/zv_w)$ is

$$H(t) = e^{j[(\omega_1+\omega_s)t + (\mu/2)t^2]} \int_x \int_{x'} F(x)\,G(x')\, e^{-\dfrac{j\omega_1 x}{v}}$$

$$\times\, e^{\dfrac{j\mu x}{v}(t - \dfrac{x'}{v})} dx\,dx' . \quad (15)$$

For the present purposes, we represent the response of the elements by the sampling function and write

$$F(x) = F(x_n)\sum \delta(x - x_n) . \qquad (16)$$

If again, we choose for simplicity the condition $\omega_1 x_n/v = 2n\pi$, which is equivalent to choosing a particular time for the start of the chirp, we can write the output in the form

$$H(t) = S(t) \int_{x'} e^{j\frac{\mu}{2}(\frac{x'}{2})^2} G(x')\,dx'$$

$$\times \sum F(x_n)\, e^{\dfrac{j\mu x_n}{v}(t - \dfrac{x'}{v})} , \qquad (17)$$

where $S(t) = \exp j(\omega t + \mu t^2/2)$, and is a function of unit magnitude. The signal from a point source at $x' = 0[G(x') = \delta(x)]$ is of the form

$$H(t) = S(t)\sum F(x_n)\, e^{\dfrac{j\mu x_n t}{v}}$$

$$= S(t) \int F(x)\, e^{\dfrac{j\mu xt}{v}} dx . \qquad (18)$$

Thus, the output from a focused array is the <u>Fourier transform of the response of the array</u>. All we need to know is

the response of the array to know the form of the output
from the focused system.

Consider now, a uniform array $[F(x_n) = 1]$ with N
elements a distance ℓ apart, where $L = N\ell$ is the length
of the array. The output from this array will be

$$H(t) = S(t) \frac{\sin\left(\frac{\mu N\ell}{2v}\right)t}{\sin\left(\frac{\mu \ell t}{2v}\right)} \qquad (19)$$

So the maximum output amplitude is at $t = 0$ and is of
amplitude N, or N times the output from a single element. There are minor lobes at times $\mu N\ell t/2v = (2n+1)\pi/2$,
the first sidelobe being 13 dB reduced in amplitude from the
main lobe. In addition there are grating lobes with their
own sidelobes at $\mu \ell t/v = 2\pi m$, where m and n are integers.

A simple way of carrying out analog computations and
illustrating this effect is to use a word generator to
modulate an rf signal, then take the Fourier transform of
the resultant signal in a spectrum analyzer. The result
for a 32 element unapodized system is shown in Fig. 10(a).
The 13 dB sidelobe level and the grating lobes can clearly
be seen.

As a second example we can consider the use of Hamming
weighting.[5] In this case, taking $x = 0$ at the center of
the array, $F(x_n)$ is chosen so that

$$F(x_n) = k + (1-k) \cos^2 \frac{\pi x_n}{L} \qquad (20)$$

where L is the length of the array. Ideally for $k = 0.08$
the sidelobe level should be reduced by 43 dB from the main
lobe. The maximum output for N elements will be

$$H(0) = \frac{N(1+k)}{2} \qquad (21)$$

$$= 0.54N \qquad (k = 0.08)$$

For $k = 0.08$ the main lobe is 1.5 times as wide at the

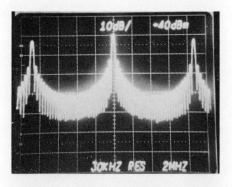

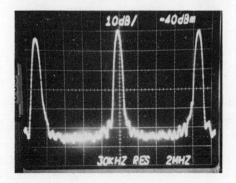

FIG. 10--(a) Top. Point response function for a 32 element
                   uniform array. Note -13 dB sidelobes
                   near the central lobe and the far grating
                   sidelobes.
        (b) Bottom. Point response function for a 32 ele-
                     ment Hamming weighted array. Note -43 dB
                     sidelobes near in.

3 dB points as for the uniformly apodized array. An illustration of Hamming weighting, taken with the help of the
analog technique, in a 32 element system, is shown in Fig.
10(b). It is clear that at the expense of a slight loss in
definition, the use of apodization should be of great help
in improving the sidelobe level.

    We now consider the effect of errors. It is apparent
that by using the formulae already given, we can calculate
the effect of both amplitude and phase errors on the output

response. We have done this on a computer, and also employed our analog technique for the purpose. However, it is helpful to try and obtain some kind of analytic formulae with which we can estimate errors and find how many elements we can afford to have missing.

Suppose that the mth element is missing. This is equivalent to subtracting an error term $e(t)$ from the output of magnitude

$$e(t) = S(t) F(x_m) e^{\frac{j\mu x_m t}{v}} \qquad (22)$$

So the total output is

$$H(t) = H_0(t) - e(t) \qquad (23)$$

where $H_0(t)$ corresponds to the output when there are no missing elements.

The period of the function $e(t)$ depends on the position x_m of the error. If it is in the center $e(t)/S(t)$ has virtually no phase change with time. If we suppose that the level of $e(t)$ is much larger than the sidelobe level in the error free array, we see that the sidelobe amplitude relative to the main lobe due to one missing element will be

$$R_m = \frac{F(x_m)}{\sum F(x_n) - F(x_m)} . \qquad (24)$$

In a Hamming weighted system, this corresponds to

$$R_m = \frac{2F(x_m)}{N(1+k) - F(x_m)} . \qquad (25)$$

Thus for a missing element at the center of the array, with $k = 0.08$

$$R_m = \frac{1}{0.54N - 1} . \qquad (26)$$

For $N = 32$ elements this corresponds to $R_m = 0.061$ or -24 dB.

A more exact calculation would take account of the sidelobe level already present, before the errors were introduced, i.e., 0.007 down from the main lobe. The true sidelobe level in the presence of errors would then be 0.061 +
0.007 or -23 dB. The result agrees fairly well with the
analog computer result shown in Fig. 11, and is almost
exactly equal to the results of the computer calculation.

When there are several missing elements, the question
is whether the effects are additive or tend to add only
randomly, i.e., if there are M missing elements, is the
amplitude error proportional to M or $M^{1/2}$. Unfortunately the former proposition is the more accurate, for
if we suppose that there are several missing elements, the
effect on the error signal is like that of an array made up
of the missing elements. This array produces a signal with
a main lobe in which the effect of all the elements is additive, and the main lobe tends to repeat itself in a distance,
if the element spacing is periodic, corresponding to that of

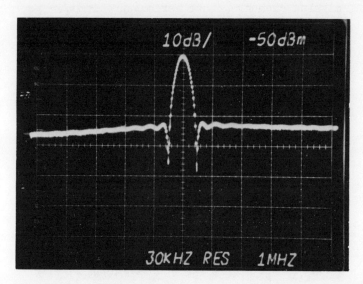

FIG. 11--Point response function for a 32 element Hamming
weighted array with one central element missing
[31 elements present]. Note that the absence of
this single element raises the sidelobe level from
-43 dB to -23 dB.

the grating lobes of the error array. If most of the elements that are missing are near the center of the array, the
main lobe of the error array will be wider than that of the
full array, so that there will be sidelobes of full amplitude near to the main lobe. More than likely, there will
be other sidelobes of similar amplitude further out from the
main lobe due to the quasiperiodicity of the error signal.
We can therefore make a good estimate of the relative maximum sidelobe amplitude due to errors from the following
formula:

$$R = \frac{\sum\limits_{error\ array} F(x_n)}{\sum\limits_{full\ array} F(x_n) - \sum\limits_{error\ array} F(x_n)} \cdot \qquad (27)$$

A picture of what occurs with 2 elements missing is shown in
Fig. 12. The following table gives computed results for
several cases, results taken with the analog system, and
results obtained from the "quick" formula. The "quick"
formula is seen to be accurate, and indicates that the
errors due to missing elements tend to be directly additive,
so that missing elements cause serious problems.

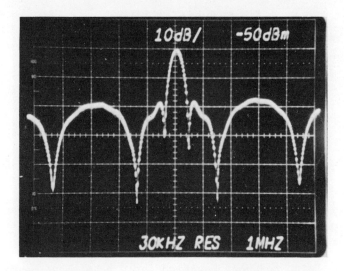

FIG. 12--Point response function for a 32 element Hamming
         weighted array with two elements missing [30 ele-
         elements present, central element and third from
         center element missing]. Note sidelobe level is
         -18 dB.

TABLE I: Comparison of Sidelobe Level Calculation using
Three Techniques. In all cases there are 32
elements [numbered -16, -15, ...-1, +1, ...15,
16]. Hamming Weighted.

Missing Elements	Sidelobe Level		
	"Quick" Calculation dB	Computer Calculation dB	Analog Measurement dB
1	-24	-23	-22
1 4	-18	-23	-18
1 12	-22	-22	-21
-10 1 12	-19.5	-19.5	-20
- 1 + 1	-17.5	-17.5	-16
- 2 2	-17.6	-18.5	-18
- 3 3	-18	-17	-17
- 4 4	-18.5	-18	-18
-10 10	-26.5	-25.5	-26
- 2 1 12	-16.5	-17	-16
-10 1	-16	-17	-16

We have examined possible ways of compensating for missing elements. One method is to increase the amplitude of the elements on each side of a missing element by a factor of 1.5. This does indeed lower the sidelobe level to the correct background level in the neighborhood of the main lobe. But as can be seen in the analog result of Fig. 13 and from the error theory, the amplitude at other times will vary as

$$e(t) \approx S(t) F(x_m) \left[1 - \cos \frac{\mu \ell t}{v} \right] . \qquad (28)$$

Thus, at a point halfway along to the main grating lobe, $\mu \ell t / v = \pi$, and the error is now 3 times the original error signal. So the method is not a useful one. It is therefore vital to carry out the engineering on these phased arrays very carefully.

A further situation of interest is that associated with the use of an extended source, and with obstructions in the path of the acoustic beam, such as a rib in medical diagnostics. We deal with these problems by considering the error signal picked up at one element. Suppose first that this element is missing and we are concerned with the error signal from a uniform extended source, $[G(x') = 1]$ at plane z . It may be shown from Eq. (17) that the amplitude of the signal picked up by a single element is approximately

$$e(t) = \frac{1}{\ell} \left(\frac{\lambda z}{2} \right)^{1/2}_{full} F(x_m) S(t) , \qquad (29)$$

whereas the signal picked up by the full array has a maximum amplitude $H_0(t)_{max} = N(1+k)/2$.

If R is the relative error signal from a point source we see that the output from an extended source now varies between

$$1 - \left(\frac{\lambda z}{2 \ell^2} \right)^{1/2} R < |H| < 1 + \left(\frac{\lambda z}{2 \ell^2} \right)^{1/2} R . \qquad (30)$$

If the sidelobe level were 25 dB, i.e., $R = 0.56$, $\lambda = 0.6$ mm, $\ell = 1.2$ mm, $z = 15$ cm , this formula would

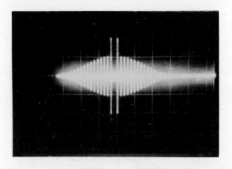

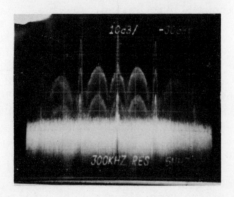

FIG. 13--Point response function for a 32 element Hamming
        weighted array with elements adjacent to the one
        missing element increased by 1.5 in amplitude.
        Top - Compensated array with missing element.
        Bottom - Point response function.
        Comparing with Fig. 11, near sidelobes reduced
        but far sidelobes increased.

indicate a variation of amplitude from 0.7 to 1.3, i.e., an
amplitude variation of over 5 dB and hence dark bands in
observing a uniform extended source, a problem which we
often encountered in our experimental systems. The basic
reason for this effect is, of course, that the sidelobes
from a uniform array of sources tend to add, while the main
lobes do not.

On the other hand, this effect is helpful as far as
obstructions in the path of the beam are concerned. As
long as the obstruction subtends an angle at individual

elements far less than their acceptance angles, it will
only obscure a small part of the signal reaching an individual element in the array, and have a very small effect
on the sidelobe level. The calculation can be carried out
mathematically by treating the plane including the object
as an extended source. The conclusions reached correspond
to this simple physical picture.

V. CONCLUSIONS

We have shown how to design one- and two-dimensional
electronic focusing systems, and demonstrated that the
principles of design are valid. A major problem is that
associated with sidelobe reduction. Apodization is of
great help for this purpose, but can fail in its intent if
there are several missing elements. It should however be
possible to obtain at least a-30 dB sidelobe level, and
theoretically -43 dB by careful apodization and engineering
design.

The image of an extended source will tend to have large
variations in amplitude even with relatively low sidelobe
levels, i.e., there will be dark bands in the picture. This
is because of diffraction effects. By using a diffuse wideband (noise modulated) source so eliminating coherence in
the source, it should be possible to radically decrease this
effect without loss of resolution. This approach needs further investigation.

An alternative approach with many additional advantages
is to use a scanned pulsed source as well as a receiver.
This makes it possible to range gate the system, i.e., obtain good resolution in the z direction as well as in the
x direction. In such a system, if we pulse the source, we
will obtain essentially the same resolution or better than
that in a CW system, provided that the time difference between rays reaching the array from a point is less than the
time for scan of one spot $t_s = d_s/v$ where $d_s = z\lambda/L$ is
the resolution. We have carried out such experiments using
a pulsed point source, as shown in Fig. 14. As the pulse
length of the source is decreased, the received signal
width decreases slightly. More important still, the sidelobes are cut off. In addition, the range resolution becomes equal in time to the transverse resolution t_s when
pulse length is equal to t_s . It is possible to extend

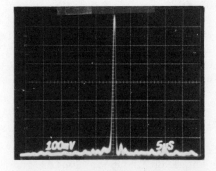

GATED TRANSMITTER

PULSE WIDTH ≈ 50 μs

Z ≃ 42 cm

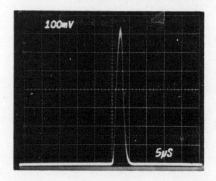

GATED TRANSMITTER

PULSE WIDTH ≈ 2 μs

Z ≃ 16.5" = 41.9 cm

FIG. 14--Point response function obtained with an 84 element receiver system. (Top): Response with a
long transmitter pulse. (Bottom): Response with
a short transmitter pulse. Note elimination of
sidelobes in the short pulse case.

this concept to envisage a system in which the array is
used as a fast scan focused transmitter as well as receiver.
Such an arrangement gives a moving point source at the
object, should eliminate most of the sidelobe problems and
give good range resolution for a B scan device.

The authors would like to acknowledge the fine technical assistance of D.J. Walsh, C.R. Hall, and L.C. Goddard.

REFERENCES

The work reported in this paper was sponsored partially
by the Office of Naval Research, under Contract N00014-67-
A-0112-0039; by the Naval Undersea Center, under Contract
N00123-72-C-0866, and by Rockwell International Science
Center under Grant RISC 74-20773.

1.   J. Fraser, J. Havlice, G. Kino, W. Leung, H. Shaw,
     K. Toda, T. Waugh, D. Winslow, and L. Zitelli,
     "A Two-Dimensional Electronically Focused Imaging
     System," presented at the IEEE Group on Sonics and
     Ultrasonics Symposium, November 11-14, 1974.

2.   F. L. Thurston and O. T. von Ramm, "A New Ultrasound
     Imaging Technique Employing Two-Dimensional Electronic
     Beam Steering," in Acoustical Holography, Vol. 5,
     edited by P.S. Green, Plenum Press, New York, 1974,
     p. 249.

3.   C. E. Cook and M. Bernfeld, Radar Signals; An Intro-
     duction to Theory and Application, Academic Press,
     New York, 1967.

ACOUSTIC MICROSCOPY - A TOOL FOR MEDICAL AND BIOLOGICAL RESEARCH

R. A. Lemons and C. F. Quate

Stanford University

Stanford, California 94305

INTRODUCTION

During the past few years the field of acoustic microscopy has changed dramatically. It has moved from a concentration on development into the realm of application. This is not to say that the process of development has ended. There will undoubtedly be major improvements and innovations in the future. But the presently existing instruments are beginning to open a few doors. The most important function of these prototype instruments is to show what can be learned at present and to indicate what may be learned in the future.

The widest spectrum of applications for the acoustic microscope lies in the fields of biology and medicine. Here the diversity is almost unlimited and there should be an opportunity for new discovery. The potential advantages of microscopic ultrasonic investigations in this area were recognized by the early workers in the field.[1,2,3] They anticipated that the acoustic properties of a biological specimen could vary substantially over microscopic dimensions. This variation should enable high contrast acoustic micrographs to be made of untreated samples. By comparison the optical properties of most biological materials show little variation. This frequently requires artificial staining agents for viewing.

The expectation of large intrinsic acoustic contrast
has been born out completely. In every specimen we have
examined there is ample contrast to provide a sharp, well-
defined image. This gives the acoustic microscope a
particular advantage in an investigation of living material.
Generally, contrast producing stains can not be used in the
preparations without endangering the vitality of the
biological system.

The fundamental difference in the interactions of
sound and light with matter makes the acoustic microscope
a unique tool. With high spatial resolution the acoustic
microscope can probe the elastic properties of a specimen
at the cellular level without damaging the natural
structural relationships.

In this paper we shall concentrate on the results
which have been obtained with a scanning acoustic microscope
developed at Stanford. We have previously described the
detailed design of this instrument and have given some
results.[4,5,6] Basically the instrument consists of a con-
focal pair of acoustic lenses. Each lens is simply a
spherical interface between a sapphire crystal and a drop
of water. These lenses typically have a numerical
aperture of .7 and a focal length of .15 mm. Such small
focal lengths are essential to compensate for the enormous
absorption of water at the operating frequency. Thin film
piezoelectric transducers are applied to the end of each
crystal opposite the lens surface. One transducer
generates the ultrasonic power while the second serves as
the detector. Similarly, one lens focuses the acoustic
beam in the water drop while the second lens collects and
collimates the diverging sound.

By mechanically scanning the object of interest
through the focused acoustic beam, the attenuation and
phase shift generated by each point on the specimen can be
measured. This information is then presented on a scan
synchronized display to produce the acoustic micrograph.
The essential components of this design are shown
schematically in Fig. 1. The lenses have been separated
in this diagram as they are when the specimen is mounted.
The specimen ring shown between the lenses is connected to
a loudspeaker movement which provides the fast scan.

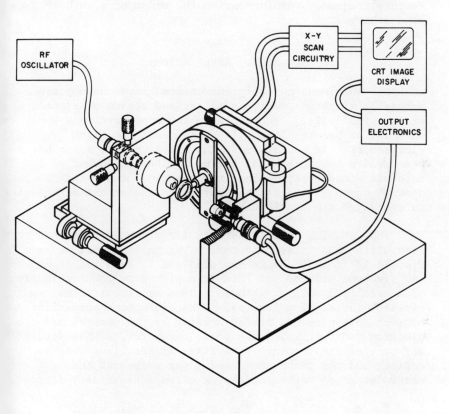

Fig. 1   Generalized diagram of the scanning acoustic
          microscope.   The micrometers on the left lens
          support provide for the precise alignment required
          by this confocal geometry.

This entire assembly is then scanned from line to line by a hydraulic system.

The advantage of this instrument is its resolution. It can be operated at acoustic frequencies up to 1 GHz. At this frequency resolutions on the order of 1 μm have been achieved.

DIAGNOSTIC APPLICATIONS

The specimens normally encountered in pathology are primarily microtome sections of excised tissue or spreads of isolated cells. In examining these preparations the pathologist has traditionally relied upon the light microscope. In recent years the electron microscope has been used on an increasing scale. With the advent of the acoustic microscope, we face the question of how its special capabilities can be applied to assist the pathologist in the diagnosis of disease.

This problem of diagnosis is largely one of pattern recognition and interpretation. The pattern must be made visible if it is to be interpreted and in both the light and electron microscopes this is achieved by the application of specialized stains. These stains are specific to certain classes of molecules. Therefore, concentrations of such molecules are differentiated by the optical or electron contrast which the stain generates. These stains are part of the tremendous heritage of the light micros- copist, and the techniques used today represent the accumulation of patient research extended over many decades.

For the pathologist the acoustic microscope may, therefore, be analogous to a new kind of stain. The acoustic image enables materials and structures with differing elastic properties to be distinguished in a recognizable pattern. As such, the acoustic microscope should be used to augment the information available with conventional stains.

Diagnostically the acoustic microscope may be important at two levels. One possibility is the case in which it is used to reveal abnormalities which go undetected with existing techniques.

The second possibility is that the acoustic instrument
may possess an advantage over conventional stains in
providing a diagnosis either more quickly, or more
accurately.   Our preliminary results indicate that this
type of role for the acoustic microscope will be realized.
In Fig. 2 we show a comparison between the acoustic and
optical micrographs of a section from a malignant breast
tumor.   This sample was prepared by first fixing the
tissue in formalin and then embedding it in paraffin.
The paraffin block was microtomed to produce a section
with a nominal 5 μm thickness.   The section was then
picked up and fastened with gelatin to the 2 μm mylar
support membrane used in the acoustic microscope.   After
the paraffin was removed this specimen was imaged with the
acoustic microscope operating at a frequency of 600 MHz.
At this frequency the instrument resolution is approximately
2 μm.   A portion of the resulting image is shown at the
top of Fig. 2.   To obtain the comparable light micrograph,
this section was stained with Hematoxylin-Eosin and cover-
slipped in the standard way.   The Hematoxylin-Eosin stains
the nuclei of each cell a deep blue so that they stand in
sharp contrast with the cytoplasm which stains pink.
This stain is the most widely used in the preparation of
tissue sections for light microscopy since it enables
individual cells to be clearly differentiated from one
another.   The optical image of the stained section taken
with a top quality light microscope is shown at the bottom
of the figure.

The differences between the two images are very
distinct.   In the acoustic image, for example, a large
irregular island of attenuating material is sharply
differentiated from surrounding material.   The distinction
in the optical image is extremely subtle.   A number of
smaller areas of large acoustic attenuation can be seen
scattered across the acoustic image.   Evidently the
elastic properties of the material in these areas
distinguishes them from surrounding areas.   It was the
opinion of several pathologists who inspected these two
images that the regions of high acoustic attenuation
represent localizations of increased collagen content.
Low frequency studies on the elastic properties of collagen[7]
have shown that fibers of this material can have an acoustic
impedance more than thirty times greater than other common
substances found in the body.   Finding distinctively high

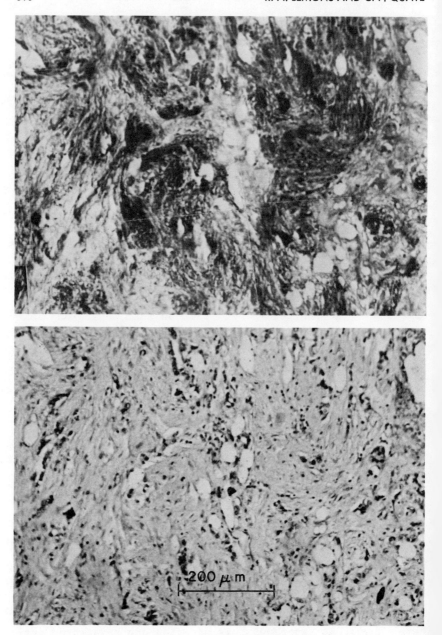

Fig. 2   Comparison of the acoustic (top) and optical images
         (bottom) of a small region of malignant breast tumor

attenuation in collagenous regions might therefore have
been anticipated in an acoustic micrograph.

The diagnostic importance of this result stems from
the relationship between the collagen and the presence of
the neoplasm. Fibroblasts in the vicinity of neoplastic
cells are stimulated to produce collagen as an inflammatory
response. In certain tumors the quantity and distribution
of collagen can be an indication of the presence or the
extent of the neoplastic growth.

In light microscopy the presence of collagen can be
indicated by using specialized stains such as the tri-
chrome stains. Such stains can give collagenous areas a
greatly increased optical contrast. The difficulty with
this technique is that the staining procedure requires
several hours. In some situations such as mastectomy
this time delay is prohibitive. The diagnosis is then
made with a much quicker stain like Hematoxylin-Eosin.
As seen in Fig. 2, this stain does not distinguish collagen
deposits in a clear way. By comparison, an acoustic
micrograph could reveal this information in a matter of
minutes. In addition, minute quantities of collagen
should be detectable. This offers the potential for a
more accurate diagnosis at an early stage in the disease.

Detection of these collagen deposits is just one
example of the potential diagnostic utility of the acoustic
microscope.

APPLICATIONS TO LIVING SYSTEMS

In living systems there is a large class of activity
relating to the elastic properties of the biological
material which is not completely understood. The changes
that occur during contraction or during fluid flow are
difficult to detect with the optical instrument. But there
is evidence that viscosity changes are involved in many of
these contractile systems and the acoustic instrument
should in principle be capable of recording some of the
important changes.

In this area of biology some of the phenomena are
poorly understood because techniques necessary to measure
some of the most basic physical parameters are unavailable.
For example, how does a single celled animal like Amoeba
proteus move from one place to another ?   When the cell
begins to move the cytoplasm flows in that direction pushing
the cell membrane ahead of it.   What drives the flow ?
What initiates it and what controls it ?   Clearly, some
physical properties of the cell must change in this process,
and it is to be expected that the elastic properties will
be prominent among them.   Since the acoustic microscope is
sensitive to changes in these properties it should be used
to measure these on a microscopic scale.   For example, a
marked change in the cytoplasmic viscosity may accompany
the initiation of movement.   This could be revealed as a
change in the absorption of that area.   Understanding the
mechanism of movement in the case of a single cell like the
amoeba could greatly extend our understanding of cellular
movement in general.   We have not as yet carried out a
complete study in this area but we can report here our
preliminary work that shows that we can work with living
systems.

Figure 3 is composed of a temporal sequence of acoustic
images of a single Amoeba proteus.   Each frame was made
with approximately a 4 sec. exposure.   Adjacent frames are
separated in time by 15 sec. with a much larger interval
between the images in the top and bottom halves of the
figure.   The acoustic frequency used was 500 MHz giving
nearly 2 μm resolution.   A constant focal plane was
maintained at approximately the mid-plane of the cell,
leaving some portions of the cell out of focus.   In all,
this cell was observed for over an hour as it moved
generally to the right across the display.   The cell
appeared to move normally, unaffected by either the scan
or the acoustic power.

In order to mount the amoeba in the acoustic microscope
the cell was sandwiched between two sheets of mylar separated
by a tiny drop of water.   This provided a miniature
aquarium which could be scanned through the acoustic cell.

Inspection of the acoustic images shows that in the
stationary pseudopodia the fine particulate structure of
the cell is easily resolved.   Of greater interest are the
localized areas of large attenuation which are seen to move

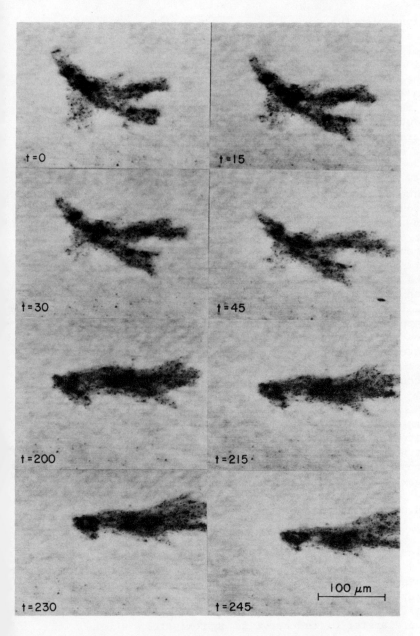

Fig. 3   Temporal sequence of acoustic images of an Amoeba
proteus.   Times are in seconds.

along with the cytoplasm.   These may be organelles such as
the nucleus or vacuoles.   Variations in the transmission
across a given pseudopod can also be seen to change with
time.   It is clearly important to observe a specimen in
the living state.   Being a dynamic situation, the changes
in the acoustic properties can be studied as a function of
time.   This is a much more accurate and reliable measurement than can be achieved by comparing separate samples.
Moreover, there is no concern that the observed detail is
an artifact of the preparation technique.

As yet we have only demonstrated that a living cell
such as the amoeba can be readily observed with the
acoustic microscope.   To ascertain whether the variations
in transmission represent changes in the physical properties
or only changes in thickness will require a much more
thorough investigation.   Measurements such as the ratio
of phase shift to attenuation must be made.   By improving
the mounting technique the thickness of the cell can be
constrained, and by controlling the environment the motion
of the cell could be started and stopped at will, enabling
careful measurements to be carried out.   With such
controls the acoustic microscope may provide us with very
interesting new information.   Progress in this direction
will open the entire field of rheology to study with the
acoustic microscope.

Within this area mitotic cell division may be a
particularly interesting subject.   During the process of
mitosis radical changes take place in the internal
structure of the cell.   Modification of the acoustic
properties in the region of the spindle may be extensive.
Again, measurement of viscosity can be important.   Studies
by other techniques have indicated that substantial changes
in cytoplasmic viscosity accompany cell division.[8]

As a second example of the application of the acoustic
microscope to a living system, Fig. 4 compares the acoustic
and optical images of a segment of a fine nerve fiber.
This fiber activates one of the stabilizer muscles on the
crayfish.   It is known to contain six separate axons and
is frequently used in the study of simple nerve networks.
The diameter of the nerve is approximately 40 μm.

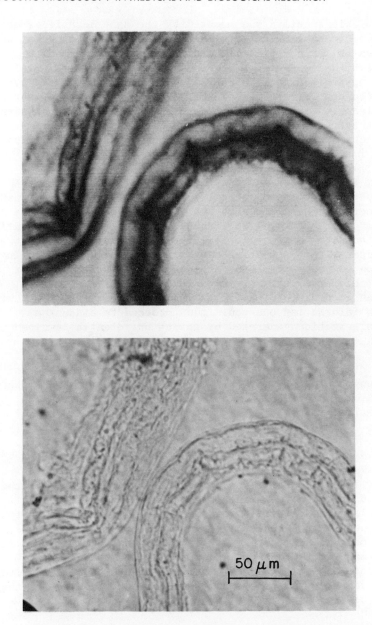

Fig. 4 Comparison of the acoustic (top) and optical images
(bottom) of a segment of crayfish nerve.

This specimen was mounted by the same techniques used to mount the amoeba. Precautions were taken to keep the nerve in the proper saline solution throughout the investigation to insure that it would remain vital. In the optical image this unstained preparation shows very little contrast and the internal structure of the fiber is unclear. By comparison the acoustic image shows abundant contrast. Moreover, some of the longitudinal axon structure can be clearly seen in the upper left portion of the image. This acoustic image was made at 500 MHz.

The interest in studying a nerve fiber with the acoustic microscope stems from the total lack of information concerning the changes in the elastic properties of the fiber which accompanies a nerve impulse or action potential. The acoustic microscope is a tool which could measure these changes with the spatial resolution required to determine which component is being affected. The starting point for this study will be an investigation of the changes which accompany a chemically induced depolarization. With this information one could then proceed to study the depolarization generated by an action potential excited with microelectrodes.

In addition to this work we have recorded acoustic micrographs of living cell cultures. Two types have been used, the heart cell of the embryonic chick and the mammalian fibroblasts.[6]

CONCLUSION

In this paper we have tried to give a glimpse of the areas in which the acoustic microscope may advance our knowledge in the fields of biology and medicine. Improved diagnostic techniques employing the acoustic microscope may be realized in the near future. On the broader horizon a more complete understanding of the detailed elastic properties of a living system may be reached with the acoustic microscope and with the new measurements that are now available.

ACKNOWLEDGEMENTS

This research was funded by a grant from The John A. Hartford Foundation, Inc., and we express our gratitude for their support. We also thank Drs. R. D. Allen, M. Billingham, R. Kempson, and J. Wine for the help and guidance they have given us with the biological preparations.

REFERENCES

1. Sokolov, S., "The Ultrasonic Microscope", Akademia Nauk SSSR, Doklady, 64, 333 (1949).

2. Dunn, F., and Fry, W. J., "Ultrasonic Absorption Microscope", J. Acoust. Soc. Am., 31, 632 (1959).

3. Kessler, L. W., "Review of Progress and Applications of Acoustic Microscopy", J. Acoust. Soc. Am., 55, 909 (1974).

4. Lemons, R. A., and Quate, C. F., "Acoustic Microscope - Scanning Version", Appl. Phys. Letters, 24, 163 (15 February 1974).

5. Lemons, R. A., and Quate, C. F., "A Scanning Acoustic Microscope", 1974 Ultrasonics Symposium Proceedings, IEEE Cat. #74CHO896-1SU, 41-44 (1974).

6. Lemons, R. A., and Quate, C. F., "Acoustic Microscopy - The Biomedical Applications", Science (in press).

7. Fields, S., and Dunn, F., "Correlation of Echographic Visualizability of Tissue with Biological Composition and Physiological State", J. Acoust. Soc. Am., 54, 809 (1973).

8. Carlson, J. Gordon, "Protoplasmic Viscosity Changes in Different Regions of the Grasshopper Neuroblast living Mitosis", Biol. Bull., 90, 109 (1946).

MICROANATOMY OF A HISTOLOGICALLY UNSTAINED

EMBRYO AS REVEALED BY ACOUSTIC MICROSCOPY

Mahfuz Ahmed
Zenith Radio Corporation, Chicago, IL. 60639

Lawrence W. Kessler
Sonoscope, Inc. , Bensenville, IL. 60106

Abstract: The microanatomy of a fixed but unstained,
optically opaque 72-hr. chick embryo is nearly completely revealed with an acoustic microscope. This
new technique differentiates soft tissue structures and
provides information on elasticity and density at the
microscopic level. Comparison of the "acoustic micrograph" with the known optically revealed microanatomy
of the chick embryo indicates that a rather complete
visualization of organ premordia has been achieved
without clearing and staining the specimen.

The study of embryos in their natural state (i. e.
unfixed and unstained) is important to embryologists.
The main reason is that this study provides information
on the processes of cell duplication and regeneration as
occurs, for example, in cancer and in aging. Furthermore, the sensitive embryonic tissue makes it a good
indicator of the effects of pollutants.

Optical visualization of whole-mount embryos is
usually very difficult, since the specimens are thick and
optically opaque. Therefore, it is necessary to employ
fixing, clearing and staining techniques which alter the
natural state of the tissue. The research reported here
was a preliminary investigation to determine the extent
to which acoustic microscopy can be employed to

circumvent optical opacity in an unprepared embryo and
to assess the degree of soft tissue differentiation. The
significant difference between this new type of imaging
process and optical imaging stems from the fact that the
physical properties governing the transmission of light
are quite different from those governing the transmission
of sound. In particular, acoustic wave propagation is
dependent directly upon the elasticity, density and viscosity of the medium, and acoustic microscopy provides a
tool for investigating these parameters on a microscopic
scale. At a typical sound frequency of 100 MHz, resolution can be attained on the order of one wavelength of
sound, i. e. 15 μm in most biological materials.

A nominally 72-hr. chick embryo was selected for
the investigation. This choice was guided by the fact
that the specimen is readily available and has already
been extensively studied. Furthermore, at the 72-hr.
stage, the organ system is sufficiently differentiated to
make visible organ primordia (initial stages of organ
formation) which possess the typical characteristics of
all early vertebrate embryos. Although the specimen,
which was obtained from a local biological supply house
was fixed in ethanol, the technique is applicable to fresh
specimens as well, as has been demonstrated by previous related investigations[1,2,3].

Since the acoustic microscope has been dealt with in
more detail elsewhere[1], its operation will only be briefly
described here. The high frequency acoustic energy is
produced by an electrically driven piezoelectric element
and is applied to the specimen through an optically transparent substrate. The sound transmitted through the
specimen strikes a nearby optical mirror which becomes
minutely distorted by the impingent sound. The distortions cause small changes in the direction of a focused
laser beam reflected from the mirror. The beam is
made to scan an area of the mirror in synchronism with
a TV monitor. The information transferred to the laser
light is recovered in the form of an electrical signal

when the light strikes an appropriate photodetector
placed behind a demodulator such as a knife edge. This
electrical signal when amplified and fed to a TV monitor
displays a real time acoustic image on the screen. In
actual practice, the mirror referred to above is made
slightly transparent to allow a portion of the probing light
beam to pass through the specimen. This light is col -
lected by a second photodetector mounted below the specimen and the electrical signal from it is fed to a second
TV monitor. Thus, a conventional optical image is produced simultaneously with the acoustic image for purposes of comparison.

Figure 1 shows an optical image of a fixed but unstained specimen. Figure 2 is an acoustic micrograph
of the same specimen. The optical image was obtained
by backlighting the embryo and photographing it with a
copying camera. It is evident that due to opacity, most
of the structures within the embryo cannot be visualized
optically without clearing and straining the specimen.
Only the large structures within the embryo such as the
optic cups, the leg and wing buds and certain regions of
the brain can be distinguished optically. The extra-
embryonic circulatory system can also be seen because
it is very thin and does not absorb much light. The
acoustic micrograph, on the other hand, demonstrates
that without any clearing or staining procedures, a great
deal more structure may be visualized. In fact, nearly
all the structures seen in the optical image of a stained
embryo preparation are clearly brought out in the acoustic image except those limited by resolution. This may
be verified by referring to Figure 3 which is a photomicrograph of a whole mount preparation that has been
bleached and stained with hematoxylin and eosin. The
numbers in Figure 2 refer to the corresponding structures identified in the caption of Figure 3. It should be
noted that all types of structures, i. e. those of known
ectodermal, endodermal and mesodermal origin are differentiated acoustically without staining. The differentiation occurs by virtue of slight differences in the

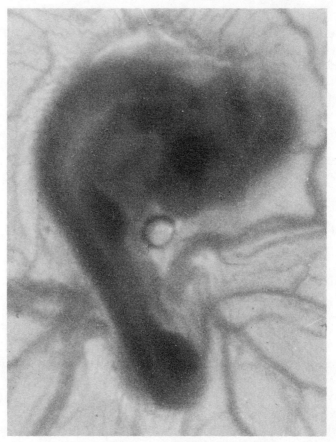

Fig. 1: Optical image of an unstained specimen

mechanical properties of the tissues, such as its elasticity, density or viscosity. Although sometimes optical and
mechanical properties may be intimately coupled, the
acoustic image in some cases shows greater contrast. A
good example is a water-air interface which reflects only
2% of light but reflects virtually 100% of the sound. Contrast in an optical image depends upon changes in the
absorption of light and changes in refractive indices of
adjacent structures. In an analogous manner, the contrast in an acoustic image depends upon changes in the
absorption of sound and changes in the acoustic refractive
index, the latter being a simple function of compliance

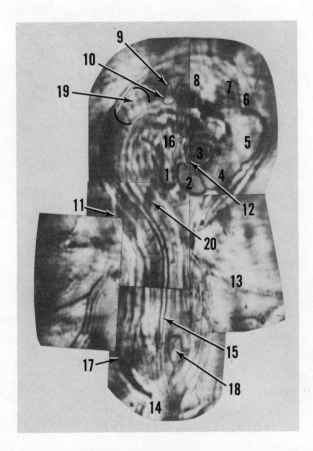

Fig. 2: Acoustic image of the specimen shown in
Fig. 1 (composite of several pictures).
The numbers refer to corresponding
structures shown in Fig. 3.

and mass density. Interfaces between different types of
tissue are visualized by acoustic wave scattering caused
by different acoustic refractive indices. Structures
composed of similar types of tissue are visualized by
the amount and closeness of the cell condensations
(changes in absorption, density and compliance).

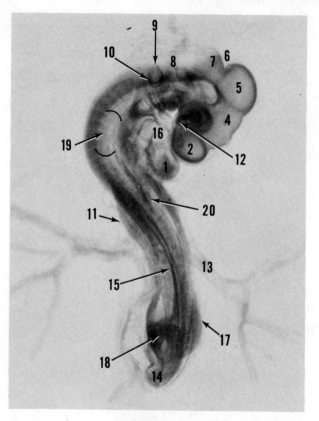

Fig. 3: Photomicrograph of a stained whole mount.

1. Ventricle
2. Telecephalon
   (cerebral hemisphere)
3. Optic cup
4. Diencephalon
5. Mesencephalon
6. Isthmus
7. Metencephalon
8. Myelencephalon
9. Endolymphatic duct
10. Auditory vesicle
11. Wing bud
12. Optic stalk
13. Vitelline artery
14. Tail
15. Spinal cord
16. Bulbus cordis
17. Leg bud
18. Allantois
19. Somites (bracketed area)
20. Anterior intestinal portal

In conclusion, acoustic microscopy has been shown to be useful in circumventing optical opacity and in differentiating tissue structure on the basis of small differences in their characteristic elastic and mechanical properties. Furthermore, inasmuch as it is possible to selectively alter the elastic properties of tissue by histological procedures, this concept may be employed to further enhance acoustic differentiation of structures.

REFERENCES

1. L. W. Kessler, A. Korpel and P. R. Palermo, "Simultaneous Acoustic and Optical Microscopy of Biological Specimens", Nature, 239, 111-112 (1972).

2. L. W. Kessler, "High Resolution Visualization of Tissue with Acoustic Microscopy", in Proceedings of the Second World Congress on Ultrasonics in Medicine, 1973, M. de Vlieger ed., Excerpta Medica, Amsterdam, Netherlands 1974.

3. L. W. Kessler, "VHF Ultrasonic Attenuation in Mammalian Tissue", J. Acoust. Soc. Amer. 53, 1759-1760 (1973).

ASSESSMENT OF BRAGG IMAGING AND THE IMPORTANCE
OF ITS VARIOUS COMPONENTS BY COMPARISON WITH
RADIOGRAPHIC IMAGING

Neal Tobochnik, Peter Spiegler, Richard Stern,
and Moses A. Greenfield
University of California, Los Angeles
Department of Radiological Sciences and School
of Engineering and Applied Sciences
Los Angeles, CA 90024

INTRODUCTION

Ultrasonic imaging by Bragg diffraction of light,
first demonstrated by Korpel (1966), may have applications
in medical diagnosis. With this in mind, a group of
researchers at the University of California, Santa Barbara
have performed various investigations using Bragg imaging
at the low frequencies (1 MHz to 5 MHz) required for
clinical use (see Landry, et al., 1972 and Hormozdyar, et
al., 1974).

In order to investigate further the clinical usefulness of this system we decided to evaluate the Bragg
system's performance using techniques similar to those
used to evaluate the performance of typical radiological
imaging systems. The performance of radiological systems
is usually measured by evaluating the ability to produce
levels of resolutions and contrast. Resolution is usually
measured by analyzing the image of a narrow slit providing
a line source of x-rays while contrast is measured with
aluminum stepwedges. Such data, plus geometric optics and
theoretical considerations about the attenuation of x-rays
through matter, is then used to predict the effect of the
various components of the x-ray system on the final
radiograph image.

The performance of a Bragg system can also be evaluated by using narrow slits for resolution measurements and
appropriate stepwedges for contrast measurements. The
effect of various system components, such as the

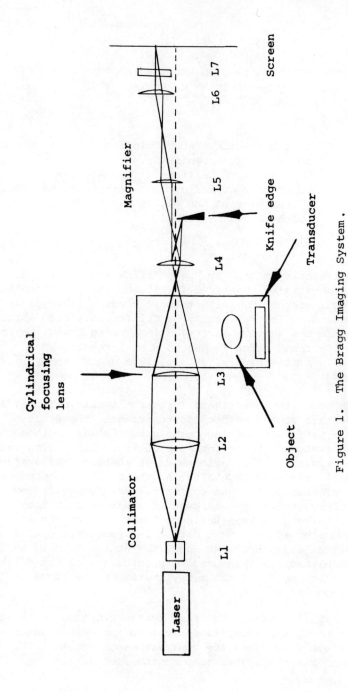

Figure 1. The Bragg Imaging System.

cylindrical focusing lens, can then be evaluated. In this study the results of such measurements are reported.

Theoretical work by Smith, et al. (1971) on resolution and Kessler (1972) on depth of field indicate that Bragg imaging is more appropriately compared to x-ray tomography or focused radioisotope scanners rather than conventional radiography. A simple theoretical model was therefore constructed to relate the resolution and depth of field properties of the system. It is believed that this model and the experimental data reported here will be useful in the consideration of the applicability of Bragg imaging to specific medical diagnostic investigations.

EXPERIMENTAL SETUP

Our system was similar in design to those of Korpel (1966), Smith, et al. (1971) and others. It is illustrated in Fig. 1.

The components were arranged along a three meter optical bench. A 5 mW helium neon laser was used as the source of light. The height of all optical components was 5 cm, a separated doublet with a focal length of about 211 mm and a width of 50 mm was used to produce the light wedge through the water tank. This lens provides considerable improvement in spatial resolution over the resolution obtainable with simple plano convex lenses of similar focal length.

A 5 cm by 5 cm and a 10 cm by 10 cm rectangular PZT-4 transducer with fundamental frequencies of 1 MHz were used as the source of ultrasound. The transducers were operated on their odd order harmonics as well as their fundamental frequency.

Korpel (1971) and Smith (1971) have pointed out the astigmatic nature of the Bragg image. At high acoustic frequencies, a cylindrical lens, L7 in Fig. 1, can be used to correct the astigmatism. However, at the acoustic frequencies of medical interest the distance between the anamorphic image and the orthoscopic image is tens of meters. This makes astigmatism correction quite difficult. The correction was therefore not made in most of our experiments. The effects of the lack of astigmatism

correction will be shown later. It is hoped that a suitable optical configuration can be constructed to allow
for astigmatism correction.

SIMPLE IMAGING MODEL

The formation of two images by the Bragg system is
analogous to the saggital and tangential images of any
astigmatic imaging device. Consider a Bragg imaging
system corrected for astigmatism such that the final
image occurs in the orthoscopic plane. Further, assume
that the horizontal and vertical numerical apertures, NA,
have the same value and the final size of the image is
the same as the size of the object. Such a system can be
conceptually illustrated as in Fig. 2. An object point S
emits sound rays which are converted by the acousto-optic
interaction into light rays which are focused to an image
point S_- at a distance $\lambda X_0/\Lambda$ downfield where λ and Λ are
the wavelengths of light and sound respectively and X_0 is
the distance between the object and the optical areas
of the system. In the process of focusing the light
rays the acousto-optic interaction acts as a lens. This
lens will have a depth of field.

Consider the imaging of two point sources A and B as
illustrated in Fig. 3. The imaging screen is placed at
the focus of point A. Point B will be imaged without
significant distortion as long as the spread of the focus
of point B labelled ΔH on the screen is less than the
resolvable distance D given by Smith as:

$$D = \Lambda/NA \qquad\qquad\qquad \text{Eq. 1}$$

The depth of field is defined as the distance in the
object plane over which the image will remain in focus.
Kessler (1972) has shown that this occurs when:

$$DF = 2\Lambda/(NA)^2 \qquad\qquad \text{Eq. 2}$$

When the horizontal and vertical numerical apertures are
not identical there will be different values for the
horizontal and vertical depth of field. The smallest
value should generally be quoted. Since Bragg imaging has
a depth of focus which under some circumstances may be
quite small, it is better to compare it with tomography
which is the radiographic procedure with a depth of field.
Conventional radiographic systems do not exhibit a depth

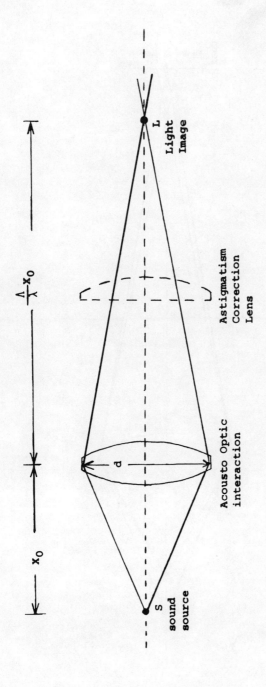

Figure 2. The Bragg System as a Lens.

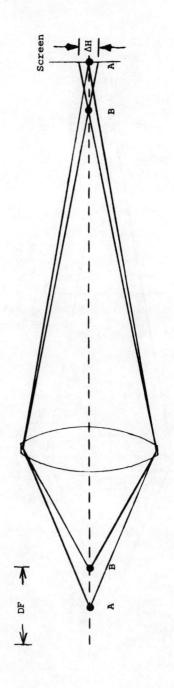

Figure 3. Depth of Focus Derivation.

of field. If it is desired to compare Bragg imaging
with a nuclear medicine device then a good choice would
be a linear motion scanner. The linear motion scanner
has a focused collimator which provides the depth of
field. In tomography or in the use of a linear motion
scanner, experience shows that optimum contrast is
obtained when the thickness of cut (equivalent to the
depth of field) is equal to the dimension of the object
to be viewed. The same rule will probably apply to
Bragg imaging.

If we write Eq. 4 in terms of the resolvable distance
D (using Eq. 1) we find:

$$DF = 2D^2/\Lambda \qquad \qquad \text{Eq. 3}$$

This equation shows that the depth of field depends on the
spatial resolution of the system. Thus, if a thick cut
is desired to be viewed through the object then resolution
will have to be compromised. If a thin cut is desired
good resolution will be possible but contrast, which is
proportional to the thickness of cut will be compromised.
A clinically useful Bragg imaging system, like a tomographic device or a linear motion scanner, should be
constructed so that the numerical aperture of the system
can be caried. This can be achieved either by having
several interchangeable lenses of different F number for
location L3 (see Fig. 1), or by having a variable aperture
in front of lens L3 and a laser with variable power.

IMAGING OF LINE SOURCES

In describing imaging systems particular attention is
given to the light intensity distribution in the images
of point and line sources. The shape and width of these
distributions are indices of the resolution capabilities
of the system. In the case of light optics it is common
to describe the imaging process as follows: The source
of light sets up a field described mathematically by a
quantity often referred to as the field amplitude. The
light intensity is then defined as the square of the field
amplitude. The imaging system in question is then represented by a spread function. When coherent light is being
used the response of the system is specified by a spread
function that operates on the field amplitude. When incoherent light is used the imaging system response is

specified by a spread function that operates on the field
intensity. The coherent and incoherent spread functions
are not the same but one can be obtained from the other.

The most commonly measured spread function is the
line spread function and intensity function which describes
the imaging system's ability to image a line of energy.
All Bragg imaging systems built to date use coherent light
and sound. Since light detectors respond only to light
intensity, it is not possible to measure an amplitude
spread function directly. The light intensity distribution
across the image of the line that can be measured may be
used to calculate an amplitude spread function if desired.

Light intensity distributions across the images of
line sources of ultrasound were measured in the following
manner: The line sources were generated by placing a
wooden plate with a narrow slit in the path of the system
ultrasonic field. As wood is acoustically opaque, sound
is transmitted only through the slit. The light
intensity profile in the optical image of the line source
was obtained by scanning across the image with a photodiode light detector mounted on an X-Y translation stage.
Since the photodiode has an active diameter of 0.5 mm
the scanning was done at 0.5 mm increments across the
image. The relative light intensity at each position
was read with a digital voltmeter connected to the output
of the photodiode preamp. As the resolution capabilities
of the system are different in the horizontal and vertical
directions measurements were taken with both orientations.
Theoretical considerations based on the work of Carter
(1972) suggest that the experimentally determined curves
should be fitted by a $sinc^2(x)$ function. The resolution
of the system, D, using the Rayleigh criterion can then
be stated as the distance from the maximum to the first
zero of this function. Smith (1971) has shown that this
resolution of the Bragg system corrected for astigmatism
should be:

$$D_i = \Lambda/(NA_i) \qquad\qquad\qquad Eq. 4$$

where D_i is the resolution of horizontal or vertical line
sources, NA_i is the numerical aperture of the system for
the appropriate line source orientation, and Λ is the
wavelength of ultrasound.

TABLE 1. Lens Characteristics

Lens	Focal Length	Measured NA	Theoretical NA
Plano convex	100	.05	.5
Plano convex	250	.04	.2
Doublet	211	.09	.24

The numerical aperture for viewing vertically oriented line sources is equal to the numerical aperture of the system cylindrical focusing lens (the diameter of the lens divided by its focal length). Spherical aberrations in the lens can seriously reduce the effective lens numerical aperture. In order to minimize the aberration effects a separated doublet cylindrical lens was constructed. Fig. 4 shows the light intensity distributions obtained with a vertically oriented slit using the doublet, 250 mm focal length plano convex lens and a 100 mm focal length plano convex lens. These measurements were performed at 5 MHz. Note that the doublet shows a much narrower distribution especially when compared to the 250 mm lens, a lens of similar focal length. Table 1 summarizes the characteristics of these three lenses. Note that even the doublet has an effective numerical aperture below theoretical expectations. This is due to the fact that the lens has not completely reduced the aberrations and that the laser beam has a gaussian intensity distribution across its diameter. A multi-element lens would reduce the aberrations further.

Fig. 5 illustrates an LID obtained at 3.2 MHz. Medical imaging of certain structures may be possible at this frequency. The resolution by the Rayleigh criterion is about .66 cm or .76 lp/cm.

The numerical aperture for viewing horizontally oriented line sources is governed by the maximum angle a sound ray can make between the optical axis of the system and the top of the wedge of focused laser light. The

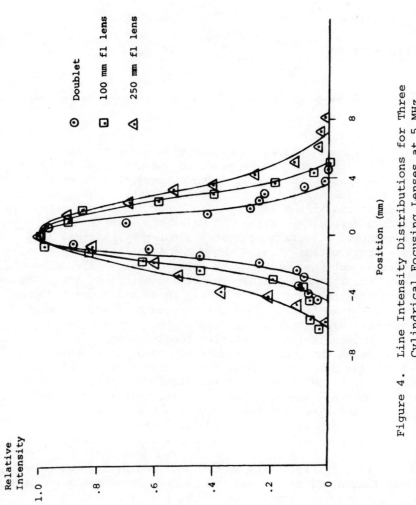

Figure 4. Line Intensity Distributions for Three Cylindrical Focusing Lenses at 5 MHz.

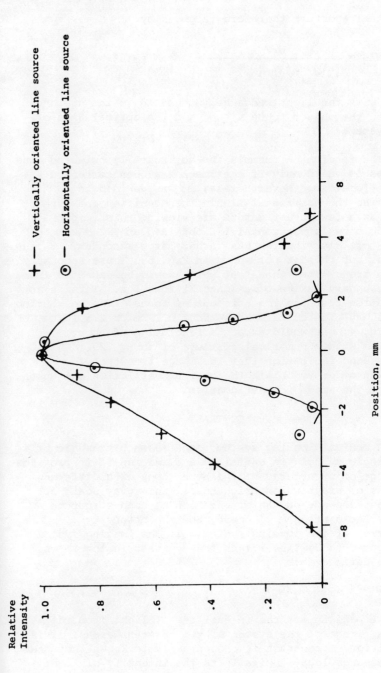

Figure 5. Line Intensity Distributions for the System at 3.2 MHz.

numerical aperture therefore is given by:

$$NA_H = \frac{W/2}{[\ (W/2)^2 + (X_0)^2\ ]^{1/2}} \qquad \text{Eq. 5}$$

where w is the light wedge height and X_0 is the distance
between the object to be viewed and the optical axis of
the system.

The resolution formula for horizontally oriented line
sources is valid only if the image has been corrected
for astigmatism. Nevertheless, it was decided to
determine the horizontal LID in the magnified anamorphic
image as a few investigators are viewing this image.
Fig. 5 shows a horizontal LID obtained at 3.2 MHz. In
fact, this distribution has rather large secondary
maxima, but this is not shown in Fig. 5. These secondary
maxima are greater than that which would be expected from
the secondary maxima of a $sinc^2(X)$ function. This is an
indication that the image is out of focus. The resolution
by the Rayleigh criterion is about 0.2 cm or 2.5 lp/cm.
From Eq. 4, one would expect a resolution of about
15 lp/cm with a numerical aperture of about .25. This
is a second indication that the image is out of focus.
Thus, in order to obtain the optimum resolution one must
correct the image for astigmatism.

CONTRAST

A medical imaging system, such as an ultrasonic or
radiologic system, is useful to a clinician if it provides
an image of structures of the human body with differing
degrees of grayness or brightness. This gray scale or
contrast depends on the attenuation by each structure
of the incident beam of energy and the intensity
linearity of the imaging system. If the imaging system
is assumed to be linear then the contrast in the image
can be defined as:

$$C = [\ \log\ (L_1/L_2)\ ] \qquad \text{Eq. 6}$$

where L_1 and L_2 are the intensities of light from the
optical image of the system of two adjacent areas. This
definition of contrast is chosen as it is known that the
eye responds logarithmically to the intensity of light

falling on it (Meredith and Massey, 1972). If we assume
an exponential model for the absorption of ultrasound in
matter (neglecting reflection for the moment) of the form:

$$I/I_0 = \exp(-2aX) \qquad \text{Eq. 7}$$

where a is the absorption coefficient and X the thickness
of material assumed to be smaller than the thickness of
cut or depth of focus. The contrast relation can then be
written:

$$C = 0.8686 \ (a_1X_1 - a_2X_2) \qquad \text{Eq. 8}$$

for adjacent areas 1 and 2.

A polyethylene stepwedge shown in Fig. 6 can be used
to measure the contrast capabilities of the system. Polyethylene was chosen for the stepwedge material as it is
one of the few solids that offers a good impedence match
to water. Thus, reflection attenuation is small compared
to absorption attenuation. The linear attenuation
coefficient of polyethylene is reported at 0.54 nepers/cm-
MHz. This value is rather high compared to a value of
about 0.15 nepers/cm-MHz for soft tissue. These values
are from Wells (1969). However, as the ultrasonic field
used in the Bragg system is of a single frequency, the
results of experiments with a polyethylene stepwedge
can be used to predict the results of experiments with a
tissue equivalent stepwedge. The equivalent tissue
thickness of polyethylene can be found by equating the
exponential absorption relations for the two materials:

$$I/I_0 = \exp(-2a_pX_p) = \exp(-2a_tX_t)$$

where a_t and a_p are the absorption coefficients and X_t and
X_p are the absorption coefficients and thickness of tissue
and polyethylene respectively. Therefore:

$$X_t = (a_p/a_t)X_p \qquad \text{Eq. 9}$$

Fig. 7 shows a plot on a log scale of the relative
brightness of the image of various steps of the stepwedge
as a function of stepwedge thickness and ultrasonic frequency. The brightness was measured with the photodiode
assembly used for the line intensity distribution measurements. The data fit straight lines supporting an

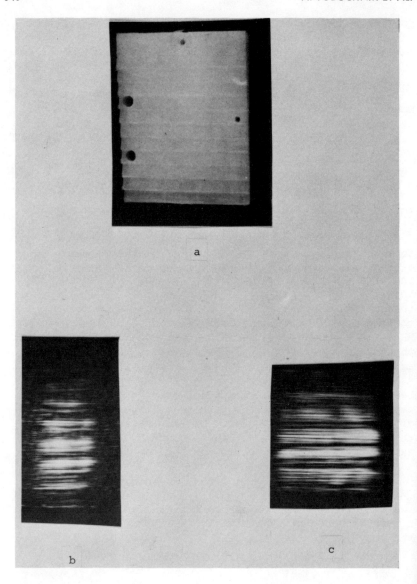

Figure 6.   The Polyethylene Stepwedge.
             (a)   Optical Image.
             (b)   Acoustical Image at 3.2 MHz.
             (c)   Acoustical Image at 5.3 MHz.

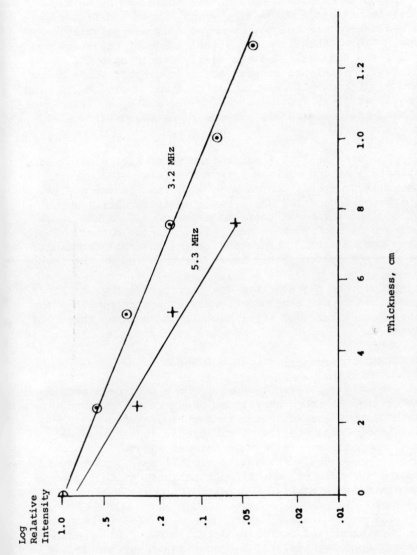

Figure 7. Intensity as a Function of Stepwedge Thickness.

exponential model of absorption. The slope of the two
lines, determined by a least squares fit of the data,
indicate that a is about 0.35 nepers/cm-MHz as compared
to the value of 0.54 nepers/cm-MHz indicated by Wells.
Possible reasons for the discrepancy include differences
in measurement technique, beam geometry and polyethylene
type. Fig. 6 shows the acoustical image of the stepwedge
at 3.2 and 5.3 MHz. One can easily distinguish the
intensity difference between adjacent steps. These steps
have a thickness differential of 2.5 mm. Using Eq. 9, and
the values of a_t = 0.15 nepers/cm-MHz and a_p = 0.35 nepers/
cm-MHz, the corresponding tissue thickness differential
is 5.8 mm.

The data in Fig. 7 and the image of Fig. 6 allow an
estimation of the minimum detectable level of contrast
(threshold contrast) to be made with the present system.
At a frequency of 3.2 MHz a one step differential using
Eq. 8 gives a subject contrast value of 0.2. This rough
estimate is much higher than values one can expect using
x-ray system data. Theoretically the threshold contrast
for a Bragg system should be at least as low as the level
for an x-ray system for a noise free situation. Optical
and acoustical degredations due to imperfections in the
lenses and transducer nonuniformities are responsible for
most of the increased experimentally determined threshold
contrast.

CONCLUSION

In this paper two techniques have been presented for
evaluating the image quality of an ultrasonic imaging
system. They are analogous to techniques used for
evaluating image quality of radiologic imaging systems.
Thus, comparisons can be made between the two types of
imaging processes. Areas of potential use of ultrasonic
imaging systems for medical applications can therefore be
pinpointed more easily.

The two evaluation techniques were used to measure
the resolution and contrast capabilities of an experimental
Bragg imaging system. If a Bragg imaging system is to
achieve its optimum resolution, it must be corrected for
astigmatism. A simple imaging model indicated that one
must balance resolution and depth of field considerations
in the design of an ultrasonic system.

The techniques for image evaluation and the model relating depth of field and resolution are applicable for use with many real time ultrasonic imaging systems.

REFERENCES

1. Korpel, A. "Visualization of the Cross Section of a Sound Beam by Bragg Diffraction of Light," Appl. Phys. Let. 9:425 (1966).

2. Landry, J., Keyone, H., and Wade, G. "Bragg Diffraction Imaging: A Potential Technique for Medical Diagnoses and Material Inspection," Acoustical Holography, Vol. 4, Plenum Press, New York (1972).

3. Hormozdyar, K., Landry, J., and Wade, G. "Bragg Diffraction Imaging: A Potential Technique for Medical Diagnosis and Material Inspection: Part II," Acoustical Holography, Vol. 5, Plenum Press, New York (1974).

4. Smith, R., Wade, G., Powers, J., and Landry, J. "Studies of Resolution in a Bragg Imaging System," J. Acoust. Soc. Amer. 49:1062 (1971).

5. Kessler,L. "A Pulsed Bragg Diffraction Method of Ultrasonic Visualization," IEEE Trans. Sonics Ultrasonics SU-19:425 (1972).

6. Korpel, A. "Astigmatic Imaging Properties of Bragg Diffraction," J. Acoust. Soc. Amer. 49:1059 (1971).

7. Meredith, W. and Massey, J. Fundamental Physics of Radiology, John Wright, Bristol (1972).

8. Wells, P. Physical Principles of Ultrasonic Diagnosis, Academic Press, New York (1969).

ACOUSTIC LENSES AND LOW-VELOCITY FLUIDS FOR IMPROVING BRAGG-
DIFFRACTION IMAGES

G. Wade, A. Coello-Vera, L. Schlussler, S.C. Pei

University of California

Department of Electrical Engineering and

Computer Science

Santa Barbara, California 93106

ABSTRACT

Horizontal resolution in a Bragg-diffraction imaging
system can be substantially improved by inserting a cylindrical acoustic lens into the sound cell between the object
and the region of acousto-optic interaction and by utilizing
a liquid of low acoustic velocity in the interaction region.
It is convenient, particularly for systems in which live
objects are being imaged, to employ water as the medium
into which the objects are placed. In all previous experiments we have used water both in the object region and in
the interaction region. By isolating the interaction region with a small Lucite cell and filling it with a transparent fluid of low velocity, not only will the horizontal
resolution be increased but also the efficiency of the
Bragg diffraction and the magnitude of the Bragg angle. An
increased Bragg angle permits lower noise operation since
the amount of unwanted zero-order light that spreads into
the image region is diminished. The insertion of an acoustic lens can also increase the resolution but only at the
expense of reducing the field of view.

Experimental results are in close agreement with the
analysis. For example, when we use Fluorinert in the interaction region instead of water, the horizontal resolution

and the Bragg angle both increase by a factor of 2.5 in
accordance with the theory.

INTRODUCTION

Bragg-diffraction imaging is a relatively simple
approach for optically imaging an acoustic field. Although
this approach is still in the laboratory stage of development, it nevertheless has produced well-resolved, real-
time images of both organic and inorganic objects in which
substantial internal detail can be seen.[1] The Bragg-
diffraction technique appears to have potential application
in several different areas, particularly those for which
sensitivity is not a factor and for which overall simplicity is important such as in non-destructive testing. This
paper describes two means for improving the system performance and for making the system more practical and useful. One means involves the use of an acoustic lens in the
sound cell between the object and the region of acousto-
optic interaction and the other, the use of a low-velocity
liquid in the interaction region. In this latter case,
experimental results have been obtained which are in close
agreement with the analysis.

CYLINDRICAL ACOUSTIC LENS TO INCREASE HORIZONTAL RESOLUTION

In the conventional Bragg-diffraction system, where no
acoustic lens is employed, the horizontal resolution is
determined by the spatial filtering action of the wedge of
light used to probe the scattered sound beam.[2] Assume that
the object is a line source of sound, the line extending in
the same direction as the apex of the wedge. The spatial
spectrum of acoustic rays emanating from the line source
enter into the interaction region and encounter the laser
light in the wedge. If the acoustic spectrum is wide and
the wedge angle narrow, the light-wave spectrum will not be
wide enough to provide the light rays needed for meeting
the Bragg condition with respect to all the acoustic rays.
Thus only a portion of the sound rays will be able to scatter light (via Bragg diffraction) into the image beam.
Consequently the resolution will be degraded, the light
wedge probing only a finite portion of the spectrum of the
line source of sound. The diffraction-limited resolution
therefore depends on the extent of the spectral window[3] and

and hence on the numerical aperture (N.A.) of the wedge.
Thus

$$\xi_h = \frac{\Lambda}{2(\text{N.A. of light wedge})} = \frac{\Lambda}{2 \sin \alpha} \tag{1}$$

where ξ_h is the resolution for object structure which
varies in the horizontal direction (the x direction of
Fig. 1), Λ is the acoustic wavelength, and α is the semi-
apex angle of the light wedge.

The nature of the horizontal resolution can be changed
with the insertion of a cylindrical acoustic lens in the
sound cell as shown in Fig. 1. To illustrate the operation,
assume that a line source is located at the front focal
plane of the acoustic lens. This lens collects and collimates acoustic rays originating from the line source. The
filtering of the acoustic spectrum in this case occurs at
the acoustic lens, not at the light sound interaction region.

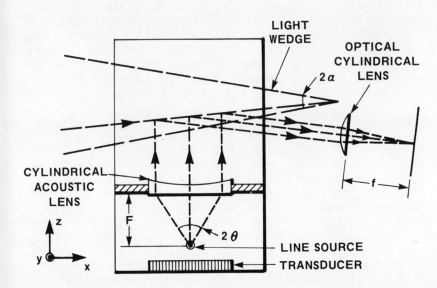

Figure 1 - Schematic diagram of a Bragg-diffraction
imaging system with an acoustic lens.

Hence the diffraction limited resolution is determined by
the numerical aperature of the acoustic lens. According to
the Rayleigh resolution criterion

$$\xi_h = \frac{\Lambda}{2(\text{N.A. of acoustic lens})} = \frac{\Lambda}{2 \sin(\theta)} \qquad (2)$$

where 2θ is the angular aperture of the lens; i.e., the
angle subtended by the two extreme rays in the figure.

The insertion of the acoustic lens must be accompanied
by the addition of a processor component in the form of an
optical cylindrical lens. This latter lens is needed to
perform an inverse transformation for viewing the image.
We can see why such a transformation is necessary if we regard the acoustic lens as a Fourier-transforming device.
The transform is then obviously needed to produce the image
in its proper form.

Thus with the acoustic lens in place, the numerical
aperature of the light wedge no longer affects the horizontal resolution. However, that numerical aperture does play
an important role in defining the field of view. This is
illustrated in Fig. 2. In the conventional system, a wide
field of view is rather easily obtained by making the sound
cell with its transducer wide and the wedge long. By adding
the acoustic lens to the system, the field of view may well
be reduced sharply because it will then be given by

$$\text{Field of view} = F \cdot 2\alpha = 2 \frac{r_\ell}{\sin \theta} \alpha \qquad (3)$$

where r_ℓ is half the width of the acoustic lens. Note
that there is a trade-off relationship between horizontal
resolution and field of view. To make ξ_h in Eq. (2) small
(for good horizontal resolution), $\sin \theta$ must be large.
However, to make the field of view in Eq. (3) large, $\sin \theta$
must be small.

We have recently built an acoustic lens and are in the
process of testing the quality of the system performance
with and without the lens in the system. However, at this
writing only preliminary experiments have been performed
and the results are not definitive.

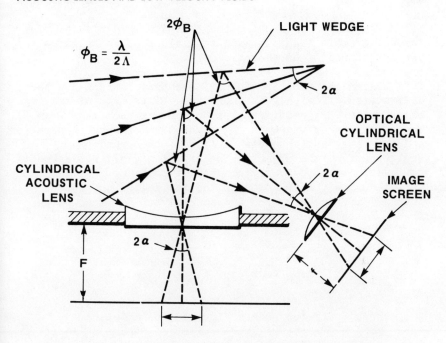

Figure 2 - Field of view is limited by the light
wedge in this new scheme.

LOW VELOCITY FLUIDS TO INCREASE DIFFRACTION EFFICIENCY,
BRAGG ANGLE AND HORIZONTAL RESOLUTION.

Another means for improving a Bragg-diffraction imaging
system is to utilize a liquid of low acoustic velocity in
the acousto-optic interaction region.

It is frequently convenient, particularly for systems
in which live objects are being imaged, to employ water as
the medium into which the objects are placed. In all of our
previous work, we have used water in both the object region
and in the interaction region. By isolating the interaction
region with a thin Lucite wall and filling it with the
proper transparent fluid of low velocity, not only will the
horizontal resolution be increased but also the efficiency
of the Bragg-diffraction and the magnitude of the Bragg angle.
These latter two effects both contribute to improving the
image signal-to-noise ratio. In the case of an increased
Bragg angle for example, the amount of unwanted zero-order

light that spreads into the image region is dimished. The
removal of this light is important since it constitutes the
largest source of unremovable noise.[4]

We can see from the following considerations how the
diffraction efficiency is increased. This type of imaging
makes use of Bragg-diffraction to produce the image beam.
The effect takes place in the light-sound interaction region
and generates a light beam replica of the sound beam
scattered from the object. The efficiency of this acousto-
optic effect is obviously of great importance. We define
efficiency as the fraction of the incident light intensity
that is diffracted by the acoustic wave. For Bragg diffraction, the power ratio of the diffracted light to incident He-Ne laser light is:[5]

$$\frac{I_1}{I_0} = \sin^2 (1.4\ell \sqrt{M P_s}) \qquad\qquad (4)$$

where ℓ is the interaction length in meters
      M is the relative numerical figure of merit for
         the liquid with respect to the water.
      P_s is the acoustic power density in watt/m^2.

If

$$\frac{I_1}{I_0} \ll 1,$$

$$\frac{I_1}{I_0} \simeq 1.96 \; \ell^2 \; M \; P_s \qquad\qquad (5)$$

Thus we see that, at low sound levels, the intensity
of the diffracted light increases linearly with acoustic
power P_s and is proportional to the relative figure of merit
M for the liquid in use. Now [5]

$$M = \frac{\dfrac{n^6 \cdot p^2}{\rho \cdot v^3}}{M \text{ water}} \qquad\qquad (6)$$

where n is the optical index of refraction

p is the elasto-optic coefficient

ρ is the density of the interaction medium

V is the acoustic phase velocity in the medium

M_{water} is the actual figure of merit for the water.

In a Bragg-diffraction imaging system, the four parameters in the above equation are properties of the liquid in the interaction region. The proper combination of medium parameters will strongly influence the quality of the device performance. The elasto-optic coefficient p does not vary much from liquid to liquid. It is about one-third for most usable liquids, being 0.31 for water and 0.35 for xylene, etc.[6] Hence, the variations in acousto-optic performance are primarily determined by the density ρ, the sound velocity V and the index of refraction n. By far the greatest variation is caused by the terms n^6 and v^3 in Eq. (6). As we can see, a liquid with a low sound velocity is highly desirable in a Bragg-diffraction imaging system.

In the previous systems we have built, we have always used water for the interaction medium. For our recent experiments we tried a different transparent liquid, known commercially as Fluorinert, FC-75, with a sound velocity of only 600 meter/sec. This is 2.5 times lower than that for water. However, Flourinert has other properties which are detrimental to imaging. Streaming in Flourinert is especially severe. It also has a large attenuation for sound which prevents the high diffraction efficiency which otherwise would be possible.

Nevertheless, two of the benefits we expected to be able to obtain from the low velocity did show up in the experimental results. The horizontal resolution and the Bragg angle were both improved by a factor of 2.5 from using Fluorinert instead of water.

The expression for the Bragg angle is given by Eq. (7)[5]

$$\emptyset_B = \sin^{-1} \frac{\lambda}{2\Lambda} = \sin^{-1} \frac{\lambda \cdot \Omega}{2 \cdot V} \qquad (7)$$

where λ if the optical wavelength of the laser light
       Λ is the acoustic wavelength in the interaction
       medium
       Ω is the acoustic frequency
       V is the acoustic velocity in the interaction
       medium.

Thus the diffraction angle is inversely porportional to the
sound velocity.

A large Bragg angle is very important in low-frequency,
Bragg-diffraction imaging.   At low frequencies, the Bragg
angle is very small. For example, it is less than 2 milli-
radian in water at 3.6 MHz.   Under these circumstances,
noise due to the zero-order light is especially serious in
degrading the image.[7] By using Flourinert, the situation
was greatly improved. Due to the larger Bragg angle, it
was easier to separate the first sideband image from the
zero-order light.

As previously stated, the diffraction-limited horizontal resolution in Bragg-diffraction imaging depends on the
numerical aperature of the light wedge. This was shown in
Eq. (1), which can be rewritten as follows:

$$\xi_h = \frac{\Lambda}{2 \sin \alpha} = \frac{V}{2 \cdot \sin \alpha \cdot \Omega} \tag{8}$$

With a lower sound velocity in the liquid, we get a smaller
acoustic wavelength in the field of the scattered sound.
Thus we can image more detail in the horizontal structure
of the object.

We can also explain this effect from another point of
view. According to Snell's law:

$$\frac{\sin \theta_2}{\sin \theta_1} = \frac{n_1}{n_2} = \frac{V_2}{V_1} = \frac{\Lambda_2}{\Lambda_1} \tag{9}$$

This is illustrated in Fig. 3 when a ray passes from a
medium of fast velocity into one of slow velocity, the rays
are always bent toward the normal. Because of the bending
of the rays at the interface between two liquids, the lobes
of the spatial spectrum of the sound beam scattered from the
object are narrowed as the sound passes into the acousto-

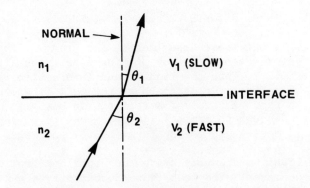

Figure 3 - Refraction of a ray.

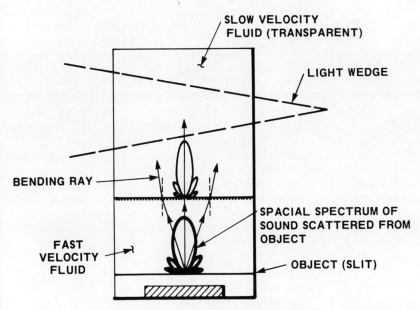

Figure 4 - Using a slow-velocity liquid in the interaction region, more of the spatial spectrum
in the sound scattered from the object can
be accepted by a given light wedge.

optic interaction region. This effect is illustrated in
Fig. 4. A given light wedge will therefore accept more of
the spatial spectrum of the scattered sound and provide
more detail in the image.

EXPERIMENTAL RESULTS USING A LOW-VELOCITY FLUID

To carry out these experiments, our Bragg-diffraction imaging system was modified by placing a cell containing Fluorinert, FC-75 into the light-sound interaction region. See Figs. 5 and 6. A photograph of the cell is shown in Fig. 7. It has the shape of a rectangular box with dimensions large enough to admit all the components of sound which would contribute to the formation of an image.

The light used in the system was obtained from a Helium-Neon laser. The sound was generated by a quartz transducer with a fundamental frequency of 3.58 MHz and a diameter of 7.62 cm. The transducer was used at its third harmonic, 10.74 MHz. The cylindrical light wedge in our system was formed by an f/7 cylindrical lens whose diameter was 2.25 inches. The light window in the low-velocity cell was 8 cm wide and 10 cm high. The sound window measured 8 cm high by 10 cm wide. The sound window was constructed of Lucite with a thickness of one-sixteenth inch. Lucite was chosen because it provides a fairly good impedance match between the water and the Fluorinert. A thin membrance was not used for the window because a degree of stiffness was desired. The density difference between the Fluorinert and the water causes an outward pushing pressure on the walls of the cell. A membrane would have a tendency to bow and cause distortion of the incoming sound field. (Note that if controlled properly, this effect could cause a focusing action and thus permit combining the effects of an acoustic lens with those of the low-velocity cell.) The window opposite the transducer was also covered with one-sixteenth inch Lucite to minimize reflections off the cell walls.

The attenuation at 10.74 MHz in Fluorinert was measured and found to be 4.92 db/cm. To make this measurement, a pencil beam of laser light was admitted into the cell containing the Fluorinert, the light coming in at the Bragg angle. The distance from the laser beam to the acoustic window was varied and the intensity of the diffracted light was measured at these various locations. The sound attenuation through the Fluorinert was then calculated from this data. To prevent streaming during this experiment the acoustic power into the transducer was kept fairly small, approximately one watt. When streaming does occur, it can cause large temporal and spatial variations in the intensity of the diffracted light.

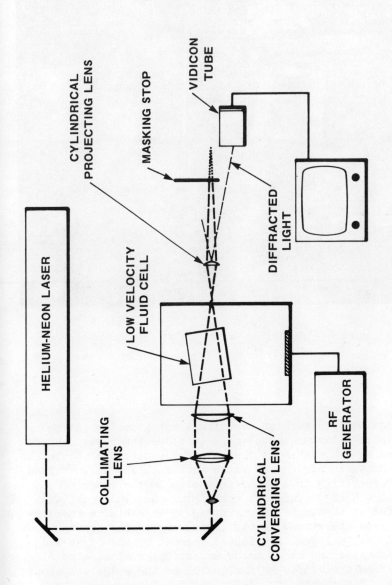

HELIUM-NEON LASER

COLLIMATING LENS

CYLINDRICAL CONVERGING LENS

CYLINDRICAL PROJECTING LENS

MASKING STOP

VIDICON TUBE

DIFFRACTED LIGHT

LOW VELOCITY FLUID CELL

RF GENERATOR

Figure 5 — Schematic diagram of the Bragg-diffraction imaging system.

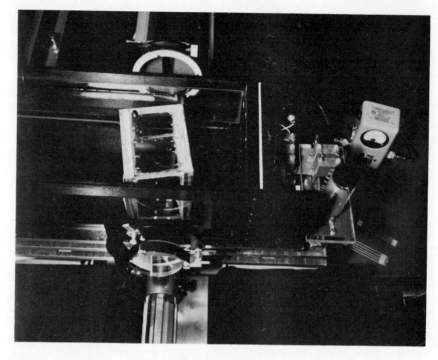

Figure 6 - Experimental setup of the Bragg-
diffraction imaging system.

Because of the high acoustical attenuation in the
Fluorinert a larger amount of acoustical power was needed
to form an image than is the case for water. To obtain the
image shown in Fig. 8, 25 watts were supplied to the transducer, resulting in a sound intensity in the water of
about $1/4$ W/cm^2. The cell containing the Fluorinert was
oriented at the angle shown in Fig. 5 to minimize the distance from the window to the light wedge. The construction
of the cell allowed the closest segment of the light wedge
to be located one cm from the acoustical window. The distance from the window to the center of the wedge was about
3.5 cm. The components of low spatial frequency in the
image were therefore attenuated by about 17 dB before the
image-forming light was diffracted off them. The higher
spatial frequencies, of course, were attenuated by varying

Figure 7 - Cell containing Fluorinert.

amounts depending on where in the light wedge they were diffracted.

The streaming velocity in a fluid is proportional to the acoustical power and attenuation of the sound in the fluid. Both of these quantities are fairly large in Fluorinert and streaming causes severe problems in image formation. When 25 watts were fed into the transducer, the image would be stable for several seconds and then break up into wavy lines. The image shown in Fig. 8 was taken immediately after the acoustic power was turned on. The wavy lines were caused by the television monitor.

The shadowgrams shown in Figs. 9 and 10 illustrate the streaming problem in the Fluorinert. Fig. 9 shows the streaming when the acoustical power is initially turned on and Fig. 10 shows the streaming after steady state conditions

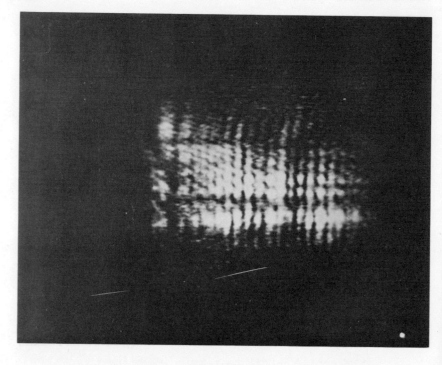

Figure 8 - Image with Flourinert of a set of
vertical wires having a diameter
and spacing of .72 mm.

have been reached. It can be seen that for steady-state
conditions, the cell is filled with circulating currents
of fluid. Once this has happened, it takes several minutes
after the acoustical power has been turned off for the motion to subside to a point which is acceptable for image
formation. With 25 watts into the transducer, streaming
could also be observed in the water section of the Bragg
cell. Problems involving streaming have previously been
solved by employing a pulsed mode of operation.[8,9] Using
the cell containing Fluorinert and an f/7 cylindrical converging lens the maximum resolution observed for a set of
vertically oriented wires was about 0.7 mm with the acoustic frequency at 10.74 MHz. The image obtained for a
spacing of 0.72 mm between wires was shown in Figure 8.
Our theoretical calculation shows that we should have

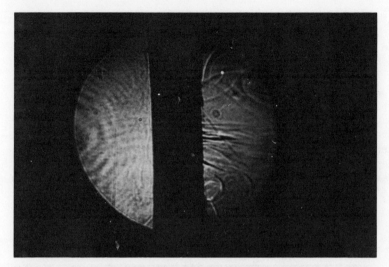

Figure 9 - Shadowgrams of streaming in water and
Flourinert shortly after the acoustical power is turned on. Flourinert is
on the right hand side and water on the
left. Black line is the aluminum frame
of the cell.

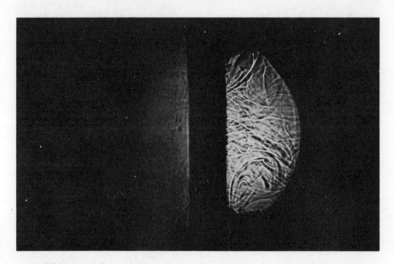

Figure 10 - After steady state conditions were
reached.

attained a resolution of 0.4 mm. Imaging of this same set
of wires was attempted after removing the low-velocity from
the Bragg cell. The wires were quite unresolvable. The
image obtained is shown in Figure 11.

There were nine wires in the set which were imaged.
Notice that more than 9 vertical lines can be seen in the
image. We have often observed these additional lines when
wires are imaged. The lines are caused by an effect, frequently called ringing, due to interference between diffracted components of the coherent sound scattered from the
evenly-spaced wires. The spacing of these additional lines
in the image is the same as the wire spacing.

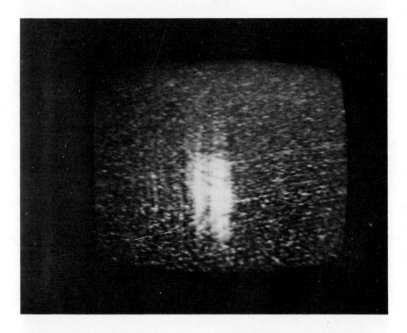

Figure 11 - Same set of wires as in Figure 8
            imaged in the conventional fashion
            (that is, with the cell containing
            the Flourinert removed). The image
            of the set of wires, if resolvable,
            would have appeared in the bright area.

CONCLUSION

Our analysis shows that the insertion of the appropriate cylindrical acoustic lens into the system will increase the horizontal resolution but reduce the field of view. Further experiments are being undertaken to check the value of such a modification to the conventional system.

Such a theory also predicts that the use of a low-velocity fluid can improve a Bragg-diffraction imaging system in terms of the efficiency of the diffraction, the magnitude of the Bragg angle and the horizontal resolution. The experimental results are in close agreement with analysis relative to the latter two effects. For example, when we used Flourinert in the interaction region instead of water, the horizontal resolution and the Bragg angle both improved by a factor of 2.5 in accordance with theory.

Fluid streaming however, is a serious problem when Flourinert is employed in CW operation. Previous experience has indicated that streaming can be greatly reduced by pulsing the acoustic power at a sufficiently low duty cycle.[8,9]

ACKNOWLEDGEMENTS

This work was supported by the National Institutes of Health of the U.S. Public Health Service. One of the authors, A. Cöello-Vera, acknowledges the support received from the Del Amo Foundation during part of this work.

REFERENCES

1. J. Landry, J. Powers, and G. Wade, "Ultrasonic imaging of internal structure by Bragg diffraction, "Applied Physics Letters 15 (6): 186-188 (1969).

2. J. Powers, "Spatial Filtering Considerations in Bragg-Diffraction Imaging," Acoustical Holography, Vol. 4, Ed. G. Wade, Plenum Press, 1972, pp. 533-567.

3. A. Korpel, "Visualization of the cross section of a sound beam by Bragg diffraction of light," Applied Physics Letters 9 (12): 425-426 (1966).

4. K. Wang and G. Wade, "Comparison of Ideal Performance of Some Real-Time Acoustic Imaging Systems," J. Acoust. Soc. Am., Vol. 56, No. 3, Sept. 1974, pp. 922-928.

5. R. Adler, "Interaction between light and sound," IEEE Spectrum, vol. 4, pp. 42-54, May 1967.

6. T. M. Smith and A. Korpel, "Measurement of light-sound interaction efficiencies in solids," IEEE J. Quantum Electronics (correspondence), vol. QE-1, pp. 283-284, September 1965.

7. J. Landry, R. Smith and G. Wade, "Optical Hetero-dyne detection in Bragg imaging," Acoustical Holography, vol. 3, pp. 47-70, Plenum Press (1971).

8. L. W. Kessler, "A pulsed Bragg diffraction method of ultrasonic visualization," IEEE Trans. on SU, vol. SU-19, No. 4, pp. 425-427, Oct. 1972.

9. H. Keyani and G. Wade, "Recent Experiments in Bragg-Diffraction Imaging," Digest International Optical Computing Conference, Zurich, Switzerland, April 1974, pp. 87-90.

RAMAN-NATH IMAGING

R. A. Smith

TRW Systems Group
One Space Park
Redondo Beach, California 90278

ABSTRACT

Analysis and experiments reported here show that Bragg
imaging systems continue to operate at frequencies too low
to support Bragg diffraction. At sound frequencies of a few
megahertz and lower, Raman-Nath theory determines light
diffraction by sound if the interaction length is no longer
than about 35 cm. Imaging rules under Raman-Nath conditions
are nearly the same as corresponding rules for imaging under
Bragg conditions. It is shown that system sensitivity under
Raman-Nath conditions is identical to system sensitivity
under Bragg conditions if spill over from central order
light is made negligible.

A major part of this paper is given to a derivation of
theoretical sensitivity for a new dual frequency optical
heterodyning system which is used to sense the image. The
presented analysis takes practical limitations into account.

INTRODUCTION

Several acoustic imaging applications require the sound
frequency be as low as possible and the imaging apparatus as
small as possible. In any acoustic imaging system, this
results in images of reduced quality. There is considerable
interest in such images where there is no alternative to the
use of low frequency sound but consequently poor resolution.
As the sound frequency is reduced in Bragg type[1] [2] imaging systems, the number of wavelengths in the light-sound

interaction region becomes too few to consider the interaction to be Bragg diffraction in the classical sense. This
paper is devoted to image formation at a sound frequency
which is too low and interaction cell dimensions too small
for Bragg diffraction, but sufficient to satisfy Raman-Nath
conditions. The system forming visual images of a sound
field under Raman-Nath conditions can be identical in form
to the Bragg imaging system (except for sound frequency)[3].

The region of operation commonly known as Raman-Nath
diffraction corresponds to light sound interaction at low
acoustic frequencies. Raman-Nath theory explains diffraction when the curvature of light rays is negligible. When
the sound intensity is low, the condition for Raman-Nath
diffraction is defined by[4]

$$d < \Lambda^2/16\lambda. \tag{1}$$

Where

    d = length of the light-sound interaction measured
        along a light ray

    λ = the wavelength of light

    Λ = the wavelength of sound.

Experiments reported in the preceding paper were done
with an Argon-Ion Laser operating with λ = .396 μm (in
water)[3]. At a sound wavelength of .53 cm (280 KHz in
water) Equation 1 limits the length of the interaction
region for Raman-Nath theory to less than 3.6 m. Images
reproduced in the preceding paper were formed under conditions well within the Raman-Nath region.

The low frequency limit for Bragg diffraction follows
from a previously published definition of Bragg diffraction
due to ultrasound[4]. It reduces to the requirement that a
light ray in the interaction region must cross at least one
wavelength of sound as illustrated in Figure 1. Both the
incident and diffracted light rays are assumed oriented to
satisfy the angular relation specified by the Bragg condition (as shown in Figure 1).

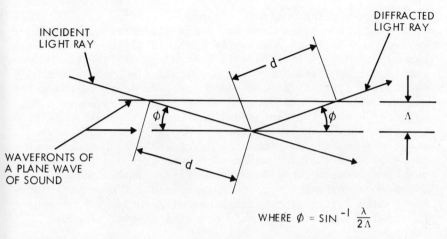

$$\text{WHERE } \phi = \text{SIN}^{-1} \frac{\lambda}{2\Lambda}$$

Figure 1. Diffraction of a light ray due to a plane wave of sound involving only two wave fronts.

Inspection of Figure 1 reveals that

$$d \sin \phi = \frac{d\lambda}{2\Lambda} = \Lambda. \tag{2}$$

It is evident from Figure 1 that the width of inter-action must be at least as large as d. A requirement for Bragg diffraction is therefore (from Equation 2)

$$\Lambda < \sqrt{d \ \lambda/2.} \tag{3}$$

For example, assuming cell illumination light is at a wavelength of .54 μm in air and an interaction length of 1.2 m, the longest sound wavelength that can be considered to diffract light in the Bragg manner is .05 cm in water. This corresponds to a minimum sound frequency of 3.0 MHz in water. It is then surprising that recognizable images shown in the preceding paper have been formed in a Bragg imaging system at a sound frequency as low as 500 KHz using a "Bragg" cell which is 1.2 m in width.

A ray-tracing analysis of image formation is essentially the same for Raman-Nath and Bragg diffraction. In either
regime, the angle through which light is diffracted is the
same. The major difference is that Bragg diffraction only
occurs when the angle between propagation vectors describing
the incident light and the sound field is equal to the Bragg
angle, while the Raman-Nath diffraction is maximum with 90
degrees between incident sound and light propagation vectors.
The significance of this difference disappears as the sound
frequency is reduced since the Bragg angle is proportional
to sound frequency. A major phenomenological difference is
that two first order sidebands, either side of incident
light, must appear simultaneously at equal intensities for
Raman-Nath diffraction while Bragg diffraction can be made
to produce a sideband on only one side of central order
light.

INTENSITY OF DIFFRACTED LIGHT

The relation between diffracted light intensity and the
intensity of a sound wave in the form of a plane wave has
been investigated by several authors. A summary of work
published prior to 1953 is included in a basic reference by
Bhatia and Noble[5] along with an extensive theory determining diffraction of light into many possible orders assuming
a wide range in the angles of incident of illuminating light.
Raman-Nath[6] showed that in the Raman-Nath regime, the maximum intensity of diffracted light occurred when an incident
plane wave of sound has a propagation vector at right angles
to the propagation vector for a plane wave of light. This
condition will be called the right angle condition.

Under the right angle condition, the intensity of light
diffracted into the mth order is given by[6]

$$I_m = I_o J_m^{\,2} \left[\frac{2\pi}{\lambda} \left(\frac{\partial n}{\partial P} \right)_s Pd \right] \tag{4}$$

where

I_o = intensity of incident light (in the form of a plane wave)

$J_m[\ ]$ = the ordinary Bessel function of order m

$\left(\dfrac{\partial n}{\partial p}\right)$ = adiabatic piezo-optic coefficient

P = peak pressure of the acoustic wave in the medium

d = length of the interaction measured in the direction of light propagation

n = is the index of refraction of the medium.

It is well-known that the intensity of diffracted light rapidly approaches zero as the incidence angle deviates from the direction given by either the Bragg condition or the right angle condition for Raman-Nath diffraction. A given spatial spectral component (a plane wave) in the sound field will therefore produce a unique related spatial spectral component in the diffracted light for both the Raman-Nath and Bragg regimes. It is interesting to note that the intensity of diffracted light under both Raman-Nath and Bragg conditions is given by the same expression when the sound level is near threshold conditions. To see this, consider the case for Raman-Nath conditions. In this case, the diffracted light intensity follows from the small argument approximation of the Bessel function which is $J_1(Z) = Z/2$ for $Z \ll 1$. At threshold conditions the intensity of light diffracted into the first order is then (from Equation (4)).

$$I_1 = \frac{\pi^2}{\lambda^2}\left[\left(\frac{\partial n}{\partial P}\right)_s\right]^2 P^2 d^2 I_o \qquad (5)$$

It can be shown that the variation in reflective index with density, ρ, at constant entropy is given by[7]

$$\left(\frac{\partial n}{\partial \rho}\right)_s = (n^2 + 2)(n^2 - 1)/(6\rho n).$$

where

ρ = density of the medium.

Using

$$\left(\frac{\partial n}{\partial P}\right)_s = \left(\frac{\partial n}{\partial \rho}\right)_s \left(\frac{\partial \rho}{\partial P}\right)_s = \left(\frac{\partial n}{\partial \rho}\right)_s \frac{1}{v_s^2} \tag{6}$$

where

v_s = longitudinal sound wave velocity

Equation (6) becomes

$$\left(\frac{\partial n}{\partial P}\right)_s = (n^2 + 2)(n^2 - 1)/(6n\rho v_s^2). \tag{7}$$

The average power of a plane sound wave in a liquid is (8)

$$I_s = P^2/(2Z_a) \tag{8}$$

where

Z_a = acoustic impedance

= ρv_s for a liquid

The first order diffracted light intensity is then (using Equations (5) through (8)

$$I_1 = \Pi^2 \left(\frac{d}{\lambda}\right)^2 \sigma^2 \frac{I_s I_o}{2\rho v_s^3} \tag{9}$$

where

$$\sigma = (n^2 + 2)(n^2 - 1)/(3n).$$

Although the analysis of Raman-Nath also covered the case where diffraction can be considered to be Bragg diffraction ($\lambda_1 d > 2\Lambda^2$), the result from analysis published by Bhatia[1] and Noble for Bragg diffraction will be compared with the result from Raman-Nath's analysis. Bhatia and Noble[16] show that the first order diffracted plane wave will have an intensity given by

$$I_{-1} = \frac{1}{4} \frac{I_o \delta^2 \sigma^2}{\left(\xi - \frac{1}{2}\right)^2 + \frac{1}{4}\delta^2\sigma^2} \sin^2\left(\beta d \left\{\left(\xi - \frac{1}{2}\right)^2 + \frac{1}{4}\delta^2\sigma^2\right\}^{1/2}\right) \quad (10)$$

where

$\delta = \Delta(\Lambda/\lambda)^2$

Δ = peak fractional change in the density of the medium

$\xi = (\Lambda/\lambda) \sin\theta$

θ = angle between the propagation vectors of the incident light and sound wave (considered plane), respectively

$\beta = \pi\lambda/(\Lambda^2)$

This equation holds only when $\xi \simeq \frac{1}{2}$ and $\delta \ll 1$. Note that the Bragg condition is defined by $\xi = \frac{1}{2}$. The condition that $\delta \ll 1$ is consistent with low intensities of sound that are of interest in this threshold study. The argument of the sine function in Equation (3) is small at $\xi = \frac{1}{2}$ when threshold conditions apply ($\beta d\delta\sigma \ll 1$). The form of Equation (10) of interst in this study then follows from the small angle approximation. For threshold analysis, Equation (10) becomes

$$I_{-1} = \frac{I_o}{4} \beta^2 d^2 \sigma^2 \delta^2. \quad (11)$$

The peak fractional change in density follows from Equation (8). It is

$$\Delta = \left[2I_s/(\rho v_s^3)\right]^{1/2} \quad (12)$$

so that

$$\delta^2 = \frac{2I_s}{\rho v_s^3} \left(\frac{\Lambda}{\lambda} \right)^4 . \tag{13}$$

Substitution of applicable definitions into Equation (11) shows that the intensity of light in a first order sideband under Bragg conditions for low amplitude sound, is given by the same expression as that for Raman-Nath conditions, Equation (9). Equation (9) will therefore be used to relate the intensities of the spatial spectrum of the distribution describing the sound field.

Equation (9) applies to incident light in the form of a uniform plane wave. The system considered here is illuminated by a cylindrically convergent wedge of light with the convergence axis directed vertically. The interacting intensity of plane wave components within incident light must be determined in order to use Equation (9). The interacting plane wave of sound on which Equation (9) is based is of limited extend (d units wide). It is the width of the interaction region (d) which determines the incident light intensity available for interaction with an infinite plane wave component of the sound field.

The x-axis is taken normal to a vertical plane which contains the axis of convergence and which bisects the wedge of light. The system considered uses illumination with a uniform spatial spectrum between x-directed spatial frequency limits given by

$$f_{xm} = \frac{\pm \sin\alpha_m}{\lambda} \tag{14}$$

where α_m = one-half the angular width of the wedge of illuminating light.

The Bragg imaging system would, ideally, be illuminated with light distributed uniformly over spatial frequencies within limits corresponding to the angular width of the "wedge" of illuminating light. We shall consider this to be the case since any variation from a uniform distribution would introduce amplitude aberrations.

The question to be answered now is "how much incident light power is involved in producing an image (neglecting noise) by a plane wave of sound which is d units in width?" Observe that a uniform plane wave of sound of width d in the horizontal direction (and infinite height) will produce an image (without any lens) of width equal to $\lambda d/(\Lambda)$. Now the horizontal directed spatial spectrum of light forming the width pattern is the same as the spatial spectrum within interacting incident light[9]. The (spatial) spectral amplitude density of interacting incident light involved in forming an image of width, $\lambda d/\Lambda$), can be written

$$u'(f_h) = \left(\frac{d\lambda}{\Lambda}\right) \frac{\sin\pi f_h\left(\frac{\lambda d}{\Lambda}\right)}{\pi f_h\left(\frac{\lambda d}{\Lambda}\right)} u_o \qquad (15)$$

where

u_o = amplitude of the incident plane wave which will produce $u'(f_h)$ if the incident plane wave of amplitude u_o were incident on a slit of width $\lambda d/(n\Lambda)$

f_h = the horizontal directed spatial frequency directed so as to yield a symmetrical distribution. The reference direction for f_h will vary with the particular sound field component considered.

The intensity of an incident plane wave of amplitude u_o is

$$I_o = |u_o|^2/2. \qquad (16)$$

To determine I_o, (the intensity available for interaction) expand the entire uniform spatial spectrum contained within incident light into equally spaced components of the same form as Equation (15). Thus, we can write (10) for the total amplitude spatial spectrum in cell illumination

$$L(f) = \frac{u_o \lambda d}{\Lambda} \sum_m \text{Sinc } (X_s f - m)$$

$$m = \ldots -2, -1, 0, 1, 2, \ldots$$

where

$$|m| = \text{an integer} \le N/2$$

$$X_s = \frac{\lambda d}{\Lambda}$$

$$\text{Sinc}(t) = \frac{\sin \pi t}{\pi t}$$

N is defined by Equation (17).

The interval between neighboring sinc functions is $1/X_s = \Lambda/(\lambda d)$. The total number of "samples" within the spatial bandspread of incident light, $(2 \sin \alpha_m)/\lambda$, is

$$N = \frac{2 \sin \alpha_m}{\lambda} \frac{\lambda d}{\Lambda} = \frac{2 d \sin \alpha_m}{\Lambda} \tag{17}$$

where

$$\frac{\sin \alpha_m}{\lambda} = \text{the maximum spatial frequency in incident light.}$$

The power in each "sample" per unit height is just the integral of the square of Equation (15) over all spatial frequencies. It is

$$W_1 = \int_{-\infty}^{\infty} \left| \frac{u'(f_h)}{2} \right|^2 df_h = \frac{u_o^2 \lambda^2 d^2}{n^2 \Lambda^2} \int_{-\infty}^{\infty} \frac{\sin^2 \pi f \frac{\lambda d}{n \Lambda}}{\left[\pi f \left(\frac{\lambda d}{n \Lambda} \right) \right]^2} df .$$

The completed integral gives

$$W_1 = \frac{u_o^2}{2} \frac{\lambda d}{\Lambda} = I_o \frac{\lambda d}{\Lambda}. \tag{18}$$

Note that the sinc function is orthogonal under the applicable integral so that the total power is given by the sum of the power in each sample.

The total power per unit height of the wedge of incident light is (from Equations (17) and (18))

$$P_e = NW_1 = \frac{2d^2 \lambda I_o \sin\alpha_m}{\Lambda^2}. \tag{19}$$

An incident plane wave of light of proper orientation and intensity, I_o, will produce the same image intensity as that which will appear when the cell is illuminated by a wedge of light with total power per unit height, P_e, if the sound field is uniform and of width d.

The equivalent intensity of each plane wave component in incident light which is available for interaction with a plane wave component of the sound field is (from Equation (19))

$$I_o = \frac{\Lambda^2 P_e}{2\lambda d^2 \sin\alpha_m}. \tag{20}$$

The intensity of light in the image due to a uniform source of sound of intensity, I, is therefore (from Equations (9) and (20))

$$I_{-1} = \frac{\pi^2 \sigma^2}{4 \rho v_s^3} \frac{\Lambda^2}{\lambda^3} \frac{I_s P_e}{\sin\alpha_m}. \tag{21}$$

Note that the intensity in an image which is magnified
sufficiently to give the correct aspect ratio, which also
is the same size as the object, will be reduced in intensity
by the factor (λ/Λ). The intensity in the orthoscopic plane
due to a uniform sound source of intensity I_s will then be

$$I_{-1f} = \frac{\pi^2}{4} \frac{\sigma^2}{\rho v_s^3} \frac{\Lambda}{\lambda^2} \frac{I_s P_e}{\sin\alpha_m}. \tag{22}$$

In the case where the object is not a uniform distribution of sound, Equation (22) expresses the idensity due to
sound passing through only that part of the object which
scatters sound into angles far less than the maximum angular
aperture of interaction. In this case, the spatial frequency spectral intensity in scattered sound will be related
to the spatial spectral intensity of light falling on the
image in such a way that Equation (22) will give the intensity of light at the applicable resolution cell in the image,
if I_s is the sound intensity passing through the corresponding point of the object.

IMAGE PICKUP BY OPTICAL HETERODYNING

Performance of this imaging system can be enhanced by
using a television camera tube or other photo sensor system
together with a television monitor to display the image.
As with all electronic devices, there is an internal source
of noise within the image pickup device which, in some
cases, will determine system sensitivity. It is well-known
that optical heterodyne detection is capable of providing
greater sensitivity than direct forms when coherent light
is to be detected. This happens because gain is associated
with the heterodyne process making the signal at the photodiode output sufficiently large to squelch other noise
terms. A convenient way to implement heterodyne detection
is to use the two frequency imaging system shown in Figure
2(11).

In this system, a reference transducer produces a uniform sound field which is subsequently imaged, producing a

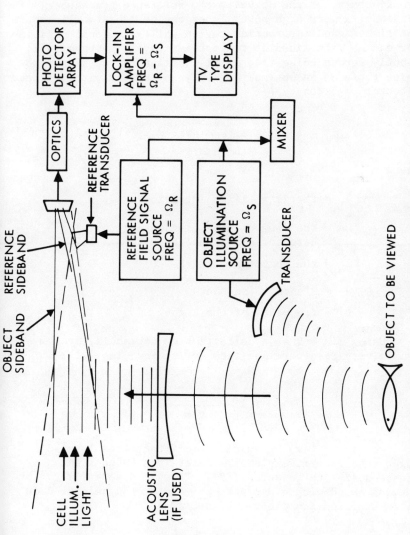

Figure 2. A double frequency heterodyning acousto-optical imaging system.

uniform "local oscillator" field[11]. Sound reflected from the object viewed produces an image that is superimposed upon the local oscillator field. The intensity of the local oscillator field can be adjusted to the optimum level by adjustment of the drive to the reference transducer. In this way, a very high power laser can be used to illuminate the light/sound interaction region while the local oscillator can be maintained at the optimum level. With a sufficiently strong noise free local oscillator, the signal to noise ratio of an optical heterodyning system with a linear detector is given by[12]

$$\frac{S}{N} = \frac{q \ I_1 \ A}{2 \ h\nu \ B_o} \tag{23}$$

where

I_1 = the image light intensity

A = the area of a resolution cell or detector area, whichever is smaller

B_o = the output bandwidth

h = Planck's constant

ν = frequency of the light.

Using $S/N = 2$ as a definition of threshold, system sensitivity is determined from (using Equation (23))

$$I_1^* = \frac{4 \ h\nu \ B_o}{q \ A}. \tag{24}$$

The sound level required to diffract this level of light onto the image plane is determined by equating Equation (24) to Equation (22). Threshold sensitivity for imaging sound with a perfect optical heterodyne system is given by (from Equations (22) and (24).

$$I_s = \frac{16 \text{ hc } B_o \, \rho \lambda_s 3 \, \lambda \, \alpha_m}{\pi^2 \, q \, A\sigma^2 \, \Lambda \, P_e}. \tag{25}$$

The last result is based upon an image which is full size. A reasonable value for the resolution area, A, is four sound wavelengths in the vertical direction by Λ/α_m in the horizontal direction or $A \cong 4\Lambda^2/\alpha m$. The value of A, applicable to any specific system, may vary and will depend upon acoustic reflectors or lenses which may be present. For example, in laboratory experiments reported in the preceding paper[3], the detector area was smaller than a resolution cell size, but magnification was not changed as the frequency was changed. As a result, observed sensitivity should be proportional to the square of frequency (inversely proportional to wavelength squared).

The frequency squared proportionality was observed experimentally, as shown in Figure 3. Figure 3 shows the results of experiments made using the system shown in Figure 2. In making these measurements, a small sound source was placed at a distance of 15 feet to allow sound to become more planar over most of the 4 foot length of the light/sound interaction region. A cylindrical acoustic lens was used to confine all sound to arrival within the 6-in. high column of cell illumination light.

In order to obtain maximum sensitivity estimates, consider an optical heterodyning system which converges all light within a resolution cell onto a detector. The collection area for a full size image will be assumed to be

$$A = 4\Lambda^2/\alpha_m.$$

Using this assumption and $\alpha_m = 1/12$ with λ .5145 µm, Equation (25) becomes

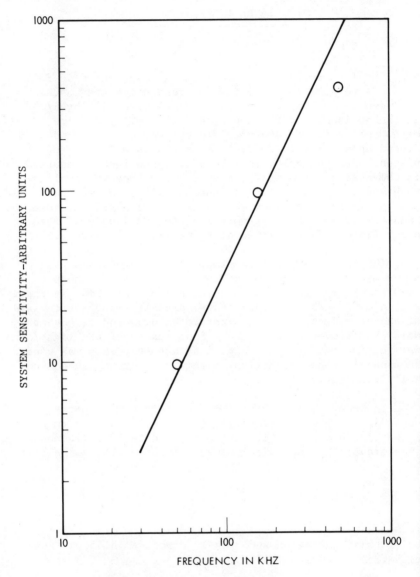

Figure 3.   Measured Sensitivity of the Dual Frequency
            Optical Heterodyning System Shown in Figure 2.

$$I_{s1}^{*} = 1.7 \times 10^{-21} \, \frac{B_o}{q \, P_e \, \Lambda^3} \; watts/cm^2 \tag{26}$$

where

B_o = system bandwidth Hz

P_e = laser power per unit height of the wedge (watts/cm)

Λ = sound wavelength in centimeters

q = quantum efficiency.

As an example of a typical case, let P_e = .1 watt/cm, q = .75 and B_o = 1 Hz. At a sound frequency of 500 KHz, Λ = .3 cm, a threshold sensitivity of 8.4×10^{-19} watts/cm^2 is theoretically possible. This corresponds to about -59 db relative to one microbar.

Recent measurements of sensitivity have been completed using a system having the dimensions and wavelength of the example given immediately above. Threshold of sensitivity was observed[13] to be 6.7×10^{-15} watt/cm^2. This result implies that the experimental system was four orders of magnitude less sensitive than the theoretical limit given immediately above.

There are several mechanisms tending to limit sensitivity to less than the theoretical limit. These include:

(1) Amplitude noise on the laser beam used to illuminate the interaction region.

(2) Saturation of the optical sensor and, consequently, a "local oscillator" intensity less than the level required to swamp thermal and shot noise generated in the sensor.

(3)   Noise introduced in the sensor used to pickup the
      image (such as shot and thermal noise).

(4)   The photosensitive illuminated area must be small-
      er than the dimensions of one fringe formed by
      super-imposed beams (for heterodyning) to prevent
      cancellation of the heterodyning signal produced
      on one part of a detector by an out of phase signal
      generated at other parts.

The measured noise level was 8×10^3 times the value
given by Equation (25) at the time laser power was on the
order of .1 watt per centimeter of height. Equivalent noise
power for each resolution cell of image light for the experiment described is then

$$P_n = \frac{4h\nu B_o}{q} \times 8 \times 10^3 = 1.6 \times 10^{-14} \text{ watt} \qquad (27$$

where

$$B_o = 1 \text{ Hz.}$$

A United Detector Technology PIN-20B photodiode was
used to make the previously reported sensitivity measurement. The noise equivalent power (NEP) specification for
this diode is 2×10^{-14} watt for a bandwidth of 1 Hz at
1000 Hz. The approximate correspondence of this NEP specification with the value determined above from measurements
is more coincidental than astounding since measurements were
made at 500 KHz and threshold determination was made in an
approximate manner.

Since system sensitivity did not significantly exceed
the value determined from the NEP of the photodiode, there
is good reason to assume that the local oscillator was of
insufficient intensity to swamp sensor noise. That is,
gain possible by optical heterodyning was either not great
enough or alternatively amplitude noise on the laser beam
was at the same effective level as intrinsic noise of the
photodiode. (Work on the experimental system is continuing

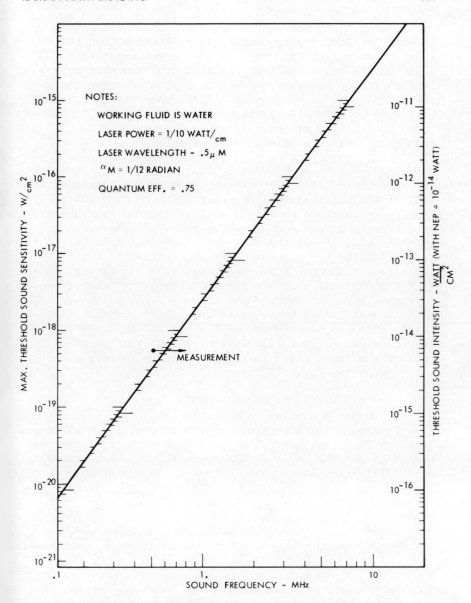

Figure 4.   Threshold Sound Intensity

It should be noted that any stray light (including
laser light scatter from particulate matter within the
light sound interaction region) will produce additional
noise in the same way dark current introduces shot noise.
Sensitivity of the acousto-optical system has been consis-
tently observed to be greatly reduced when particulate
matter is in evidence by visual inspection of the light
sound interaction region. (Just as theory predicts when
the local oscillator field is insufficient.)

When system sensitivity is limited by sensor sensitivity, threshold sensitivity is obtained by equating Equation
(22) to the noise equivalent power (NEP) of the sensor.
Figure 4 presents a numerical evaluation of a case where
threshold is determined by sensor noise together with a
second evaluation of the theoretical limit. A signal-to-
noise ratio of 2.0 was used to define threshold using a
photodiode with an NEP of 10^{-14} watt (for a 1 Hz bandwidth).
(Results for this case are shown on the right hand ordinate
in Figure 4.) Threshold sound intensity, shown on the right
hand side of Figure 4, is based upon the assumption that
local oscillator intensity is insufficient to swamp optical
sensor noise and that all light within only one resolution
cell on the image plane is conducted to the sensor. The
effect of stray light is assumed to be negligible.

ACKNOWLEDGMENT

This work was supported in part by the Defense Advanced
Research Projects Agency. Sensitive measurements were per-
formed by James H. Cole of TRW Systems Group. My thanks to
Helen Miller for her effort in typing the manuscript.

REFERENCES

1. Korpel, A., "Visualization of the Cross-Section of a
    Sound Beam for Bragg Diffraction of Light," Appl. Phys.
    Letters, 9:425-427 (15 December 1966).

2. Wade, G., C. J. Landry and A. A. deSouza, "Acoustic
    Transparencies for Optical Imaging and Ultrasonic
    Diffraction," presented at the First International
    Symposium on Acoustical Holography, Huntington Beach,

Calif., 1967 [Subsequently published in <u>Acoustical Holo-graphy</u>, edited by A. F. Metherell, H.M.A. El-Sum, and Lewis Larmore (Plenum Press, Inc., New York, 1969), Vol. I].

3. Cole, J. H., et al., Previous paper in this volume.

4. Pierce, D. T. and R. L. Byer, "Experiments on the Inter-action of Light and Sound for the Advanced Laboratory," Am. J. Phys., Vol. 41, p. 314 (1973).

5. Bhatia A. B. and W. J. Noble, "Diffraction of Light by Ultrasonic Waves: I. General Theory and II. Approximate Expression for the Intensities and Comparison with Experi-ment," <u>Proc. Royal Soc.</u>, Vol. A220, pp. 356-385, 1953.

6. Raman, C. V. and N. S. Nagendra Nath, Proc. Indian Acad. Sci., Vol. 2, 413 (1935).

7. Smith, R. A. and G. Wade, "Noise Characteristics of Bragg Imaging," Chapter 6 in <u>Acoustical Holography</u>, Vol. 3, A. F. Metherell Editor, Plenum Press, 1971.

8. Urick, R. J., <u>Principles of Underwater Sound for Engi-neers</u>, McGraw-Hill Book Co., p. 20, 1967.

9. Smith, R. A., et al., "Studies of Resolution in a Bragg Imaging System," J. Acoust. Soc. Am., Vol. 40, pp. 1062-1068, March 1971.

10. C.f., Carlson, Bruce A., <u>Communication Systems</u>, McGraw-Hill Book Co., New York, 1968, p. 279.

11. Smith, R. A., "Optical Imaging of Sound Fields by Heterodyning," U.S. Patent No. 3,831,135 filed Sept. 1973.

12. Pratt, W. K., Laser Communication Systems, John Wiley and Sons, Inc., New York, 1969.

13. Smith, R. A. and J. H. Cole, "Underwater Applications of Acousto-Optical Imaging," TRW Report AT-SVD-TR-7, 29 March 1974.

OPTICAL IMAGING SONAR*

J.H. Cole, R.A. Smith, R.L. Johnson and P.G. Bhuta

TRW Systems Group

Redondo Beach, California

ABSTRACT

Prior work dealing with optical imaging of sound waves was devoted to nondestructive testing applications at sound frequencies of 1 MHz and higher. Bragg diffraction of light by sound waves permits one to obtain optical images of sound fields on a real time basis. This paper describes the formation of optical images of sound fields which are at frequencies of approximately 100 KHz and higher for sonar applications. A new dual frequency optical heterodyning technique, necessary for image formation at low sound intensity and low frequencies, was devised for sonar applications. Optical images of objects insonified at a distance of 5 meters from the Bragg cell are presented. Experimental results pertaining to sensitivity measurements are given and compared with sensitivity reported for other imaging systems.

INTRODUCTION

The acousto-optical system described here has been under investigation at TRW for various sonar applications for several years. In this system, an acoustical image formed by reflected sound waves diffracts light from a laser

---

*Work supported in part by Defense Advanced Projects Agency (ARPA Order No. 1951) under contract with the Office of Naval Research (Contract N00014-72-C-0182).

beam in the light/sound interaction region. The configuration considered is in the form of a Bragg imaging system.[1,2]
This system operates with an expanded beam of laser light
which is brought to a line of focus just beyond the water
filled region which receives sound reflected from the object
viewed. Anamorphic images are formed in the side bands of
light (either side of cell illumination) due to the interaction of sound in the water filled volume which also receives both light and sound. A good image of an object in
the media is formed on a viewing screen by magnifying a
side-band image in one direction by just the amount sufficient to correct the anamorphism.

The configuration considered (Figure 1) is quite simple. In its most elementary form, it consists of only a
laser, laser beam forming optics and a cylindrical lens in
front of a viewing screen. There is also a water volume
between lenses which, for best performance, must be kept
free of particles which would scatter light onto the image
plane.

IMAGING PRINCIPLES

Objects illuminated with sound produce (by reflection
or scatter) a wave pattern completely analogous to the wave
pattern reflected from visually viewed objects. Just as
objects visually observed may be described by a distribution
of plane waves of light, it is also convenient to describe
insonified objects by a distribution of plane waves of
sound. This viewpoint simplifies the analysis of acousto-
optical imaging systems since individual plane waves can be
considered separately. In order to understand acousto-
optical imaging it is only necessary to observe the relation between each individual plane wave in the sound field
and the resulting plane wave forming the image in the light
field. At frequencies and system dimensions of interest
for sonar applications, a plane wave of sound moving at
right angles to a plane wave of light will diffract light.
Diffraction will not occur when light and sound do not have
propagation vectors satisfying either the Bragg or Raman-
Nath condition.[3] Both conditions are quite close to a right
angle between respective vectors at frequencies of interest
here.

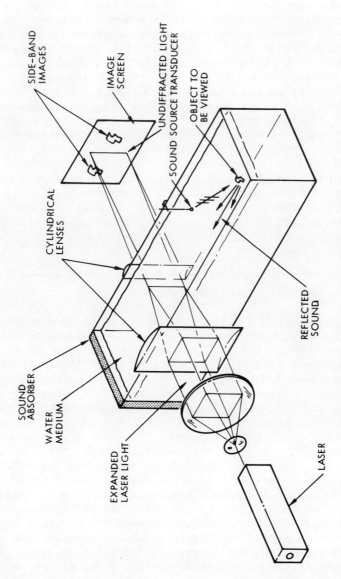

SIDE-BAND IMAGES

IMAGE SCREEN

UNDIFFRACTED LIGHT

SOUND SOURCE TRANSDUCER

OBJECT TO BE VIEWED

CYLINDRICAL LENSES

REFLECTED SOUND

SOUND ABSORBER

WATER MEDIUM

EXPANDED LASER LIGHT

LASER

Figure 1: Elementary form of the acousto-optical imaging system.

Formation of an image from a sound field depends on
the existence of light over a range of directions so that
each plane wave component in the sound field will be replicated. Plane wave components in a sound field which are
absent, of course, do not affect the light, and components
in the sound field which find no light at right angles to
the direction of this component would also be absent from
diffracted light. An undistorted image is formed only if
all components in the sound field defining the insonified
image meet with light which satisfies the right angle condition. In order to provide light necessary for satisfying
the right angle condition for all components of interest,
the acousto-optical cell is illuminated by cylindrically
convergent light. There is then a continuum of light rays
describing the light field. An alternate scheme uses spherically convergent light to provide a continuum of rays. In
either scheme, a range of sound field vectors will encounter
light satisfying the right angle condition. The cylindrically convergent light pattern was used exclusively for all
results reported here.

LIGHT HETERODYNING

A real image is formed, as described, having sufficient
intensity to be photographed when the sound level is well
above threshold. The most sensitive system requires photo-
sensing of the side-band image rather than direct viewing
with the naked eye or photography. Optical heterodyning
can be used to make the most sensitive laser receiver[4] and
provides a way to retain doppler information in sound reflected from the object viewed.

Sensitivity of an acousto-optical system tends to increase with laser power because the diffracted light intensity at a given sound level increases with laser power.
The advantage of increased laser intensity is not always
obtained due to light scattered from dust particles and
component surfaces. In directly viewed systems this light
limits system sensitivity by the simple mechanism that the
sound level must be sufficient to diffract more light than
background light due to random scatter sources. By using
light heterodyning an image produced by very low sound
levels can be sensed even when it is completely obscured
from direct viewing. Figure 2 shows the results of an

experiment where this was done. Sound was reflected from
the object viewed in these experiments. The viewed object
is shown in Figure 2a. A direct image as displayed on a
TV monitor is shown in Figure 2b. With illuminating sound
of sufficiently low power to allow general background light
to completely obscure the image, a photosensor was scanned
over the image and the heterodyne output recorded on a
memory scope (Figure 2c). Note that the image is completely
free of background.

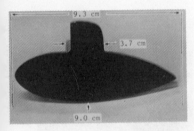

2a. Flat object used in
    1.8 MHz imaging.

2b. TV display of the direct
    image of the object at
    1.8 MHz.

2c. Heterodyned image of object
     at 1.8 MHz.

Figure 2: Dual frequency heterodyned reflection image at
           1.8 MHz at a distance of 5 meters. (Photos not
           to scale.)

OPTICAL HOLOGRAPHY

   Due to frequency and phase response, optical holography is similar in nature to optical heterodyning. Therefore, optical holography can be used to improve sensitivity
of an acousto-optical imaging system.[6] Even when an image
is completely obscured from direct viewing, optical holography can extract information. The results of an experiment
where this was done are shown in Figure 3. Sound was reflected from the object shown in Figure 3a. A direct image
of the focal plane of the system is shown in Figure 3b.
Figure 3c shows the holographically reconstructed image.

a. Object imaged at 4.5 MHz   b. Image plane of the
   using TRW's acousto-   acousto-optical im-
   optical imaging system   aging system without
                                         holographic improve-
                                         ment

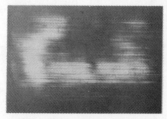

c. Holographically reconstructed acousto-
   optical image of the above object

Figure 3: Optical holographic reflection image at 4.5 MHz.

SENSITIVITY

Sensitivity of the optical heterodyning system was measured by directing image light to a photodiode. It was found that the sound level to produce a unity signal-to-noise ratio at a sound frequency of 500 KHz was 1.46×10^{-16} Watt/cm^2. In making these measurements, a small sound source was placed at a distance of 5 meters to allow sound to become more planar over most of the length of the light/sound interaction region. A cylindrical acoustic reflector was used to confine all sound to arrival within the column of cell illumination light. The acoustic reflector could have contributed up to 16 db of sensitivity gain.

Analysis presented in Reference 5 shows that theoretical minimum intensity detectable by the acousto-optical imaging system is given by

$$I = 6.92 \times 10^{-19} \frac{B_o}{q P_e \Lambda^3} \quad \text{Watts/cm}^2 \tag{1}$$

where B_o = system bandwidth, Hz

P_e = laser power per unit height of the wedge, Watts/cm

Λ = sound wavelength, cm

q = quantum efficiency of the photodetector

This equation represents the maximum threshold sensitivity based on the minimum sound intensity required to form a Raman-Nath side-band in the limit of optical heterodyning. Experimental measurements have not approached this sensitivity due to system noise consisting of laser noise, local oscillator noise and detection system noise.

RANGE CALCULATIONS

Predicted ranges for the TRW Acousto-Optical Imaging System have been calculated. This has been accomplished using a computer program[5] which solves the basic two way transmission sonar equations using system sensitivity as observed experimentally. Reverberations were neglected since inclusion would make it quite difficult to compare this system with alternatives. Additionally, two-dimensional cross-range imaging implies that sound arriving from different directions due to reflections at the same range will show up in different parts of the image and therefore, reduces the degrading effects of reverberations. Maximum operating range is determined by the sonar equation which is[7]

$$SL - 2TL + TS = EL \qquad\qquad (2)$$

where   SL = source level = 130 db (including directivity index)
        TL = 20 log R + αR db
        R = range in meters
        α = logarithmic absorption coefficient in db/meter
        TS = target strength in db
        EL = minimum detectable echo in db

The logarithmic absorption coefficient, α, was calculated using the equation of Schulkin and Marsh[8]

$$\alpha = ASf_T f^2 / (f_T^2 + f^2) + Bf^2 / f_T$$

$$f_T = 21.9 \times 10^{6 - 1520/(T + 273)} \qquad\qquad (3)$$

where   A,B = constants
        S = salinity in parts per thousand
        f = acoustic frequency in kilohertz
        T = temperature in degrees centigrade

Maximum range with sensitivity no greater than experimental observations was calculated. The minimum detectable echo was based upon the best sensitivity that has been experimentally determined from the most recent measurements using the laboratory acousto-optical imaging system. After substitution of Eq. (3) into the sonar Eq. (2) for two way transmission, the equation was solved for range. The result appears in Figure 4.

RESOLUTION

Resolution measurements were made to determine resolution characteristics of the laboratory system shown in Figure 5.

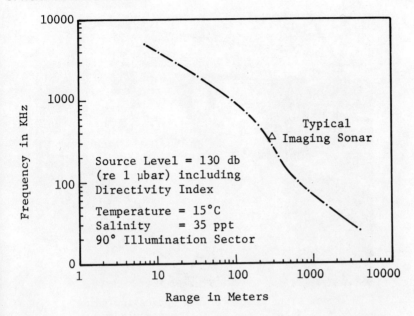

Figure 4: Acousto-optical imaging system range capability
for -20 db target strength at the surface.

Resolution measurements consisted of observing two
object transducers. The resulting two light spots in the
image were then made to approach one another by moving the
transducers together until the images were separated by the
Rayleigh criterion. The distance between the two transducers was then measured and recorded as the acousto-optical
imaging system resolution. Measurements were taken for a
frequency of 500 KHz with the object transducers 5 meters
from light/sound interaction region.

An aperture positioned between the diverging and collimating lens was adjusted to reduce the light wedge angle.
The light wedge angle was reduced until further reduction
resulted in unresolved images. The minimum light wedge
angle, $\alpha = 0.9°$, where resolved images were observed, was
measured. This result was used in the theoretical expres-
sion for nearfield resolution, $\Lambda/2\sin\alpha$, where Λ is the
acoustic wavelength. The theoretical expression for the
farfield resolution is $\Lambda R/D$ where D is the acoustic
aperture and R is the range. The cross-over point between the nearfield and farfield is defined as $R = D/2\sin\alpha^3$.

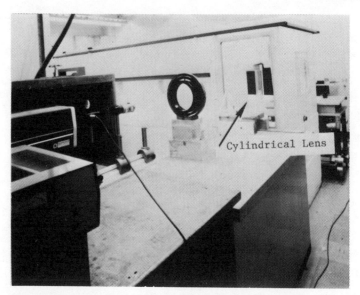

a. Light wedge forming optics including
   interaction cell entrance window.

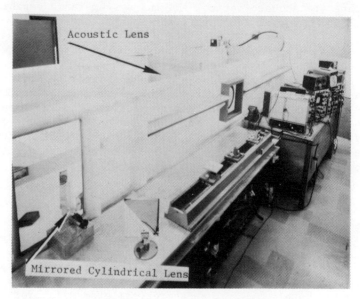

b. Imaging optics including interaction
   cell exit window.

Figure 5:   Laboratory acousto-optical imaging system.

For the laboratory configuration the nearfield/farfield cross-over point is 35 meters. Therefore, all laboratory azimuth resolution measurements were obtained in the near-field. For objects within the nearfield of the laboratory system, the nearfield resolution, $\Lambda/2\sin\alpha$, is less then $\Lambda R/D$. However, beyond the nearfield/farfield cross-over point, $\Lambda R/D$ resolution applies. The computed nearfield resolution value for each case was compared with previous resolution measurements. The results are shown in Table 1. For each experimental frequency, measured resolution is better than predicted by theory. If two mutually coherent point sources are approximately 180° out of phase, the resolution will be better than predicted by theory.[9]

Results presented in Table 1 demonstrate that a prop-erly designed acousto-optical imaging system can be expected to have maximum theoretical resolution in the azimuth direc-tion. However, the resolution in the elevation direction depends on the acoustic focusing element. Therefore, ver-tical resolution measurements were taken with the acoustic lens used in Reference 3 and with a parabolic acoustic reflector. These results are presented in Table 2 and are compared with theoretical resolution, $\Lambda R/D$.

TWO-DIMENSIONAL CROSS-RANGE CONFIGURATION

Use of an acoustic cylindrical lens or reflector to provide imaging quality in the vertical direction provides

Table 1: Comparison of Observed and Theoretical Resolution with Aberrated Light Removed

Frequency	Measured Resolution	Theory $\Lambda/2 \sin \alpha$
1.18 MHz	3.6 cm	4.1 cm
500 KHz	8.9 cm	9.6 cm
160 KHz	27.9 cm	30.0 cm
130 KHz	26.1 cm	36.9 cm

Table 2:   Vertical Resolution Measurements Determining the
           Quality of an Acoustic Lens and Acoustic Reflector
           Placed to Focus Sound into the Light/Sound Inter-
           action Region

Acoustic Processor	Frequency	Distance	Measured Resolution	Theory ΛR/D
Lens	500 KHz	2 meters	3.8 cm	.8 cm
Reflector	500 KHz	2 meters	1.9 cm	1.1 cm
Reflector	500 KHz	4.5 meters	6.3 cm	2.5 cm

two-dimensional cross-range imaging (analogous to any photographic camera). Images already presented were made with
this configuration. Figure 6 is an additional example of
images obtained with our laboratory system operating in
this configuration. Range gating techniques normally employed in sonars can be combined with two-dimensional cross-
range imaging to yield three-dimensional information.

RANGE AZIMUTH IMAGING CONFIGURATION

      To test the principle of range-azimuth imaging with
the acousto-optical system, our laboratory system was modified to generate range-azimuth images. In this experiment,
only image detail in azimuth is reproduced on the image
plane. Image detail in elevation was suppressed by simply
sensing variation in azimuth. Time of arrival of sound from
the object which is imaged was converted to image detail in
the other direction by a linear sweep in the vertical direction which was generated within the memory scope. Each horizontal position on the image plane was made to correspond
to a horizontal position of a vertical line on the memory
scope. The vertical display generated on the scope corre-
sponded to the time when sound from the object caused the
writing action on the scope face for any given vertical
line. One outgoing sound pulse was associated with one
vertical sweep in the memory scope. The actual range interval which was recorded was determined by the delay provided
between the sound pulse and occurrence of the returning
light signal due to diffraction within the Bragg cell. A
complete image was formed by successively recording vertical

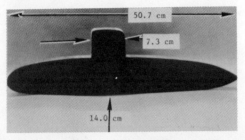

a. Object used for 500 KHz and 280 KHz reflection imaging

b. TV display of direct image   c. TV display of the direct
   of above object at 500 KHz   image of above object at
                                    280 KHz

Figure 6: Reflection images at 500 KHz and 280 KHz at a
           distance of 5 meters.

lines continuously displaced horizontally until the entire
horizontal width forming the image was scanned. The image
shown in Figure 7 was made with this system using 500 KHz
sound and a sound pulse width of 35 microseconds.

CONCLUSION

TRW's optical imaging system has several important
advantages for imaging sonars. The optical imaging system
has broadband characteristics enabling improvement in range
resolution and the use of a multiple frequency system. In
addition, two-dimensional cross-range imaging coupled with
phase and range information result in three-dimensional
image processing. Good images were formed from sound reflected by small objects located at the maximum distance
possible in the existing laboratory facility. Cross-range
resolution was checked and observed to agree with the limit

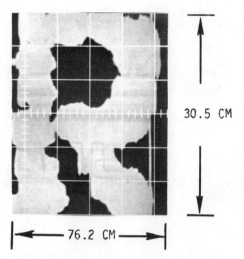

30.5 CM

|←——— 76.2 CM ———→|

Figure 7: Range-azimuth image generated by tracing a 1-inch
diameter sound transducer through a path in the
form of the letter "R" and storing images formed
by each successive position.

predicted by theory. Theory shows that further improvements
in sensitivity and projected range can be expected.

The acousto-optical imaging sonar can be configured
.to provide either range azimuth imaging or two-dimensional
cross-range imaging; that is, resolution in both elevation
and azimuth.

REFERENCES

1. Korpel, A., "Visualization of the Cross-Section of a
Sound Beam for Bragg Diffraction of Light," Appl. Phys.
Letters, 9:425-427 (15 Dec. 1966).

2. Wade, G., C. J. Landry and A. A. deSouza, "Acoustic
Transparencies for Optical Imaging and Ultrasonic Diffraction," presented at the First International Symposium on Acoustical Holography, Huntington Beach, Calif.,
1967 [Subsequently published in Acoustical Holography,
edited by A. F. Metherell, H. M. A. El-Sum, and Lewis
Larmore (Plenum Press, New York, 1969) Vol. I].

3. Smith, R. A., J. H. Cole, R. L. Johnson and P. G. Bhuta, "Feasibility Demonstration of Low Frequency Acousto-Optical Imaging for Sonar Applications," TRW Systems report 8520.2.73-110, 12 July 1973.

4. Pratt, W. K., Laser Communication Systems, John Wiley and Sons, Inc., N.Y., 1969.

5. Smith, R. A. and J. H. Cole, "Underwater Applications of Acousto-Optical Imaging," TRW report AT-SVD-TR-74-7, 29 March 1974.

6. Winter, D. C., "Noise Reduction in Acousto-Optic (Bragg) Imaging Systems by Holographic Recording," Appl. Phys. Letters, 15 Feb. 1973.

7. Urick, R. J., Principles of Underwater Sound for Engineers, McGraw-Hill Book Co., page 20, 1967.

8. Schulkin, M. and H. W. Marsh, "Sound Absorption in Sea Water," J. Acous. Soc. Am., 34:864 (1962).

9. Goodman, J. W., Introduction to Fourier Optics, McGraw-Hill, 1968, pgs. 86 and 87.

HOLOGRAPHIC PROCESSING

OF NEAR-FIELD SONAR DATA

M. Luetkemeyer   and   R. Diehl

FRIED. KRUPP GMBH
KRUPP ATLAS-ELEKTRONIK BREMEN
2800 Bremen - 44 / Postfach 448545
GERMANY

ABSTRACT

In active sonar the location of objects and the resolution
of their details is achieved by beam scanning in one plane combined with measurement of echo delay. The required signal processing can be done by holographic
means   - i.e. optical processing -   and holographic
methods   - i.e. spatial Fourier Transform and multiplication by a quadratic phase factor. Experimental results supporting this statement were obtained by a linear receiving array.

Under near-field conditions, as characterized by a
large angular field of view and high radial resolution
requirements, the delay differences between receiver elements must be taken into account.

An uncertainty condition, relating the radial and angular resolution capability to the field of view is presented.

A method essentially based on holographic processing
of subsectors of the whole angular field of view is proposed for the case, in which this condition is violated.

INTRODUCTION

The. function of active sonar is the location of objects.
An often used presentation in sonar techniques is the
PPI scan, which provides a conformal projective map-
ping by displaying the distance and the azimuthal angle.
In order to obtain more detailed information about an
object, the resolution of object details is necessary.
This for example is the case for near-field sonar.

In sonar techniques the location of objects and the resolution of their details is achieved by beam-forming
and scanning in one plane combined with measurement
of echo delay time.

It is possible to realize the beam forming by an echo
delay time compensation, which requires a large technical effort. To reduce this effort, it is expedient to
substitute   the echo delay time compensation by a mere
phase compensation. In this case beam-forming and
scanning are limited to stationary signals.

A possibility for this is beam-forming and scanning by
Fourier Transform. The required signal processing
can be done by holographic means as well, for example
by optical processing in a coherent optical processor (1).

The following deals with

o   limiting conditions of sonar mapping
         in the near-field

o   experimental investigations of one di-
         mensional holographic signal processing

         and

o   a proposal to extend the angular field of view
         by a combination of the echo delay time and
         the phase compensation.

Technological problems, especially those of optical signal processing, are not subject of this paper.

SOME ASPECTS OF NEAR-FIELD SONAR MAPPING

To realize the location of objects, beam-forming and
scanning is required. An exact solution for realizing
this beam-forming and scanning is shown in   Fig. 1.

The compensation of the different echo delay times is
achieved by variable delay lines. These delay lines will
be continuously set corresponding to the respective direction of observation. The delay times are stored in a
rotating register and will be decoded and transfered to
the delay lines.

Within a sample interval N delay times must be successively set for N beams. The addition in proper
phase of the compensated signals leads to one beam at
a time.

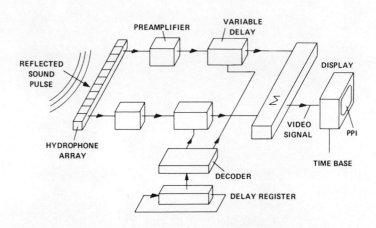

Fig. 1 :   Principle of Sonar Beam
            Scanning by Variable Delay

In the temporal sequence this sum is equivalent to the
desired video signal, which can be directly displayed
on a CRT.

If the echo delay times are chosen to be range-dependent, focusing is practicable, as it is necessary in
near-field sonar. For this purpose, further delay-time
registers must be provided for each focal range.

Near-field conditions mostly require very high sonar
frequencies and large pulse bandwidths. Under these
conditions the realization of the exact delay time compensation requires a large technical effort. It is possible to reduce this effort by selecting the parameters
such that the reflected sound pulses can be considered
quasi-monochromatic. In this case the delay time compensation can be replaced by a mere phase compensation e.g. (2) . Fig. 2 presents the scheme of compensation as realized by serial processing technique.

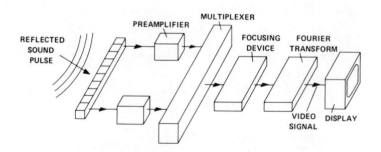

Fig. 2 :   Beam Scanning in Case of
           Quasi Continuous Echo-Signals

The electrical signals received are proportional to the sound pressure and are separately preamplified and sub-sequently scanned in a sequential mode. The serial signal will be multiplied by a range-dependent quadratic phase factor. Fig. 2 shows this operation as "Focusing Device". Subsequently the signal has to be Fourier transformed. This method has the advantage of requiring essentially less technical effort compared with the processing shown in Fig. 1 .

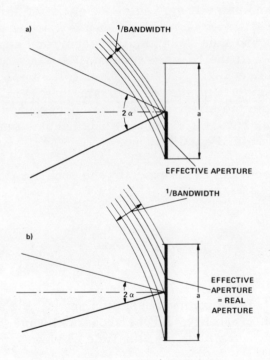

ig. 3 : Limitation of the Effective Aperture resp. of the
Field of View in Case of Phase Compensation

As mentioned before, the compensation of the echo delay
times by an appropriate phase shift of the hydrophone
signals is only possible under certain conditions. In the
following, the conditions for this form of signal processing will be investigated in more detail using Fig. 3 ,
where a is the aperture and α the maximum angle between the normal of the aperture and an object point.
In the following, this angle is referred to as angular field
of view. A partial wave coming in from the boundary of
the angular field of view can only be regarded as quasi-
monchromatic over the pulse-length i. e. the inverse
pulse-bandwidth. As can be seen from Fig. 3 a , only
part of the aperture is insonified simultaneously by the
reflected sound pulse if the angle of view is too large and
the pulse duration too short. This means that the effective aperture and thus the azimuthal resolution capability are limited. If the resolution capability is not to be
affected, it is necessary to limit the angular field of
view while maintaining the other parameters (Fig. 3 b) .
In this example the effective aperture corresponds to the
real aperture, but the field of view has become smaller
relative to that in Fig. 3 a

This can be generally expressed by the uncertainty condition

$$\Delta r \cdot \gamma = \lambda \cdot \sin \alpha$$

where Δr is the radial resolution and γ the angular resolution capability, λ the acoustical wave-length
and 2α the maximum angular field of view.

From the uncertainty condition it can for example, be
derived that in case of the mere phase compensation
with constant field of view a raise of the radial resolution capability can only be reached by a loss of azimuthal resolution capability.

SOME EXPERIMENTAL RESULTS OF ONE DIMENSIONAL HOLOGRAPHIC SIGNAL PROCESSING

A special form of phase compensation can be realized by holographic signal processing. For that purpose experimental investigations have been carried out with a test equipment in a lake. The required focusing and Fourier transforming, shown in Fig. 2 in simplified form, was done by coherent optical means as well as by digital means.

In connection with these investigations first results were published in (3).

The experiments have been carried out using a linear receiving array. This array, consisting of 64 receiving channels, is shown in Fig. 4. The ultrasonic frequency used is approximately 100 kHz, the pulse length 0.2 ms.

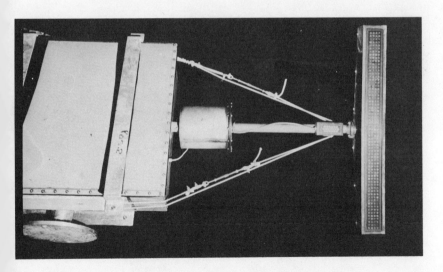

Fig. 4 :   Experimental 100 kHz-Equipment

The Fresnel region extends to about 20 m . The aperture width of about 0.8 m leads to a resolution angle of 1.2° , the radial resolution is approximately 0.15 m .

As mentioned before, a serial signal will be obtained by multiplexing the output signals of each preamplifier. This signal can be displayed as an intensity plot on a CRT . In this way a set of one-dimensional holograms with the distance R as a parameter is displayed (Fig. 5) . The hologram information was photographed and then fed into a computer via an optical equipment.

A typical one-dimensional pulse-hologram is shown in Fig. 6 . The objects were some air-filled glas-spheres with a diameter of 5 cm . The reconstruction of this hologram is shown in Fig. 7 . The theoretically predicted azimuthal resolution is achieved both in the near-field and in the far-field region.

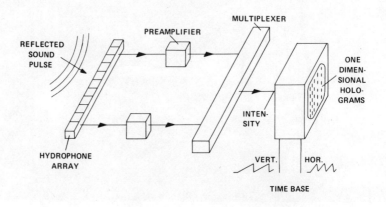

Fig. 5 :   Experimental Set Up for Holographic Trials

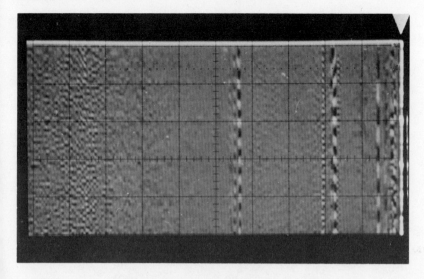

Fig. 6 :   One Dimensional Pulse Hologram
of Point Objects

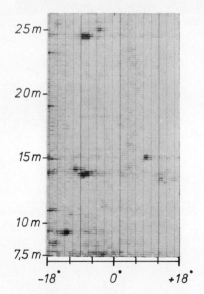

Fig. 7 :   Digital Reconstruction of an
One-Dimensional Pulse Hologram

The following picture shows the hologram and the reconstruction of a line target   (Fig. 8)   . The focused image
is compared with the result of the unfocused reconstruction,   i.e.   the quadratic phase factor is replaced by the
value one. The difference in azimuthal resolution in the
near-field can be clearly recognized. The coherent optical reconstruction has been done by SIEMENS AG   and
led essentially to the same results. As stated above, the
signal processing shown here reaches its limits if the

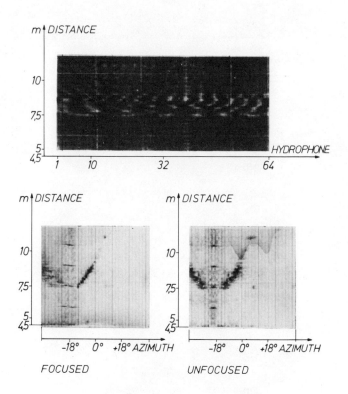

Fig. 8 :   Hologram and Reconstruction
of Line Target

field of view is to be increased and/or the radial and azimuthal resolution capability improved.

A PROPOSED COMPENSATION OF THE ECHO DELAY TIMES FOR A LARGE FIELD OF VIEW

Near-field applications require a high azimuthal and radial
resolution. As stated before the echo delay time compensation can only be replaced by a phase compensation if the
insonified or investigated angular field of view fulfills the
condition

$$sin\, \alpha \leq \frac{\Delta r \cdot \vartheta}{\lambda}$$

In the following a method will be described which allows an
application of the Fourier Transform even for angular
fields of view larger than the maximum angle given by the
unvertainty condition.

The method is based on dividing the whole angular field of
view $2\alpha_0$ in K subsectors. For these subsectors the
above mentioned condition must be fulfilled.

This division is shown in   Fig. 9.

These subsectors are processed afterwards by phase compensation. A simple method to obtain the echo delay time
compensation for K main directions is the utilization of
the linear delay effect by a sequential sampling of the received signals. If the individual transducer elements are
sampled sequentially, a linearly increasing delay of the
signals results. This delay corresponds to a definite receiving direction. By reversing the sample sequence, the
mirror inverted receiving direction will be produced.

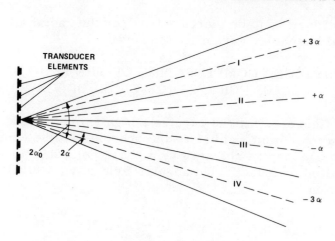

Fig. 9 :   Dividing the Field of View into Subsectors

Now a method will be shown which, within certain limits, permits a random choice of the main receiving directions β_ℓ (ℓ = 1 ... K).

According to the sampling theorem, the received signals of the individual transducer elements must be sampled with a frequency   2 B   in the real or with a frequency   B   in the complex case.

From this a frequency of

$$f_s \geq N \cdot B \qquad \text{complex}$$
$$f_s \geq 2 \cdot N \cdot B \qquad \text{real}$$

B = Bandwidth

follows for sequential processing.

The linearly increasing delay resulting from sequential sampling corresponds to a receiving direction β.

This angle is calculated from the equation

$$sin\beta = \frac{N \cdot c}{\alpha \cdot f_s}$$

where N is the number of transducer elements, c the sound velocity and a the aperture width.

If the angle to be compensated by a delay compensation is smaller than this value β , it is sufficient to sample the received signals with a proportionally higher sampling frequency f'_s and to store the respective sampled values for a short time. The stored signals will be converted into a serial signal by means of a multiplexer. This multiplexer works at the sampling frequency f_s .

This case is illustrated in Fig. 10 as a block diagram and in Fig. 11 as a sampling diagram.

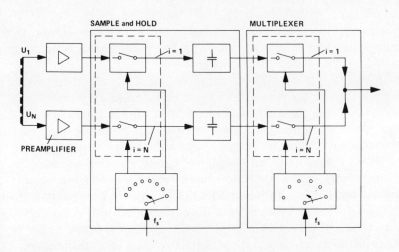

Fig. 10 :   Delay Compensation by Sampling

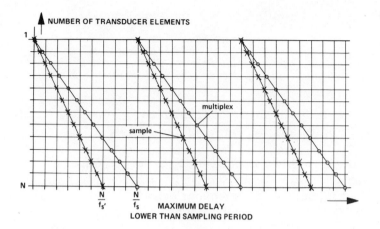

Fig. 11 :   Delay Compensation by Sampling
- Sampling Diagram -

If the angle that has to be compensated by tapped delay lines is larger than the angle β resulting from the sampling frequency, the processing is basically the same. However, this results in a sampling frequency which no longer
satisfies the sampling theorem.

In order to achieve a linearly increasing delay with the
desired slope as well as a sufficient sampling rate it is
therefore necessary to make two sample periods overlap
respectively. For example, the scanner alternates from
the first element to an element placed in the second half
of the array, then switches back to the second element
and so on. This can be seen in Fig. 12   To process the
signals for the respective subsector it is now required
to extract those echo signals inciding from other subsectors. For each single subsector this is obtained by
time-filtering of the sequential signal using a bandpass-
filter.

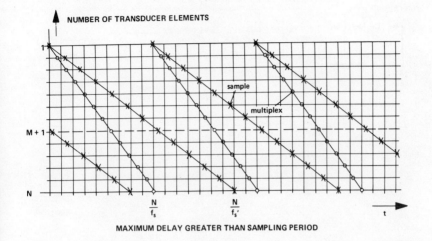

Fig. 12 : Sample Diagram

The processing of K subsectors requires K parallel channels. The increased technical effort connected with this parallel processing is avoidable if a ping-to-ping processing is selected. This method is practicable in the near-field, as the ping period is relatively short. A proposal for realization of the method described is shown in Fig. 13 . The output signals of the transducer array are separately preamplified. A clock signal corresponding to the ping period is produced in a clock-generator. The sample and hold circuit generates the delay corresponding to a subsector by sequential sampling and buffer storage. The multiplexer scans the stored signals with the sampling frequency and transforms them into a serial signal. Since it is reasonable to apply complex sampling of the sound field at the array for obtaining the phase and thus the direction of incidence, quadrature sampling is used for signal processing (4) .

M. LUETKEMEYER AND R. DIEHL

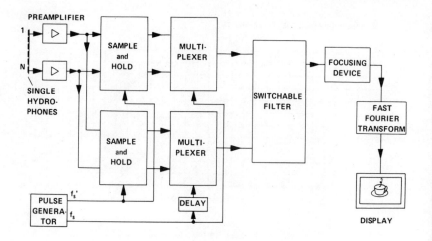

Fig. 13 :   Block Diagram of Near-Field
             Wide Angle Sonar Device

The second identical unit which is triggered with a delay
serves this purpose. At the output of both multiplexers,
therefore, the real as well as the imaginary part of the
analytical signal are present. The extraction of the non-
desired subsectors is performed by a switchable filter.
Afterwards each of the K subsectors are submitted to
a phase compensation by Fourier Transform. The reconstructed image displayed on a CRT is thus built up by a
continuous presentation of the subsectors. Presently our
research team is working on the realization of the concept
described. Problems are expected primarly in connection
with the development of a suitable line array. Fig. 14
shows an experimental model of the transducer developed
for this realization. To isolate the individual transducer
elements the so-called dicing technique has been applied.
The acoustic behaviour of this test transducer has led to
quite adequate results in first surveys.

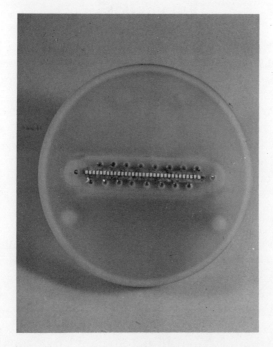

Fig. 14 :   Test   Array   Consisting
of 32 Single Hydrophones

The development of the electrical components seems to be
uncritical so that it can be expected that a functional model
will be completed in the near future.

ACKNOWLEDGMENT

The work is carried out under a contract with the BWB
(Bundesamt für Wehrtechnik und Beschaffung) , Koblenz /
West-Germany to whom graditude is expressed for the per-
mission to publish this paper. Special thanks are due to
G. K r a c h t   and   G. T u m m o s c h e i t , for their
helps in the experiments.

REFERENCES

(1)   Penn W. A. , Chovan J. L. : The Application
         of Holographic Concepts to Sonar, Acoustical
         Hol., Vol 2, (A. F. Metherell, L. Larmore),
         Plenum Press, 1970 ,
         Page   133 - 172

(2)   Welsby V. G. , Dunn J. R. : A High Resolution
         Electronic Sector-Scanning Sonar,
         Journal Brit. I. R. E. Sept. 1963,
         Page   205 - 208

(3)   Diehl R : Holographic Processing of Acoustic
         Data Obtained by a Linear Array, Signal Pro-
         cessing (J. W. R. Griffiths, P. L. Stocklin,
         C. van Schooneveld), Academic Press, 1973 ,
         Page   561 - 575

(4)   Grace O. D. , Pitt S. P. : Sampling and Interpo-
         lation of Bandlimited Signals by Quadrature
         Methods, JASA 48 , 1970 ,
         Page   1311 - 1318

SOLID PLATE ACOUSTICAL WAVE FOCUSING

Karyl-Lynn K. Stone

Visual Acoustics

Magnolia, Mass. 01930

INTRODUCTION

This paper is devoted to the operating procedures of solid plate wave focusing, the heart of the solid plate acoustical viewer, first presented at the Fifth International Symposium. The waves are the imaging device somewhat akin to the focused electronic signals in a television display, the next to last step in the viewing procedure by acoustical methods.[1] The theory behind the wave focusing is incorporated into the system's physical characteristics. How and why the wave's velocity, focusing and character are determined by these characteristics is discussed. Experimental changes from the original viewer are described. Photographs of experimental data (solid plate waves) are presented which illustrate the multitude of effects varying frequencies have on wave type and strength.

THEORY

The solid plate waves are concentrated in a small section of the system, on the acousto-solid plate (a-sp) converter, joining strip (j.s.) and the solid plate (sp) lens. The solid plate waves propagate along the a-sp converter as flexural waves, at velocity cp.[2] cp is calculated from the approximate formula for flexural waves in a plate.

$$cp = \left(\frac{4\pi a f}{\sqrt{12}}\right)^{1/2} \cdot \left(\frac{E}{\rho}\right)^{1/4}$$

a = ½ thickness of plate
f = frequency
E = Young's Modulus
ρ = density of plate

Therefore, cp, velocity of flexural waves in a plate, is determined from the characteristics of the a-sp converter plate. (3)

These waves focus because of the difference in the velocity of the waves on the two plates and the radius of curvature of the joining strip. Snell's Law explains why. If we use Snell's Law for sufficiently small angles, it takes the following form

$$\frac{\theta_1}{\theta_2} = \frac{n_1}{n_2} = \frac{v_1}{v_2} \quad \text{WHERE } \sin\theta_1 \simeq \sin\theta_2$$

θ_1 = angle diverging sp waves hit the joining strip
θ_2 = angle converging sp waves leave the joining strip
n_1 = index of refraction of the a-sp converter
n_2 = index of refraction of the sp lens
v_1 = velocity of sp waves in a-sp plate (cp)
v_2 = velocity of sp waves in sp lens (cp_1)

Now, when $v_1 = v_2$, then $\theta_1 = \theta_2$ and the rays are not affected. They are neither bent, nor refracted and consequently do not focus. Photographs comprising Figures 4 - 6 illustrate this phenomena, which exists when the joining strip is straight. However when $v_1 \neq v_2$ then $\theta_1 \neq \theta_2$ and the waves are bent or refracted. If the interface (in this case, the joining strip) between the two media is curved in an arc, then these bent rays will focus to a point. The photographs comprising Figures 7 - 9 illustrate this phenomena.

This focal length can be calculated from the general formula

$$\frac{n_1}{s_1} + \frac{n_2}{s_2} = \frac{n_2 - n_1}{r}$$

where s = arbitrary distance. Now, if $s_1 = \infty$ then $n_1/s_1 = 0$. Therefore, $n_2/s_2 = n_2 - n_1/r$. Now, if $s_2 = f$, where f is the focal length, then $n_2/f = n_2 - n_1/r$. (4). By setting $n_1 = 1$, we are using the a-sp plate as a reference. We then derive

$$f = (n_2/n_2 - 1) \; r$$

where r is the radius of curvature of the joining strip. Also from Snell's Law

$$\frac{\theta_1}{\theta_2} = \frac{\lambda_1}{\lambda_2}$$

where λ_1 = wavelength of sp wave in a-sp plate
λ_2 = wavelength of sp wave in sp lens

In order for the waves to be transmitted by the j.s., λ_1 must be greater than the j.s.; $\lambda_2 = 10^{-4}$ was experimentally determined.

THEORETICAL APPLICATION

Figure 1 schematically represents the application of Snell's Law to the operation of the Solid Plate Lens System. There are high velocity waves on the bottom plate, the a-sp converter plate, and low velocity waves on the top plate, the sp lens. In our system, the joining strip is attached to the a-sp plate, which is the bottom plate. It lightly touches the top plate, the sp lens. Therefore, imagine for a moment that the a-sp plate is cut at the joining strip and so is the sp lens. By now drawing them together, we can schematically draw the complete propagation and focusing of the sp waves. Figure 2 illustrates the continuous propagation and focusing of sp waves into sp images.

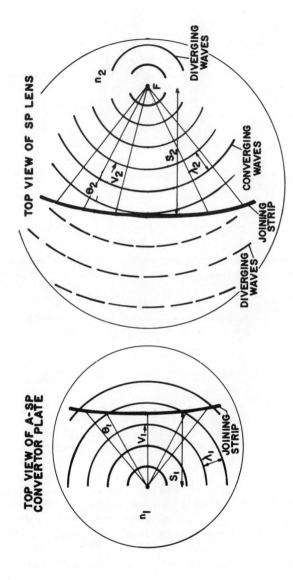

Figure 1   Schematic Representation of Snell's Law to the
Operation of the Solid Plate Lens System.

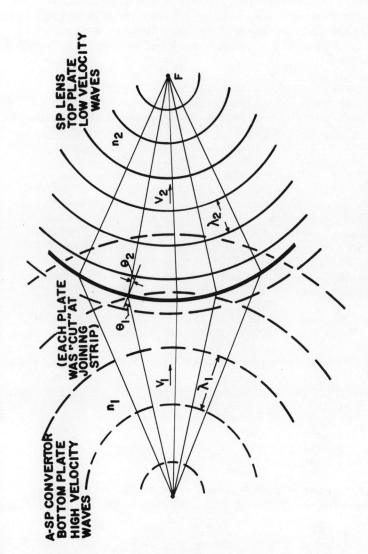

Figure 2   Continuous Propagation and Focusing of sp Waves
into sp Images.

EXPERIMENTAL CHANGES

There have been some changes in the original design
and materials used for the component parts of the Solid
Plate Lens System. These changes have better enabled us
to photograph the sp waves. Perhaps further design or
material changes might be necessary before the final system
is developed.

Firstly, for the a-sp plate, we are using piezo
electric ceramic in place of the .025" aluminum used in
the first model. Secondly, we have replaced the plastic
joining strip with a metal one. This metal joining strip
is paper thin and razor sharp. It is permanently attached
to the ceramic and is in light, even contact with the membrane of the sp lens. It's radius of curvature, as
mentioned before, determines the focusing of the image.
Thirdly, we are no longer using a 1 milliwatt laser as the
point light source but a bulb. Fourthly, in place of the
collimating lens we are using a mirror, which eliminates
the light dispersion caused by the collimating lens.
Fifthly, our new spatial filter is a black dot in the
middle of a piece of glass. Sixthly, we have a longer
focal length which improves image contrast.

EXPERIMENTAL RESULTS

The relationship between the wavelength, frequency, and
velocity of waves is given by the following equation:

$$c = \lambda f$$

where c = velocity
    f = frequency
    λ = wavelength

Since the equation defining the velocity of the membrane
waves is

$$cp_1 = \sqrt{t/M}$$

where cp_1 = velocity of waves on sp lens
    t   = tension of the membrane
    M   = area density

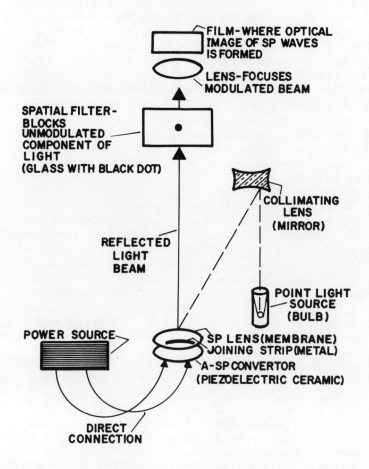

Figure 3   Diagram of System Set-Up for Photographs.

Membrane waves have a velocity independent of frequency.
Consequently, the wavelength will vary inversely with
frequency. The wavelength is measured by the distance the
lines are apart.

Figure 3 is a schematic diagram of the set-up used
for the taking of the photographs in Figures 4 - 9. All
photographs are the top view of the sp lens (membrane).
(Refer to Fig. 1). The a-sp plate is immediately underneath the sp lens but is not visible in the photographs.
The joining strip is glued to the a-sp plate and is in
direct contact with the sp lens. When the position of
the joining strip is not visible, imagine it to be in the
upper half of the sp lens. The membrane is black and the
waves are white. Figures 4 - 6 illustrate the inverse
variation of wavelength with frequency. Figures 7 - 9
illustrate the application of Snell's Law to the focusing
of the sp waves.

In Figures 4 - 6, the joining strip is straight making
the focal length infinity. The waves are neither refracted
nor bent and, therefore, not focused. These photographs
were taken at three different frequencies to illustrate
wavelength variation with frequency. The first photograph, Figure 4, was taken at 610 KHZ. The joining strip
is very distinct in this photo and the waves are propagating in both directions. These waves, however, attenuate
before the edges of the membrane are reached, resulting in
the absence of standing waves. The waves are fairly uniform and close, the unevenness resulting from the
irregular contact of the joining strip to the membrane.
The white semi-circle, visible at the end of the joining
strip is a result of the joining strip exerting too much
pressure on the membrane. As the frequency is lowered to
235 KHZ, the standing waves become visible, with waves
propagating in only one direction from the joining strip.
The patterns become more complicated and the lines get
farther apart. This is even more apparent in the third
photograph taken at 94 KHZ. There are distinct standing
waves or interference patterns caused by the waves reaching
the edge and being reflected back. This results in waves
seemingly to emanate in both directions but in fact it is
only the result of excessive reflection of the waves.

Figure 4   Straight Joining Strip at Frequency of 610 KHZ.

Figure 5 Straight Joining Strip at Frequency of 235 KHZ.

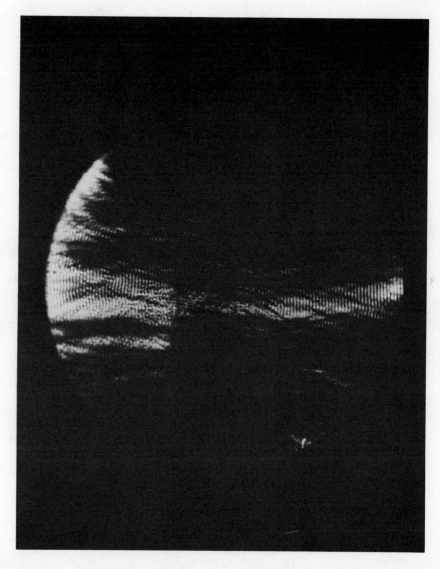

Figure 6 Straight Joining Strip at Frequency of 94 KHZ

In Figures 7 - 9 the next three photographs are presented. In these photographs the joining strip is curved.
It has a radius of curvature of 1". By using the formula
presented earlier in this paper we can calculate the focal
length of this particular Solid Plate Lens System set up.
The a-sp plate is the reference, $r = 1$, and $n_2 = 100$,
$f = 100/99$, making the focal length almost at the center
of curvature. Therefore, the waves will focus.

These photographs, applying Snell's Law to the
focusing of sp waves, show the resolution of the focal
point. These focused waves are also being shown at three
different frequencies. The first photograph (Figure 7)
is at 580 KHZ. In this photograph the waves attenuate
before reaching the edges of the membrane, resulting in
the absence of any standing waves. The waves are uniform
and close focusing to a point before the edge of the
membrane is reached. The presence of multiple focal
points is due to the uneven contact of the joining strip
to the membrane. The white circles at each end of the
joining strip is the joining strip exerting too much
pressure on the membrane. As the frequency is lowered
to 320 KHZ (Figure 8) the standing waves become more
visible causing the multiple focal points to be more
pronounced, though still distinct. The two small white
circles in the middle and edge of the membrane are the
terminal points of the joining strip, which again, is
exerting a bit more force than it should. The third
photograph (Figure 9) taken at 105 KHZ shows the multiple
focal points of Figure 8 as one, indistinct, blurred
focal point. Also, there are many standing waves or
interference patterns; this results in many "superfluous" waves i.e. ones that don't focus but just keep
reflecting back and forth, over the entire surface of the
membrane.

CONCLUSIONS

The joining strip is the essence of the Solid Plate
Lens System. The design and shape of the joining strip
determine if the waves will or will not focus. It also
can determine where on the membrane the waves will focus.
The material of the joining strip can vary. We have used
both plastic and metal joining strips. Both have focused

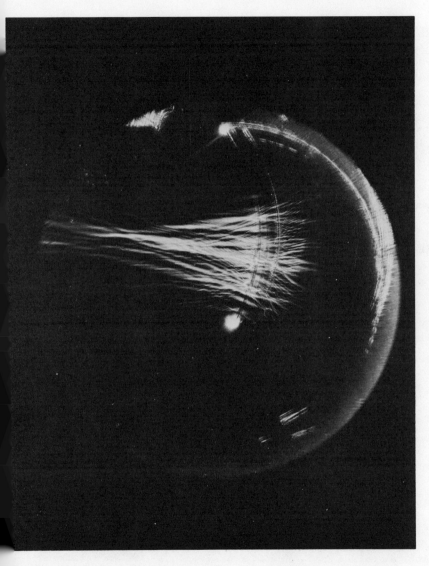

Figure 7   Curved Joining Strip at Frequency of 580 KHZ.

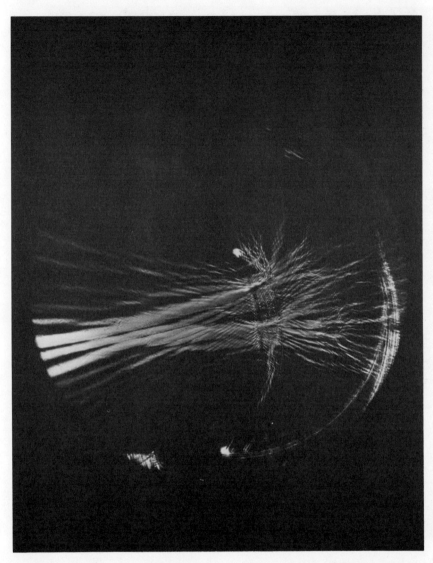

Figure 8   Curved Joining Strip at Frequency of 320 KHZ.

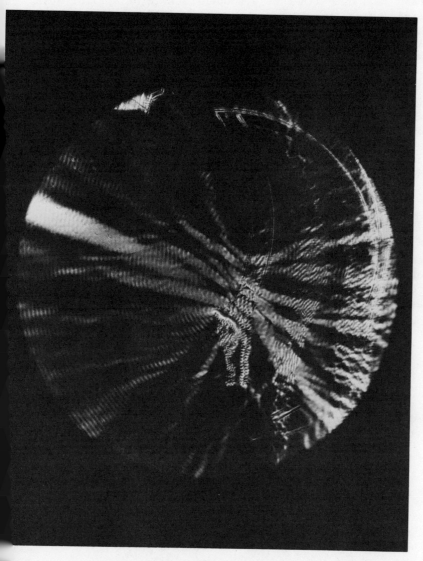

Figure 9   Curved Joining Strip at Frequency of 105 KHZ.

the waves but metal seems to have the edge. The surface
of the joining strip which is in contact with the sp membrane lens can either be a smooth plastic or a razor sharp
metal, though the latter appears to yield better images.
The materials used for the other components of the system
can also vary though we have found some materials experimentally superior to others.

The imaging of the sound source, presented in the last
six photographs, shows the varying patterns and waves obtained with different frequencies and differently shaped
joining strips. From the photographs we can see both the
inverse relationship of the wavelength to the frequency
and the application of Snell's Law to the functioning of
the Solid Plate Lens System.

ACKNOWLEDGMENTS

The author wishes to thank Dr. J. A. Clark for his
help and the use of his photographs. The research reported in this paper was supported by the Advanced Research
Projects Agency and the Department of Defense. The author
was contracted by Dr. Clark for this project.

REFERENCES

1. Clark, J.A., and Stone, K. L., "A Solid Plate
    Acoustical Viewer for Underwater Diving," Acoustical
    Holography, Vol. 5, Plenum Press, 701-714 (1973).

2. IBID

3. Clark, J. A., Durelli, A. J., and Laura, P. A.,
    "On the Effect of Initial Stress on the Propagation
    of Flexural Waves in Elastic Rectangular Bars,"
    J.A.S.A. 52, #4, 1077 - 1086 (1972)

4. See Reference 1.

EARTH HOLOGRAPHY AS A METHOD TO DELINEATE BURIED STRUCTURES

John B. Farr

Western Geophysical

Houston, Texas

Monofrequency signals were used to record holographic data over a segment of a small piercement salt dome in the Gulf of Mexico. Conventional pulse-echo seismogram sections and equivalent two-dimensional holographic reconstructions are shown to compare the effectiveness of the two methods in delineating scattering sources. The holographic reconstructions contain a number of near-vertical lineations which may be associated with faulting.

Using a synthetic aperture technique with a stationary marine cable deployed perpendicular to the source line, a hologram was generated. Reconstructions of this hologram at two depths are shown. These images have high intensity scattering points clustered in areas beyond the dome boundary. Weak scattering point lineations are observed which parallel the salt-sediment interface. No significant scattering points were noted within the dome boundary. Although the field recorded data was somewhat undersampled and the corresponding reconstructed holographic images very crude, the examples given illustrate the holographic method's ability to delineate scattering points within the earth.

435

INTRODUCTION

1.1 Utilization of Diffracted Waves

Earth holographic imaging involves the location
of subsurface scattering features, rather than the
location of the relatively smooth reflection interfaces mapped by the conventional seismic method. Evi-
dences of scatterers are familiar to most geophysicists.
They are expressed on seismograms as diffraction patterns
associated with faults, salt domes, reefs, or other
subsurface structures involving the termination of
individual beds.

The diffractions are considered by most explora-
tion geophysicists to be troublesome sources of inter-
pretive error since they mask the desired reflection
events. They have value only as indicators of subsurface faulting. Even in this role, the diffractions
are of questionable use since they appear only sporadi-
cally and can result from phenomena other than faulting.

Conventional pulse-echo seismic profiling is
directed toward the identification and location of
reflection interfaces. Generally such interfaces must
be smooth and extend for some lateral distance before
they can be clearly delineated. The procedures for
reflection mapping have been perfected to high degree
over a period of forty years.

Diffraction mapping, on the otherhand, involves
rough or broken interfaces which have very limited
lateral extent. The pulse-echo methods which are
optimized for smooth interface detection only show
diffractions on occassion and then only when mixed
with stronger reflections. The holographic method is
directed toward the utilization of diffractions and
therefore offers a fresh look at diffraction mapping
and interpretation.

1.2 Synthetic Diffractions

(1)*   The first figure shows diffraction patterns
on synthetic seismograms generated within a digital

*Parentheses at the left margin contain figure numbers and indicate the discussion of that figure.

computer for a simple model of a single truncated bed. The PP and PS diffracted waves from the 90° corner are labeled on the slide. As can be seen, the PP diffraction is less than one-half the amplitude of the conventional reflected P-wave and very much weaker than the direct wave arrivals. The lower shear wave or PS diffraction is somewhat stronger than the PP diffraction but is weaker than the corresponding PS shear-wave reflection. On this model, as in the real earth, the diffractions are always weaker than their associated reflection arrivals.

1.3 Field Diffractions

Unlike the smooth continuity seen on seismograms from synthetic or laboratory models, the appearance of diffractions on field seismograms is at best sporadic. Rarely they are seen in their entirety; more commonly, only portions of the total diffraction pattern are seen among the stronger reflected events. And in many cases, diffractions are absent entirely from the conventional seismic section. This is partially due to the almost universal practice of multifold compositing, where diffractions are for the most part attenuated along with multiple reflections and other events having normal moveouts different from the conventional reflections.

(2)   Figure 2 shows a conventional pulse-echo marine seismic record section on which the diffractions from a fault are relatively clear. Note that these diffractions are only seen where strong reflections are absent. The automatic gain control applied to the data in the display has amplified the diffractions to a level where they can be seen. This is not the common situation, and for diffractions to be useful in routine interpretation the scattering points must be reliably located, even in the presence of stronger reflections.

1.4 Diffraction Amplitudes

In the offshore experiment, which is the subject of this paper, great care was taken to measure the diffraction amplitude as a function of frequency.

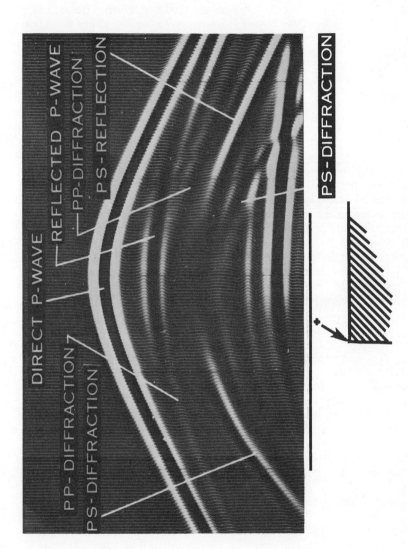

Figure 2. Offshore section with fault diffractions.

(3)   This work is summarized in the third figure.
At the offshore site of this experiment the direct-
waterwave signals (in green), reflection signals (in
red), and identifiable diffracted signals (in white)
were measured on horizontal and vertical wave tests.
These measurements were made at a number of locations
with several swept-frequency vibratory signals. The
averages of a number of observations are plotted on
this figure along with the measurements made of the
cable towing noise, shown in blue.

The amplitude of the diffracted signal is one-
half to one-tenth of that of the reflected signal.
At one particular good spot, the average reflection
signal amplitude reduced to seawater pressure was
29.6 microbars and the diffracted signal 16.5 microbars. This was the highest relative diffraction amplitude found in the entire area.

From these field measurements, it is apparent that
the weak diffractions, to be reliably mapped, require
enhancement. Utilizing monofrequency source signals
matched filtering and integration of the signal detected
during a long time duration, can improve the signal-
to-noise ratio to the point where diffractors can be
reliably mapped.

1.5 Similarity of Holographic and Conventional
Vibratory Methods

The basic differences and similarities between
the earth holographic method and the conventional
vibratory seismic reflection method should be
emphasized. Only the vibrator source time function
and the representation of the seismic waves detected
at the surface are different in the two methods.
However, the interaction of the earth with the
seismic waves in reflecting, refracting, and scattering
them back to the surface is identical in both cases.

The recording techniques are essentially the same,
with the exception that when steady-state mono-frequency
arrivals are recorded, a reference signal such as a
pilot signal or baseplate signal is recorded on an
additional channel to provide a phase reference. In
holographic recording the phase of the steady-state
signal at a point on the surface from a single point

scatterer is the equivalent information to the pulse-echo traveltime in conventional seismic recording.

A vibratory source, as used in the present experiment, can generate either steady-state monofrequency waves or swept-frequency waves, or a combination of both. The conventional vibratory sweep can be used as a monofrequency signal by appropriate filtering; however the power input to the earth at any single frequency is relatively low since a very short time is spent vibrating at any one frequency in the typical swept signal. By vibrating at a certain frequency of the sweep for a longer time duration, the power at that particular frequency can be increased, thereby enhancing the signal-to-noise ratio for holographic imaging of diffraction points at that frequency.

The experiment was conducted 75 miles offshore Louisiana in approximately 85 feet of water. The vibrator and one of the two hydrophone streamer cables used were towed behind the boat at a speed of 4 knots. The motion of the source and receivers introduced a slight smearing effect which limited the monofrequency integration time to few seconds.

The F.M. sweep signal used for the conventional work was 6 seconds long and consequently the same smearing was present on the pulse-echo recordings. This smearing is an accepted part of marine geophysical surveying and provides some useful noise attenuation even though limiting the horizontal resolution.

2.0 HOLOGRAPHIC IMAGING METHOD USED IN THE EXPERIMENT

2.1 Crosscorrelation method of imaging

The imaging method used in these experiments involved spatial crosscorrelation rather than the more conventional direct mathematical inversion of phase and amplitude data. This procedure was selected, primarily since it was relatively straightforward on the large digital computer we had at our disposal. This computer, an IBM 360/75 was equipped with an array processor which greatly speeded up the time

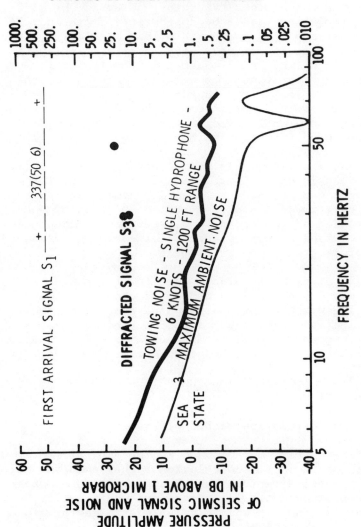

Figure 3. Graph of relative amplitudes of direct, reflected and diffracted signals.

consuming multiply and add operation needed to cross
correlate. Also the areal cross-correlation of complex
numbers was a logical extension of the linear cross-
correlation of real time series values used routinely
to compress the vibratory chirp signals in pulse-echo
work.

Cross-correlation also simplified the problems
introduced by non-standard array geometries   and
vertical variations of velocity within the earth. These
irregularities need be taken into account only one time,
when constructing the correlator rather repeatedly
when direct mathematical inversion is used.

2.2 Holographic Correlators and Spatial Correlation

The cross-correlation method of earth holographic
imaging requires a velocity model for the subsurface
of a given area. This velocity model is used to construct the areal correlator which in turn is used to
reposition the energy observed at the surface, back
to the diffracting point. In this respect, the holographic method is similar to the common-depth-point
(CDP) reflection method, which also requires a velocity model for a given area. The velocity model in the
CDP technique is used to remove the normal-moveout
of reflections due to the different source-seismometer
ranges. The objective of the CDP method is to time-
align reflections from the same subsurface point. In
earth holography, the scattered seismic waves are
space-aligned at the same diffracting point. If the
velocity of an area is not known accurately in either
method, the reflections will not be precisely time-
aligned or the diffractions precisely space-aligned.

To get the correlator needed to form the image
in earth holography, a piecewise linear velocity-depth
model is first constructed for the prospect area. This
velocity model is chosen to fit the check-shot traveltime, sonic log-integrated traveltime, or any other
traveltime information with depth determined for the
area.

(4)   An example of a piecewise linear velocity model
containing high velocity layers in a low velocity section

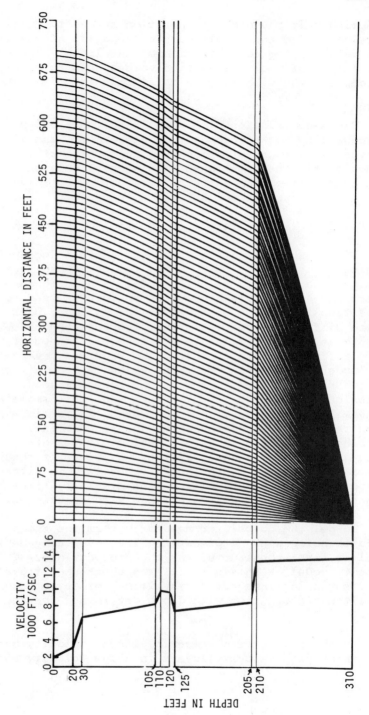

Figure 4. Velocity-earth model with ray tracing through layers.

appears on the left of the fourth figure. A plot of
the individual rays from a subsurface point, to the
surface, through a medium with this velocity layering
appears on the right-hand side of the slide.

The tracing of rays from a subsurface point to
the surface, through a medium with the piecewise
linear velocity model permits the determination of
amplitude and traveltime or phase with range. The
phase, ϕ, of a steady-state wave pattern from such
a monofrequency source with frequency, f, is simply
related to the traveltime, t, by the familiar relation, $\phi = 2\pi$ ft.

The image of a subsurface diffracting structure,
typically generated by a digital computer, is obtained
(Sondhi, 1969; Maginess, 1972) by cross-correlating the
field recorded steady-state wave pattern (or hologram)
with the complex conjugate of the correlator. The
correlator is constructed from the calculation of
traveltime and amplitude with range for the appropriate
velocity model, given the field geometry and recording
procedure. The holographic field data acquisition procedure is almost identical to that used in routine
seismic work and can be accomplished using a stationary
source and moving receiver array (or vice versa), a
moving source and moving receiver with constant
separation, as in marine work, or crosslinear source
and receiver arrays as outlined by Milder and Wells
in 1970. Once the field geometry and procedures are
specified, separate correlators are constructed for
each of the depths at which reconstructed images are
desired. Only one set of correlators is needed for a
given prospect area, assuming the velocity remains
unchanged. The correlators are in reality the response
from a single scatterer just as in Vibroseis the correlator is the response from one reflector.

(5)   Examples of correlators for a velocity model
of the experiment area, to be discussed later, appear
on the fifth figure. The vertical axis is depth and
the horizontal axis is range. For each depth, two
curves are plotted. The first curve is the real part

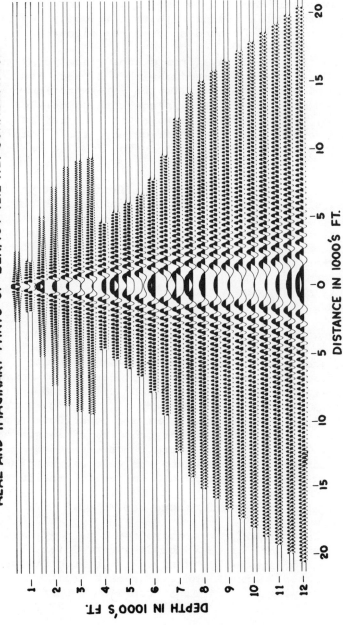

REAL AND IMAGINARY PARTS OF BLK. 184 42.2 HZ. CORRELATORS

Figure 5. Example of single-frequency correlation v. depth.

of the complex correlator, and the second the imaginary part. These correlators are spatially cross-correlated with the recorded data to obtain the image of the bedding terminations against faults as well as other subsurface scatterers.

2.3 Tests of Holographic Crosscorrelation Method on Mathematical Model

Before field testing was undertaken, a number of mathematical model studies was performed to test the holographic crosscorrelation method. These studies showed it would be possible to image scattering (6) points as a function of depth. As an example of these synthetic model investigations, figure 6 illustrates the two-dimensional holographic image produced from a buried object having three sharp diffracting corners. The exact acoustic wave field produced by this object when insonified by vertically traveling plane waves was mathematically derived. The amplitude and phase values for a series of simulated detector locations along an observation line were then tabulated, and input to the digital cross correlator program for imaging the field-recorded holographic data. As can be seen, the corner scattering points were successfully imaged at the correct locations. Note that the corners not the object itself were imaged. The three images shown on this slide represent different gain levels. On the highest gain level image, a number of faint streaks can be seen. These streaks, must be considered as undesirable noise in the holographic technique. This noise is one of the reasons for the relatively less accurate vertical resolution of the holographic method when contrasted with its lateral resolution.

2.4 Tests of Holographic Imaging Method on Physical Models

In addition to the exact mathematical model shown on figure 6, the holographic reconstruction technique was tested on a variety of physical models placed in a water tank and insonified with high-frequency (1 megahertz) acoustic waves.

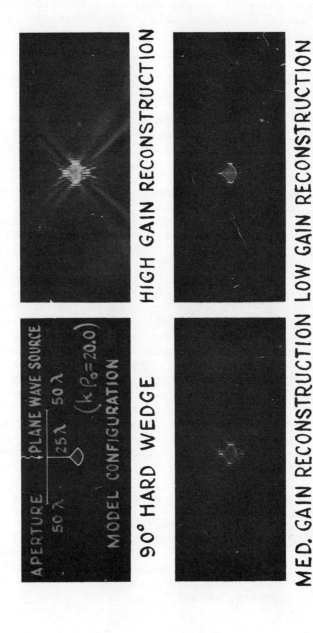

Figure 6. Holographic reconstruction of mathematical 3 pt. piewedge.

Figure 7. Water tank-scissors fault model.

(7)   A typical physical model is shown on figure 7.
It was constructed from a thick slab of aluminum which
had been milled in such a manner as to simulate a
fault whose throw increased with distance. Holographic
data was recorded using a scanning crystal receiver and
a 1-megahertz (8) scanning source. This data was then
holographically imaged and produced the results seen
on figure 8. Note that the dark dashed line on the
image delineates the fault trace. It is of interest
that each of the dashes along the fault correspond to
points where the throw has changed exactly one-half
wavelength, equivalent to a one wavelength change of
the two-way travel path. On the model, the fault started
with zero throw, linearly increasing until an eight-
wavelength (1 megahertz at water velocity) throw was
measured at the edge of the slab. Careful examination
shows there are exactly 16 dashes along the fault.
Although this is undoubtedly due to the noise-free
laboratory environment and the very simple model tested,
the dash counting technique does suggest a method for
determining changes in fault throw, given a control well
where an original fault displacement can be established.

(9)   Another water tank model consisting of a series
of Plexiglas beds was reconstructed and the image as
well as the original model are shown in figure 9. The
cutout area represents the salt dome and the corresponding image at the depth of the first bed is what
would be anticipated from beds truncating against salt.

     Before the field experiment was conducted it was
felt that the bed terminations against the rough salt
would produce the largest diffractions even though
this had not been observed on the pulse-echo recording.
Consequently the field data was expected to look more
like figure 9 than like figure 8.

3.0 FIELD TESTS OF HOLOGRAPHIC METHOD

3.1 Description of Test Site

(10)   As illustrated in the figure 10, a small
piercement salt dome located 70 miles southwest of
Morgan City, Louisiana, in Eugene Island Block 184,
was selected for a field evaluation of holographic
imaging. It was chosen for the experiment due to its

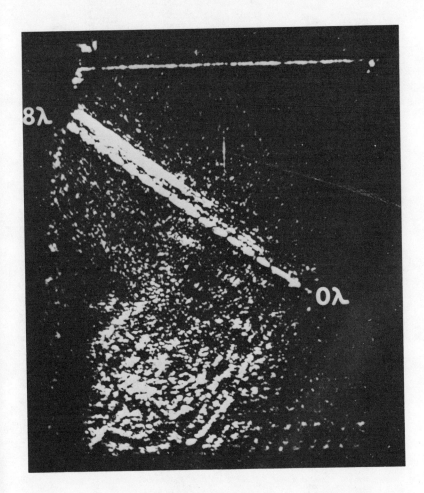

Figure 8. Reconstruction of scissors fault hologram.

PHYSICAL MODEL STUDIES
APPLICATION OF MODELING TO BLOCK 184 HOLOGRAPHIC EXPERIMENT

BLOCK 184 MODEL

IMAGE OF TERMINATION OF TOP LAYER

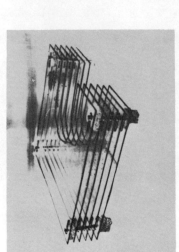

Figure 9. Plexiglas beds model with dome cutout.

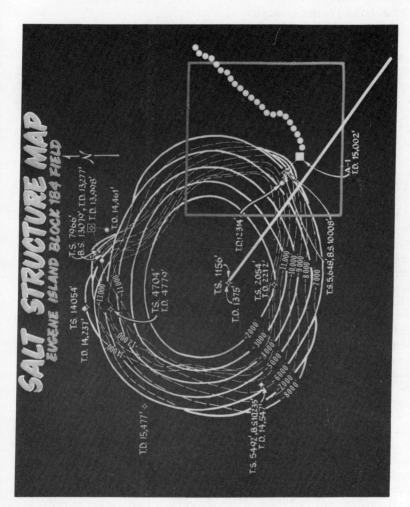

Figure 10. Salt dome contours and line index map.

small areal extent, shallow salt and wells which confirmed the existence of large salt overhangs. The water
depth of approximately 80 feet was optimum for the
single large vibrator source and conventional marine
streamer cables which were available for acquisition
of the holographic data.

The existence of a permanent gathering system
platform at the well site provided an ideal location
for the digital recording system used to acquire areal
coverage from the stationary bottom cable. The recording
geometry is shown superposed on the salt contours. The
blue rectangular area in the lower right shows the
portion of the dome over which the holographic imaging
was performed.

3.2 Crossed Array Recording Geometry Used For Areal
Hologram

(11)   Figure 11 is a pictorial representation of the
crossed array recording geometry used to acquire an
areal hologram over this image area. A conventional
marine cable and vibrator source were mounted on a
boat which traveled back and forth along the boat line.
The vibrator ran continuously using a mono-frequency
pilot signal. Signals from the streamer were recorded
in a continuous series of 10-second records. Each record
contained location information for the source and
receivers on the cable. Several traverses along this
line produced a series of two-dimensional or line
holograms. In addition to the monofrequency records,
25-75 Hertz sweeps were used to acquire conventional
swept-frequency seismic data.

A second cable, shown as red dots on the figure
10 at right angles to the boat traverse, was located
on the bottom and remained stationary throughout the
entire recording procedure. A second set of digital
recording instruments was located on the well platform and recorded the signals from the bottom cable.
The instruments aboard ship and on the platform were
synchronized by a radio link. By means of the link,
records on the platform were taken simultaneously
with those recorded aboard ship. This crossed array
technique provided inexpensive areal coverage. Using
the bottom cable and separate recorder eliminated

Crossed Array Holographic Recording Method

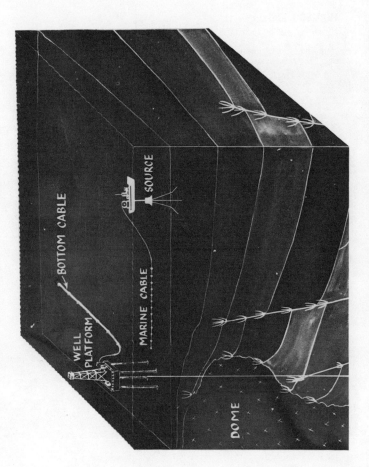

Figure 11. Pictorial diagram of hologram recording geometry.

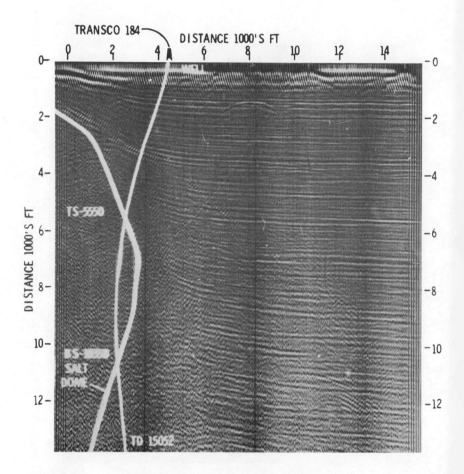

Figure 12. Conventional 25-75 Hz sweep line with well and dome.

23/24 or 96% of the vessel time which would have been
required if only the boat had been used to sample the
blue area.

3.3 Field Results from 2-D Linear Holographic Recordings

Where each conventional pulse-echo seismic section represents a vertical slice of the earth, each
holographic image represents a horizontal or constant
depth slice through the earth. If the holographic
data are recorded as a line on the surface, the image
is a constant depth line beneath the surface. If holographic data are recorded areally on the surface, a
constant depth areal image or contour map is produced
as the output. In the experiment described in this
paper, both types of recording were made simultaneously.

(12)   As a reference, figure 12 shows the conventional
seismic depth section taken along the boat traverse
using a compressed 25- to 75- Hz linear sweep. On the
section is seen a number of steeply dipping events and
diffractions along with a rather remarkable wealth of
high-frequency reflections. Note the typical shadow
zone were reflections are absent surrounding the salt
boundary as delineated by the well. These shadow zones
are typically seen around most salt domes and are
usually attributed to zones of interfering steep
reflections, refractions and diffractions.

(13)   Figure 13 shows a section consisting of a
series of holographic line images at different depths
along the same line as the pulse-echo section. Amplitudes of the scattered arrivals are plotted as a
function of distance along the traverse for a 42- Hz
monofrequency signal, which is close to the center
frequency of the 25-75 Hz Vibroseis sweep used to
record the previous section. Each variable-area trace
represents a single line reconstruction at a different
depth. Each trace is for successive 250-foot depth
increments so that a total depth of 12,000 feet is
represented over the entire section. The horizontal
and vertical scale is the same on this section as on the
pulse-echo depth section. The strongest diffraction
points are seen beyond the dome boundary and appear
to line up in a manner suggesting a large tangential fault.
Other lineations, having different orientations, are also
seen, which may likewise indicate faulting around the

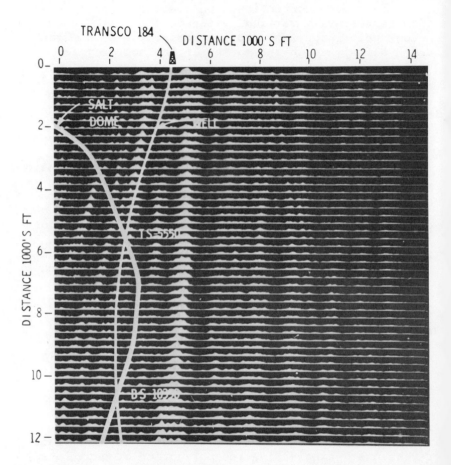

Figure 13. Holographic linear reconstruction-variable
area with dome.

dome. Some unexplained lineups are seen within the
salt itself.

3.4 Comparison of Holographic and Conventional Sections

To compare the conventional pulse-echo and holographic representations, the variable-area display of
figure 13 was replotted using the variable-density
mode and has been superposed on the conventional
seismic reflection section on figure 14.

(14)   Some interruptions and terminations of reflections can be seen at the lower portion of the
section, although the upper beds appear to be continuous where the holographic data indicate faulting
is present. It is of interest that no continuous reflections are seen at depth, on the dome side of the
holographically defined fault. It would appear the
so-called "shadow zone" of no seismic reflections lies
between the holographic fault and the salt boundary
at least to the accuracy with which that boundary
can be determined by the well. Throughout the offshore
Louisiana salt dome province many domes are typically
associated with regional depositional faults, which
may be indicated by the holographic fault image.

Other traverses along this same line in the same
direction gave nearly identical results to the image
shown, confirming the repeatability of the process.
Other lineups were noted beyond the immediate area of
the dome itself. Here it is important to note that
if only line holograms are recorded, the problem of
diffractions occurring out of the vertical plane of
the section cannot be resolved.

3.5 Areal Holographic Images Around Dome

(15)   Figure 15 shows the 3500- foot image from the
areal array data. This data was recorded on the marine
platform from the bottom cable receivers while the
source boat made several approximately parallel traverses along the radial line. Since only 24 channels
were available for this recording procedure, the
holographic data was considerably under-sampled in
one direction. The intensities below 35% which are

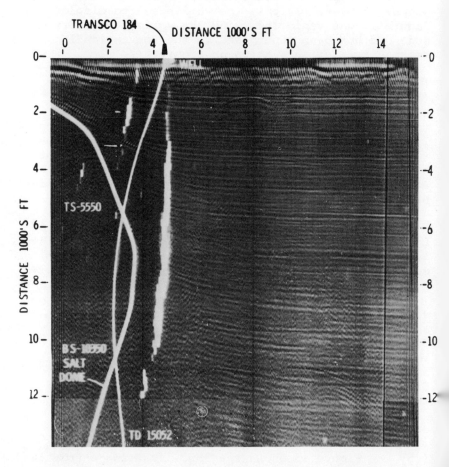

Figure 14. Superposed holographic and conventional lines
(VS mode).

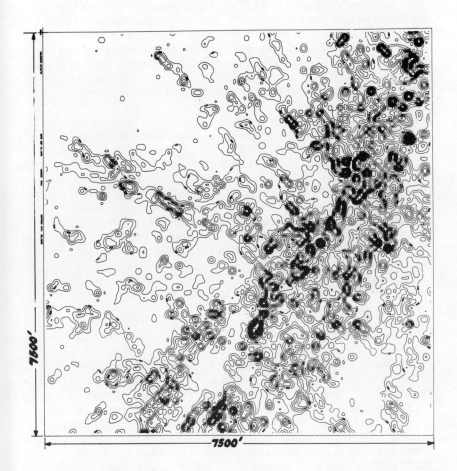

Figure 15. 3500- foot areal hologram reconstruction, con-
        toured.

not shown can be considered as ambient noise resulting
from flare patterns and the like. The contours permit
a much greater dynamic range to be seen than the
typical photographic image. Several high-intensity
lineations can be seen which appear fault-associated.
The concentration of diffracting points generally
coincides in position with the holographic fault as
seen on the two-dimensional reconstruction seen
earlier. A few smaller amplitude diffractions are
seen which fall within the salt boundary. These could
be associated with small radial faults which have
been interpreted using the well information.

(16)   Figure 16 is a variable density or photographic representation of the same data plotted in
contoured form on the previous figure. The limited
dynamic range of the variable density plots eliminates
the low intensity values, permitting only the major
features of the scattered field at the depth of 3500
feet to be seen. As in the two-dimensional case,
repeated recordings from different boat traverses over
the same area gave similar, although not identical,
images. Several strong lineations in the immediate
vicinity of the boat traverse in the lower portion of
the mapped area can be associated with the events
seen on the two-dimensional linear images shown on
the earlier slides.

        In general, the dark scatterers define the
boundary of the area which is void of conventional
continuous reflections. It is somewhat surprising
that the salt sediment contact is not better delineated.
The radial and circumferential faulting appear to provide better scattering sources than the salt trunca-
tions. The high concentration of scattering points
along the tangential fault is encouraging since on the
conventional section shown earlier few, if any,
diffractions were seen at or near the interpreted
fault interface.

(17)   Figure 17 shows an areal image of the same
holographic data at a depth of 6000 feet. Note that
the same major features (i.e., the "holographic
tangential fault and radial fault") are seen in the
same general areas, although shifted slightly with
depths as would be anticipated.

Figure 16. 3500- foot areal reconstruction variable density with map.

Figure 17. 6000- foot areal reconstruction variable density with map.

CONCLUSIONS

The holographic field results in the Block 184 area were, as might have been anticipated, far more complicated and less definitive than the noise-free laboratory models. Seismic waves cannot be easily focused on a scatterer from a source on the earth's surface as they can in a water tank. Consequently when recording a field earth hologram, not only the scattered wave but also many other undesired waves are recorded with large amplitudes. Surface waves, direct water waves, refracted waves, and multiple reflections, all corrupt the earth hologram to varying degrees.

The several hundred foot long groups of 20 to 30 interconnected hydrophones used on existing marine seismic cables prevent accurate phase measurements at the higher seismic frequencies. For improved resolution, these cables must be replaced by cables with a large number of closely spaced, independent hydrophones to provide the higher density of individual sample points required for detailed holographic imaging.

The crossed array recording geometry suggested by Milder and Wells (1970) appears practical and provides high-density areal coverage with a minimum expenditure of field time. Using modern telemetry, the need for a platform or other stationary recorder location could be eliminated and all data recorded aboard ship.

In conventional pulse-echo work, using compressed chirp signals, the F.M. sweeps are chosen to optimize the signal-to-noise ratio in particular areas. In like manner monofrequency signals of the correct frequency and duration will enhance the signal-to-noise ratio of diffractions to the extent that they are easily seen despite their somewhat lower amplitudes relative to typical reflections.

The holographic source signal can be a steady monofrequency, a shot or preferably a portion of a F.M. sweep. The power at a particular frequency in a vibratory sweep signal is a function of its duration.

By extending the duration of a monofrequency portion
of the sweep, diffraction signal amplitudes can be
enhanced to whatever degree desired.

Combination mono-swept frequency source signals
permit the recording of conventional and holographic
data simultaneously with only the addition of one
extra recording channel to provide a replica of the
outgoing source signal for phase measurements. Conventional seismic reflection data is best used to
map the structure in geologically simple provinces.
It has superior depth resolution to holographic
data. However, holographic data provides superior
lateral resolution in locating scatterers such as
faults, pinchouts, reefs, etc. Certainly the additional cost of acquiring holographic data is small
when done in conjunction with a conventional seismic
survey.

The processing of holographic data is also very
similar to the familiar techniques. The cross-correlation procedure, normally performed in time in conventional vibratory work, is merely performed in space
in the holographic technique. Thus, the two systems
are very similar in concept, data acquisition procedures, and in the data processing methods required
to produce an output display. It is the displays
themselves which differ, each with its own particular
advantages.

The holographic images of scatters require interpretation just as do the conventional seismogram sections. Each type of image or section emphasizes different information and the two together may provide the
geophysicist with a better understanding and definition
of the complexities of the subsurface.

The conventional seismic reflection method has
enjoyed forty years of development and testing. Today,
a reflection survey utilizes source and seismometer
arrays, multifold coverage, and extensive digital
processing to obtain an interpretable seismogram
section. The situation is quite different from a survey
using a single dynamite shot and recording the signals
from three individual seismometers on a single paper
record. Mathematical and physical model studies have

shown the holographic method to be a viable technique.
The initial field tests of the method, while not up
to laboratory model quality, are certainly encouraging.
The holographic method must be compared with the seismic
reflection method at an equivalent stage of development,
not present-day technology.

The author wishes to thank Amoco Production Company
for permission to present this paper, and particularly
Mr. M.E. Arnold who supervised the data acquisition
and provided the signal strength analyses, as well as
Dr. G.M. Ruckgaber who developed the digital computer
program used to produce the output images shown in this
presentation.

REFERENCES

French, W. S. (1974), "Two-dimensional and three-
dimensional migration of model-experiment reflection
profiles", Geophysics, v. 39, p. 265-277.

Milder, D. M. and W. H. Wells (1970), "Acoustic
Holography with Crosses Linear Arrays", IBM J. Res.Dev.,
p. 492-500.

Maginness, M. G. (1972), "The Reconstruction of
Elastic Wave Fields from Measurements over a Transducer Array", J. Sound Vibr., v. 20, p. 219-240.

Sondhi, M. M. (1969), "Reconstruction of Objects
from Their Sound-Diffraction Patterns", J. Acoust.
Soc. Am., 46, p. 1158-1164.

ENHANCEMENT BY NON-COHERENT SUPERPOSITION OF MICROWAVE

IMAGES FORMED WITH CROSSED, COHERENT ARRAYS

R. A. Hayward, E. L. Rope, G. Tricoles,
On-Ching Yue

General Dynamics Electronics Division, P.O. Box
81127, San Diego, California, 92138

Several potential applications motivate research in microwave holography. These applications include the determination of ionospheric properties and imaging through fog or soil.[1-3] Additional applications are to diagnostics of microwave devices such as antennas and radomes.[4,5,6]

Although progress has been made,[7-10] microwave holography is rather slow. Delays occur in acquiring data by scanning a single probe antenna over a surface to sample a wavefront. Although improved methods are being investigated,[11,12] scanning is sometimes impractical because of equipment size or because object motion during the scan can blur the image. The scanning delays can be eliminated by an areal array of receiving antennas, but the costs of an array seem rather large. Therefore, as a compromise between cost and speed we have collected data with crossed, linear arrays, which were first suggested for conical holography.[13]

Additional delays occur if reconstructions are made from optical holograms because the preparation and scale reduction of a transperency takes time.

Image quality is low because the ratio of hologram size to wavelength is limited; consequently, the speckle has dimensions comparable to those of the object. Images can also be small and distorted if they are produced with coherent light from holograms that are not reduced by the ratio of microwave and optical wavelengths. To eliminate the scale reduction we have computed images, utilizing the

formation wavelength for the computed reconstruction. To
reduce speckle we have non-coherently superimposed several
images that were formed with coherent microwaves for distinct
object orientations.

The paper describes an experimental, wavefront sampling
array that consists of two orthogonal, linear antennas.
One array transmits sequentially. The other array receives
sequentially while a particular transmitting antenna radiates. A switching arrangement generates an areal array of
phase and intensity values that are approximately equivalent
to those from an areal array of antennas with a fixed transmitting antenna.

The paper describes a numerical technique for computing
and displaying images, and it presents enhanced images.
The enhanced images were composites, formed by non-coherently
adding the intensities of several distinct images that were
formed with coherent microwaves but for distinct object
orientations.

THEORY

Diffraction Theory

Images were computed from data that were measured over
a plane with a wavefront sampling array, which is described
in the next section. The data were processed by a fast
Fourier transform following their multiplication by a quadratic phase function that simulated a lens. The computed
image intensities were displayed by an electronic printer
with 10 grey levels.

The computations were based on the Fresnel approximation to scalar diffraction theory. From the Huygens-Fresnel
principle a complex-valued scalar component of field $U(P_0)$
at position P_0 due to a distribution of known field $U(P_1)$ at
P_1 is

$$U(P_0) = \iint_{\Sigma} g\, U(P_1)\, \cos\, (\underset{\sim}{n},\, \underset{\sim}{r}_{01})\, d^S \qquad (1)$$

where $\underset{\sim}{r}_{01}$ is the unit vector connecting P_0 and P_1, $\underset{\sim}{n}$ is the
unit vector normal to the aperture, g is $(j\lambda r_{01})^{-1} \exp\,(jkr_{01})$
with $k = 2\pi/\lambda$, and λ is the wavelength. P_0 is in the measurement plane, and P_1 is in the object plane of Figure 1.

If the distance between the aperture and observation plane greatly exceeds the dimensions of the aperture and the region of interest in the observation plane, then the Fresnel approximation of Equation 1 can be used. The cosine-factor has approximately unit value, r_{01} is approximated by quadratic terms, and g is $(j\lambda z)^{-1} \exp(jkr_{01})$- where z is the distance between the planes containing P_0 and P_1; therefore,

$$U(x_0, y_0, z) = g(z_1) \, \theta_0 \iint U(x_1, y_1, 0)$$

$$\exp[-jk(x_0 x_1 + y_0 y_1)/z] \, dx_1, dy_1 \quad , \quad (2)$$

where $e_i = \exp(jk\rho_i^2/2z)$, with $\rho_i^2 = x_i^2 + y_i^2$, for i = 0 or 1.

Now suppose that a thin lens, focal length f, is simulated at the plane $z = d_1$ by a quadratic phase function. The field just after the lens is then

$$U_+(x_0, y_0, d_1) = U(x_0, y_0, d_1) \exp(-jk\rho_1^2/2f) . \quad (3)$$

By applying Equation 2 to U_+ in Equation 3, we obtain the amplitude in the image plane,

$$U(x_2, y_2, d_1 + d_2) = g(d_1) g(d_2) \, e_2 \iint U_+(x_0, y_0, d_1) e_0$$

$$\exp[jk(x_2 x_0 + y_2 y_0)/d_2] dx_0 dy_0 . \quad (4)$$

Notice that Equation 4 has the form of a Fourier transform if we assume U_+ is bounded and integrate to infinity. With the focusing condition $d_1^{-1} + d_2^{-1} = f^{-1}$, the integrals over x_0 and y_0 lead to δ-functions, and their sifting property gives

$$U(x_2, y_2, d_1 + d_2) = -k^2 (d_1 d_2)^{-1} \exp[jk(d_1 + d_2)]$$

$$\exp\left[jk(1+m)\left(x_2^2 + y_2^2\right)\right] U(mx_2, my_2) , \quad (5)$$

where m is $-d_1/d_2$. Equation 5 shows that the image amplitude is an inverted replica of the object amplitude. The phase factors are of no consequence if we plot $|U|^2$.

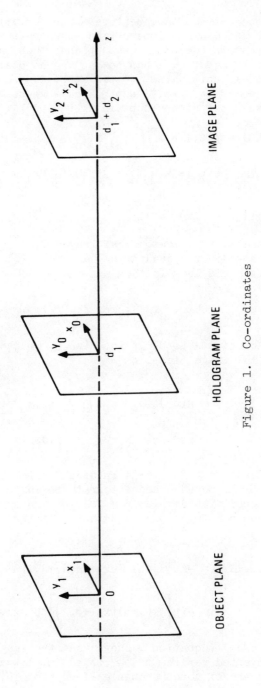

Figure 1.   Co-ordinates

Notice that the reconstruction continues the field through the simulated lens. The present method therefore differs somewhat from conical holography of Wells. Wells analyzed measurements with crossed arrays for coherent fields, but the reconstruction theory considered that the hologram was illuminated by the transmit array. In the present method the equivalent areal array of data is considered as propagating through the simulated lens.

Equation 4 suggests a convenient imaging algorithm, a fast Fourier transform. The next subsection describes the procedure.

Computational Algorithm

The object-scattered field is measured at sample points so the input data for calculations consist of sampled field values in a rectangular matrix, where the uniform sampling intervals are Δx and Δy. Let U_{mn} be the complex-valued field of position $(m\Delta x, n\Delta y, d_1)$ then the data matrix will have size M x N, with integral indices m and n such that $-M/2 + 1 \leq m \leq M/2$, and $-N/2 + 1 \leq n < N/2$. The size of the data sampling surface is $M\Delta x$ by $N\Delta x$.

The algorithm for computer implementation consisted of two steps, as follows:

(1) $V_{mn} = U_{mn} \exp \{j\pi [(m\Delta x)^2 + (n\Delta y)^2]/\lambda d\}$,

(2) $W_{ij} = $ Fourier transform $\{V_{mn}\}$,

where W_{ij} are samples of the image field, and d is object distance. The size of the image is fixed by Δx and Δy, the sampling interval is the hologram as $(\lambda d/\Delta x)$ by $\lambda d/\Delta y$.

Conical Holography Approximation of Measured Fields

The crossed arrays measure data that are approximately equivalent to those measured by an area array. The accuracy of the approximation was discussed by Wells for conical holography; see Reference 13. Here we verify that to second order the phase measured by crossed arrays of receivers & transmitters is equivalent to the phase measured by an areal array for a fixed transmitter. The equivalence restricts the equivalent areal array to have antennas centered at points on a rectangular mesh. The third order term was considered in Reference 13.

Consider a point object and ignore the directive properties of the receiving & transmitting antennas. Let the co-ordinates of the transmitter object and receiver be $(x_t, y_t, 0)$, (x_0, y_0, z_0), and $(x_r, y_r, 0)$, respectively. Then in the measurements with either an areal array or a crossed array the ray paths for transmission and reception are respectively

$$r_t = \left[(x_0-x_t)^2 + (y_0-y_t)^2 + z_0^2 \right]^{1/2}$$

and

$$r_{ra} = \left[(x_0-x_r)^2 + (y_0-y_r)^2 + z_0^2 \right]^{1/2} .$$

For an areal array, let us fix x_t at x'_t and $y_t = 0$, and let x_{ra} and y_{ra} denote explicitly the receiving locations. For crossed arrays let us fix x_t also at x'_t with y_{tc} variable and fix $y_r = 0$ with x_{rc} denoting receiving locations. With a binomial expansion to second order, for an areal array

$$r_{ra} + r_{ta} = 2z_0 + \left[(x_0-x'_t)^2 + y_0^2 + (x_0-x_{ra})^2 \right.$$
$$\left. + (y_0-y_{ra})^2 \right]/2z_0 .$$

For a conical array

$$r_{rc} + r_{tc} = 2z_0 + \left[(x_0-x'_t)^2 + (y_0-y_{tc})^2 + (x_0-x_{rc})^2 \right.$$
$$\left. + y_0^2 \right]/2z_0$$

The path difference is

$$\delta p = (r_{ra}+r_{ta})-(r_{rc}+r_{tc}) = \left[(x_0-x_{ra})^2 - (x_0-x_{rc})^2 \right.$$
$$\left. + (y_0-y_{ra})^2 - (y_0-y_{tc})^2 \right]/2z_0 ;$$

Sufficient conditions for δp equal zero are that $x_{ra} = x_{rc}$
and $y_{ra} = y_{tc}$. The first, $x_{ra} = x_{rc}$, requires that the x
co-ordinates of the antennas in the horizontal receiving
array be equal to those of the antennas in the areal array.
Thus the equivalent receiving array has vertical columns of
antennas. The second condition $y_{ra} = y_{tc}$ requires that the
y co-ordinates of the vertical transmitting array be equal
to the antennas in the areal array. Therefore the equivalent array would have vertical rows & horizontal columns.

EXPERIMENT

The apparatus was a multi-channel interferometer that
is sketched in Figure 2. A signal generator and travelling
wave tube amplifier generated coherent, 16.000 GHz microwaves. This source was connected to a linear, vertical
array of vertically-polarized horn antennas through a
branching network that was formed from waveguide and mechanical, waveguide switches. Figure 3 shows the antenna arrays
and a branching network. A second network connected the
antennas in the horizontal array to a coherent receiver, a

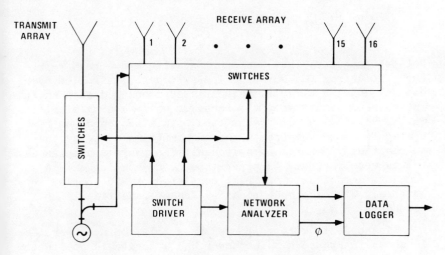

Figure 2. Arrangement of Apparatus

Figure 3. Wavefront Sampling Array

Network Analyzer, that was supplied with a coherent reference field from the source. The switches, controlled by a
digital driving circuit, sequentially connected each antenna
in the receive array to the receiver while a single antenna
in the vertical array transmitted. This sequence was repeated for each antenna in the transmitting array by the
automatically controlled switches. The phase and intensity
values were recorded on paper tape by a data logger. A
complete set of data or frame consisted of 256 complex
numbers. With the mechanical switches frame time is 24
seconds.

The objects were supported by a telescoping tower, as
is shown in Figure 4. The tower had a turntable at its top
to rotate objects about horizontal and vertical axes. A

frame of data was recorded for a fixed object position.
Frames were recorded for each of several positions that
differed by 2° in azimuth, elevation, or both. The object
in Figure 4 is a concave paraboloid with diameter 38 cm or
20 wavelengths. The squares on the paraboloid are merely
brightly colored paint.

The main sources of error were the unequal path
lengths in the array. The differences arise from mechanical tolerances, waveguide bends, and the switches. The
differences were reduced by spacers following a test that
utilized a field transmitted through the networks toward
each antenna. The waveguide near each antenna was shorted,

Figure 4.   Tower and Object (Object a 38 cm diameter,
            concave, paraboloid)

and the phase of the reflected field was measured. Residual errors were determined in a second test that utilized a transmitting antenna near each receive antenna. The transmitting antenna was moved on a line approximately parallel to and a few wavelengths from the plane of the array. A laser beam was utilized to determine variations in the spacing of the travelling antenna and the fixed antennas. Random phase errors of ±12° from a mean were accepted. Other error sources were small; for example, the receiver has ±1° phase accuracy.

RESULTS

Figure 5 shows a microwave image of the 38 cm diameter paraboloid in Figure 4. This image, formed for a fixed position of the paraboloid, has considerable speckle. The spots have dimensions approximately Rλ/D where R is the

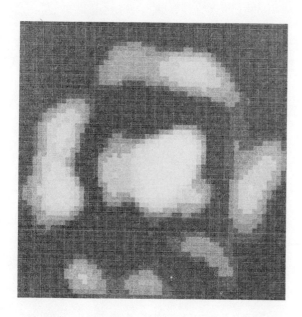

Figure 5.   Microwave of Image of Object in Figure 4

object distance (18 meters), λ is the wavelength 1.875 cm, and D is the length of an array side 3.5 meters. Rλ/D is 9 cm. For comparison, the object diameter is 38 cm, approximately four resolution cells.

A composite image was produced by non-coherently adding the intensities of nine coherent images. See Figure 6. The coherent images were formed from data measured for each of nine object orientations. These orientations correspond to two degree rotations in azimuth, elevation, or both about the central orientation, which was that for Figure 5. It is clear that the composite image has less speckle than the single coherent image. No effort was made to register the images about a common center; in fact, the summation of intensities was done by computer and only the composite picture was produced. More complex objects may give registration problems.

The images have a plausible physical explanation. The bright central region is a specular reflection from the

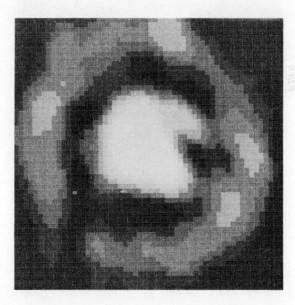

Figure 6. Enhanced Microwave Image of Object in Figure 4

vertex. The outer ring appears to be edge diffraction. We have not explored the effects of changing the radius of curvature of the paraboloid's rim.

A second composite image was formed. The object was a concave paraboloid with diameter 30 cm. It subtends approximately 3.3 resolution cells. Figure 7 shows a coherent image. Figure 8 shows a composite enhanced image formed by adding 6 coherent images that were formed for three azimuth & two elevation positions.

The images give the object dimensions accurately. This size determination requires knowing the distance between array & object. The relative sizes of images in Figures 5 and 7 are not in the ratio of the object diameters (30/38) only because of a difference in the plotting of the computed images.

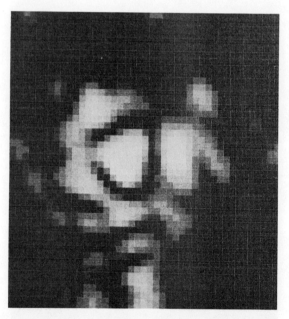

Figure 7. Microwave Image of 30 cm Diameter Paraboloid

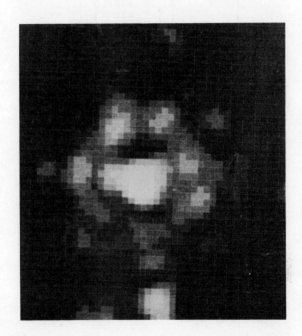

Figure 8. Enhanced Image of 30 cm Diameter Paraboloid

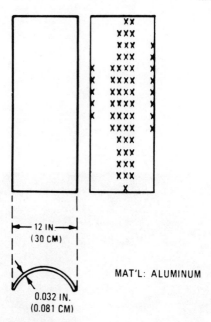

Figure 9. Hemicylinder and Image

    Several additional images have been formed but are
omitted for brevity. Flat objects such as discs or plates
produce good images. A convex hemi-cylinder gave a rather
interesting specular return and the image has two bands at
the edges. This image, shown in Figure 9 was plotted by a
computer line printer. A threshold was set at 16% of the
peak image intensity, and the printer plotted a point only
if the intensity exceeded the threshold. The data were
measured with a preliminary setup that had only one transmitting antenna, which was vertically displaced to a set of
fixed positions to synthesize the data array.

REFERENCES

1. G. L. Rogers, Jour. Atm. & Terr. Phys., 11, p. 51
    (1957).

2. N. Farhat & W. Gaurd, Proc. IEEE, 58, p. 1955 (1970).

3. O-C. Yue, E. L. Rope, G. Tricoles, IEEE Trans on Computers, to appear April, 1975.

4. P. F. Checcacci, V. Russo, A. M. Scheggi, Alta Frequenza, XXXVIII, p. 378 (1969).

5. P. F. Checcacci, V. Russo, A. M. Scheggi, Proc. IEEE, 56 p. 2165 (1968).

6. G. Tricoles & E. L. Rope, Digest IEEE G-AP Antennas & Propagation Symposium, 1971. p. 160.

7. R. P. Dooley, Proc. IEEE, 53, p. 1733 (1965).

8. G. Tricoles & E. L. Rope, J. Opt. Soc. Am. 56, p. 542A (M66); 57, p. 97 (1967).

9. Y. Aoki, Proc. IEEE, 56, p. 1402 (1968).

10. R. W. Larson, E. L. Johanson, J. S. Zelenka, Proc. IEEE, 57 p. 2162 (1969).

11. G. Papi, V. Russo, & S. Sothni, IEEE Trans. AP-19, p. 740 (1971).

12. N. Farhat, et al Acoustical Holography vol V, Plenum Press

13. W. Wells, Acoustical Holography, vol II, Plenum Press.

HOLOGRAPHIC APERTURE SYNTHESIS VIA A TRANSMITTER ARRAY*

P. N. Keating, R. F. Koppelmann, T. Sawatari,
and R. F. Steinberg
Bendix Research Laboratories, Southfield, Mich.

ABSTRACT

The problem of the synthesis of a large holographic
aperture by means of a smaller aperture and an array of
sources is addressed. Emphasis is placed on the problem of
parallel acquisition of the contributions associated with
the different sources and on a comparison between alternative superposition approaches (either superposition of holograms prior to reconstruction or superposition of reconstructed image fields). Experimental results are presented
in which synthetic-aperture images due to a 10 x 10 hologram
array and a 2 x 2 transmitter array are compared with images
due to (a) a 10 x 10 array and a single transmitter, and
(b) a 20 x 20 array and a single transmitter.

1. INTRODUCTION

Aperture synthesis[1] is a technique which can provide a
high resolution image from a small aperture, even with a
single detector, by making a tradeoff elsewhere in the system (e.g., by taking appreciable time to acquire the data,
as in side-looking radar[2]). Aperture synthesis by scanning
the source and detector is well-known, and aperture synthesis by allowing the object to move at a specific velocity

---
*Work supported in part by the Office of Naval Research under
Contract N00014-74-C-0030.

has also been investigated. In this paper, we are concerned
with aperture synthesis by source multiplexing.

Among the many synthetic-aperture configurations, the
Well's cross[3] is a convenient geometry to allow the synthesis of an N x N aperture from 2 N elements. In this arrangement, N transmitters and N receivers are aligned orthogonally
as linear arrays and no phase correction is required. Phase
corrections are required for any other configuration. From
a cost-effectiveness, however, the Well's cross configuration
is not necessarily an optimum one, for the cost of a transmitter element is generally more expensive than that of a
receiver element.

In the present paper, we discuss the implementation of
the synthetic aperture technique for an acoustic underwater-
viewing system[4] using a fixed transmitter array. In Section 2, the data processing needed to synthesize the final
images is discussed. One approach is to superpose the
separately reconstructed images while the other is to synthesize the hologram aperture before reconstruction. Mathematically, both methods are equivalent. The first method
has been used in preliminary studies to simplify compensation of phase and intensity uncertainties which are unavoidable in a real system.

In Section 3, two different approaches for simultaneous
data acquisition are analyzed. In the first approach, each
transmitter emits coherent radiation of a different frequency so that the received signals can be decomposed into
the separate sub-holograms associated with each transmitter.
The second approach is to use different, coded modulations
of a single carrier frequency from each transmitter so that
this decomposition can be effected. Comparison of the two
approaches indicates that the former method requires less
acquisition time. However, the latter method provides a
higher range resolution.

In the last section, an underwater experiment is described which demonstrates successful aperture synthesis
using a 10 x 10 receiver array and a 2 x 2 transmitter
array. The results are compared with (a) those of 20 x 20
receiver array with a single transmitter, and (b) those of
a 10 x 10 receiver array and a single transmitter.

2. SYNTHETIC APERTURE PROCESSING

In the Fresnel approximation, the hologram due to an object distribution $f(x)$ illuminated by a monochromatic source at a point $(\xi_0 \mp d_0)$ in the detector plane (see Figure 1) is

$$g_n(\underset{\sim}{u}) = \int d^2x \; f(\underset{\sim}{x}) e^{i\frac{k_0}{2R}(\underset{\sim}{x}-\underset{\sim}{\xi}_0-\underset{\sim}{d}_n)^2 + i\frac{k_0}{2R}(\underset{\sim}{x}-\underset{\sim}{u})^2} \tag{1}$$

apart from some unimportant factors. The reconstructed field for the case $d_0 = 0$ and a large aperture A can be written

$$F_0(\underset{\sim}{v}) = \int d^2x \; H(\underset{\sim}{v}-\underset{\sim}{x}) \; \mathcal{J}(\underset{\sim}{x}) \tag{2}$$

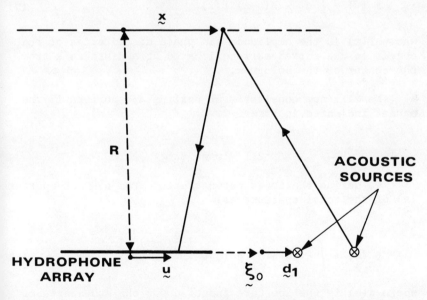

Figure 1.   Geometry of Transmitter and Receiver Arrays and Object Plane

where

$$H(\underset{\sim}{x}) = \int d^2 u \, A(\underset{\sim}{u}) e^{i\frac{k_o}{R}\underset{\sim}{u}\cdot\underset{\sim}{x}} \tag{3}$$

This is the point spread function associated with the large aperture and

$$\mathcal{H}(\underset{\sim}{x}) = f(\underset{\sim}{x}) e^{-i\frac{k_o}{R}(\underset{\sim}{v}^2-\underset{\sim}{x}^2)-i\frac{k_o}{R}\underset{\sim}{\xi}_o\cdot\underset{\sim}{x}+i\frac{k_o}{2R}\underset{\sim}{\xi}_o^2} \tag{4}$$

It is noted that the first phase of this equation is very small when $\underset{\sim}{v} = \underset{\sim}{x}$ and $H(\underset{\sim}{v}-\underset{\sim}{x})$ has a finite value only when $\underset{\sim}{v} \approx \underset{\sim}{x}$ especially for the large aperture $A(\underset{\sim}{u})$. Furthermore, ξ_o is an arbitrary constant (location of the transmitter). If we choose $\xi_o = 0$, then we obtain a well-known convolution expression for the image,

$$F_o(\underset{\sim}{v}) = \int d^2 x \, H(\underset{\sim}{v}-\underset{\sim}{x}) \, f(\underset{\sim}{x}) \tag{5}$$

where $f(\underset{\sim}{x})$ is the amplitude and phase distribution of the object on the curved surface located at a distance R from the center of the hologram.

We will now consider synthesizing an aperture in the manner indicated in Figure 2.

2.1   Image Sum

We define the image reconstructed from n^{th} sub-aperture (a) of the total aperture as:

$$f_n(\underset{\sim}{v}) = \int d^2 u \, a(\underset{\sim}{u}) \, g_n(\underset{\sim}{u}) e^{-i\frac{k_o}{2R}(\underset{\sim}{v}-\underset{\sim}{u})^2-i\frac{k_o}{2R}\underset{\sim}{v}^2} \tag{6}$$

where $a(u)$ is the aperture function for the sub-aperture. Hence

$$f_n(\underset{\sim}{v}) = \int d^2 x \, h(\underset{\sim}{v}-\underset{\sim}{x}) \, \mathcal{H}(\underset{\sim}{x}) e^{-i\frac{k_o}{R}\underset{\sim}{d}_n\cdot(\underset{\sim}{x}-\underset{\sim}{\xi}_o)+i\frac{k_o}{2R}\underset{\sim}{d}_n^2} \tag{7}$$

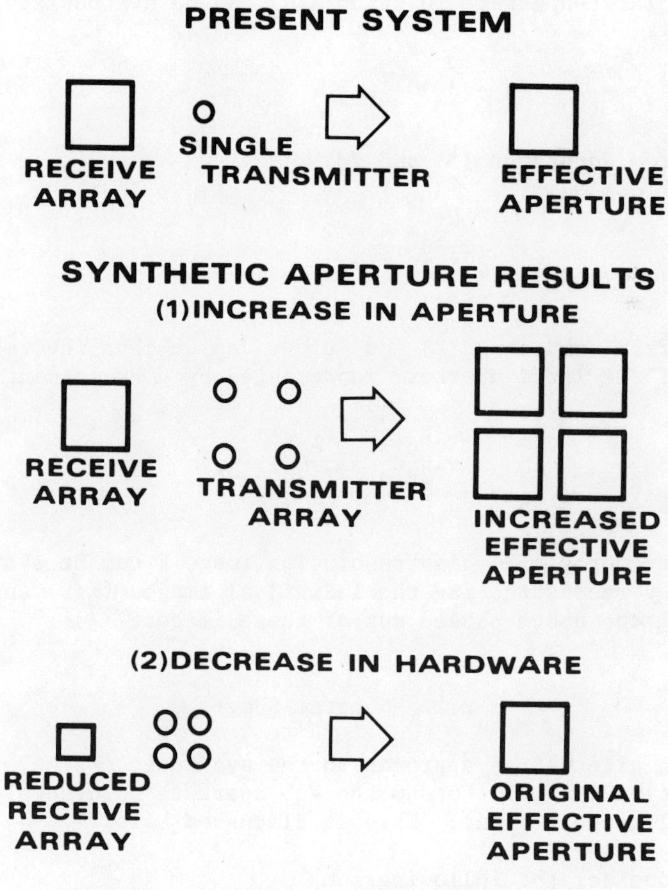

Figure 2. Synthetic Aperture Geometry.

where

$$h(\underset{\sim}{x}) = \int d^2 u \; a(\underset{\sim}{u}) e^{i\frac{k_o}{R}\underset{\sim}{u}\cdot\underset{\sim}{x}} \tag{8}$$

If the large-aperture function $A(\underset{\sim}{u})$ can be synthesized by a sum

$$A(\underset{\sim}{u}) = \sum_n a \; (\underset{\sim}{u} - \underset{\sim}{d}_n) \tag{9}$$

then from equations (3) and (8),

$$H(\underset{\sim}{x}) = h(\underset{\sim}{x}) \sum_n e^{i\frac{k_o}{R}\underset{\sim}{d}_n\cdot\underset{\sim}{x}} \tag{10}$$

Thus, from equations (2) and (6) we can achieve the resolution of the large aperture represented by H by forming the sum

$$F_1(\underset{\sim}{\nu}) = \sum_n f_n(\underset{\sim}{\nu}) e^{-i\frac{k_o}{2R}d_{\sim n}^2 + i\frac{k_o}{R}\underset{\sim}{d}_n(\underset{\sim}{\nu}-\underset{\sim}{\xi}_o)} \tag{11}$$

In other words, the high resolution image F can be synthesized by reconstructing the individual images $f_n(\underset{\sim}{\nu})$ and then forming the above phased sum of these images.

2.2 Hologram Sum

An alternative approach to the synthesis of the high resolution image is to sum the sub-aperture holograms at the holographic plane. This is discussed below.

Consider the following sum

$$F_2(\underset{\sim}{\nu}) = \sum_n \int d^2 u \; a(\underset{\sim}{u}-\underset{\sim}{d}_n) \; g_n(\underset{\sim}{u}-\underset{\sim}{d}_n) \; B_n(\underset{\sim}{u}-\underset{\sim}{d}_n)$$

$$e^{-i\frac{k_o}{2R}(\underset{\sim}{u}-\underset{\sim}{\nu})^2 - i\frac{k_o}{R}\nu^2} \tag{12}$$

where $B_n(\underset{\sim}{u})$ is a new quantity, to be determined. In order to make this expression (12) identical with that of equation (11), we have to have

$$B_n(\underset{\sim}{u}) = e^{i\frac{k_o}{R}\underset{\sim}{d}_n(\underset{\sim}{u}-\underset{\sim}{\xi}_o)} \tag{13}$$

Therefore, if we set

$$\tilde{g}_n(\underset{\sim}{u}) = a(\underset{\sim}{u}-\underset{\sim}{d}_n)\, g_n(\underset{\sim}{u}-\underset{\sim}{d}_n)e^{i\frac{k_o}{R}\underset{\sim}{d}_n(\underset{\sim}{u}-\underset{\sim}{\xi}_o-\underset{\sim}{d}_n)} \tag{14}$$

We can determine a large synthesized aperture hologram as

$$G(\underset{\sim}{u}) = \sum_n \tilde{g}_n(\underset{\sim}{u}) \tag{15}$$

The synthesized image is, therefore,

$$F_2(\underset{\sim}{v}) = \int d^2 u\, G(\underset{\sim}{u})\, e^{-i\frac{k_o}{2R}(\underset{\sim}{u}-\underset{\sim}{v})^2 - i\frac{k_o}{2R}\underset{\sim}{v}^2} \tag{16}$$

which is equivalent to equation (11).

In other words, the high resolution image F can also be synthesized by forming a larger hologram $G(\underset{\sim}{u})$ via the addition of shifted "sub-holograms" $a(\underset{\sim}{u}) \cdot g(\underset{\sim}{u})$, and reconstructing the large hologram.

2.3 Comparison of the Two Methods

In terms of intrinsic sensitivity to phase errors, there does not appear to be any important difference between the two above approaches. In terms of processing time, there is an appreciable difference since the image sum approach of 2.1 requires as many reconstructions as there are contributions, while the hologram sum of 2.2 requires only one. However, of greater importance in the preliminary studies is the question of adjusting the phases of the contributions before adding. In other words, there will probably be phase uncertainty in each of the image or hologram contributions due to uncertainties in propagation and alignment of the two arrays. The most convenient way of monitoring and removing them is of considerable importance, and for this reason, the image sum approach of 2.1 was implemented.

2.4 Coherence Requirements

Prior to any discussion of simultaneous data acquisition it is necessary to discuss the coherence requirement; i.e., the maximum bandwidth which can be used to differentiate different signals from the different transmitters.

The maximum path difference of the radiation between the source to the hologram plane is given [equation (1)] as

$$\Delta \ell = \frac{1}{2R} \left(\underset{\sim}{x}_1 - \underset{\sim}{\xi}_0 - \underset{\sim}{d}_{n1} \right)^2 + \frac{1}{2R} \left(\underset{\sim}{x}_1 - \underset{\sim}{u}_1 \right)^2 - \frac{1}{R} \underset{\sim}{x}_1^2 \qquad (17)$$

where $\underset{\sim}{x}_1$, $\underset{\sim}{u}_1$ and $\underset{\sim}{d}_{n1}$ are the values which make $\Delta \ell$ maximum within their limit, i.e., x_1 and u_1 are the coordinates of an edge of the field of view and hologram respectively. Note that the last spherical term is added since we defined the object plane as the spherical surface instead of the planar surface [see equation (4)]. Without losing generality, we can assume that $\underset{\sim}{\xi}_0 = 0$ and $\underset{\sim}{u}_1 < \underset{\sim}{d}_{n1} < \underset{\sim}{x}_1$. We can

approximate the above equation as

$$\Delta \ell = \frac{|d_{n1}| \ |x_1|}{R} \qquad (18)$$

This can be expressed in terms of the wavelength, λ : the total number of resolution elements of the sub-hologram, α, over the field of view and the number of transmitter elements, M.

$$\Delta \ell = \frac{\lambda \ \alpha \ \sqrt{M}}{4} \qquad (19)$$

In obtaining equation (18) from equation (19), the relations of $\delta = \lambda R/2\, d_{n1}$, (where δ is a resolution of the synthesized aperture) and $2\,|x_1|/\delta = \alpha\sqrt{M}$ are used. The minimum pulse length is then given by

$$T_s > \frac{2\Delta\ell}{c} = \frac{\lambda\alpha\sqrt{M}}{2c} \tag{20}$$

The corresponding maximum bandwidth Δk_o allowable is then given as

$$k_o = \frac{2\pi}{\lambda\alpha\sqrt{M}} \tag{21}$$

3. PARALLEL ACQUISITION OF INFORMATION

In the preceding section, two methods of synthesizing a high resolution image from sub-holograms "$a(\underset{\sim}{u})g_n(\underset{\sim}{u})$" corresponding to a different transmitter located at $(\underset{\sim}{\xi}_o + \underset{\sim}{d}_n)$ has been described. In the above analysis the sub-hologram $a(\underset{\sim}{u})\, g_n(\underset{\sim}{u})$ is assumed to be produced by a monochromatic wave; i.e., each transmitter radiates monochromatic wave of identical frequency k_o and width of Δk_o. It is, of course, possible to obtain such sub-holograms if the transmitters are activated sequentially. The sequential acquisition of the holographic information is, however, time-consuming and unsuitable for practical systems when moving targets are expected.

For simultaneous acquisition of the data, such transmitter signals must be coded so that the received signals can be decomposed into each sub-hologram. The selection of a proper code for simultaneous acquisition is very important for a practical application of the synthetic aperture method, since the image quality, dynamic range and other significant characteristics of the system are dependent upon the type of code we choose. Further, the cost of the system is subject to the method of the coding.

In this section two methods of coding: (1) different wavelength, and (2) modulation code on a carrier with a single frequency, are analyzed and their characteristics are compared.

3.1 Synthesis Using Frequency Coding

In this case, the different transmitter contributions are characterized by different acoustic wavelengths. If the received signals are decomposed by narrow bandpass filters, we have the holographic information corresponding to each transmitter,

$$g_n'(\underset{\sim}{u}) = \int d^2 x \; f(\underset{\sim}{x}) e^{i\frac{k_n}{2R}(\underset{\sim}{x}-\underset{\sim}{\xi}_o-\underset{\sim}{d}_n')^2 + i\frac{k_n}{2R}(\underset{\sim}{x}-\underset{\sim}{u})^2} \tag{22}$$

In a similar manner as equation (6), we define the reconstructed image from the sub-hologram as

$$f_n'(\underset{\sim}{v}) = \int d^2 u \; a(\underset{\sim}{u}) g_n'(\underset{\sim}{u}) e^{-i\frac{k_n}{2R}(\underset{\sim}{v}-\underset{\sim}{u})^2 - i\frac{k_n}{2R}\underset{\sim}{v}^2} \tag{23}$$

Note that, for the reconstruction process, we used the wave vector k_n instead of k_o.

We form a similar summation with that given in equation (11)

$$F_1'(\underset{\sim}{v}) = \sum_n f_n'(\underset{\sim}{v}) e^{-i\frac{k_n}{2R}\underset{\sim}{d}_n'^2 + i\frac{k_n}{R}\underset{\sim}{d}_n'(\underset{\sim}{v}-\underset{\sim}{\xi}_o)} \tag{24}$$

Comparison of the expressions of equation (11) and equation (24), lead to the choice of $\underset{\sim}{d}_n'$ to satisfy $k_n\underset{\sim}{d}_n' = k_o\underset{\sim}{d}_n$ and, furthermore, we can set $\underset{\sim}{\xi}_o = 0$ without losing generality. Then, in order to make the above summation [equation (24)] identical with that of equation (11) we have to have a condition given as

$$h(\underset{\sim}{v}-\underset{\sim}{x}) e^{-i\frac{k_o}{R}(\underset{\sim}{v}^2-\underset{\sim}{x}^2)} \underset{\sim}{\approx} h_n(\underset{\sim}{v}-\underset{\sim}{x}) e^{-i\frac{k_n}{R}(\underset{\sim}{v}^2-\underset{\sim}{x}^2)} \tag{25}$$

where $h(x)$ is given in equation (8) and

$$h_n(\underset{\sim}{x}) = \int d^2 u \, a(\underset{\sim}{u}) e^{i\frac{k_n}{R}\underset{\sim}{u}\cdot\underset{\sim}{x}} \qquad (26)$$

Both $h(x)$ and $h_n(x)$, the point-spread function of the sub-hologram image, have a significant value for $|x| < \delta_0$ ($; \delta_0$ is the resolution of the sub-hologram) and are small for $|x| > \delta_0$. The condition of equation (25) can, therefore, be expressed as

$$\Delta k < \frac{1}{\delta_0 \, \theta_0} \qquad (27)$$

where Δk is the bandwidth of the set $\{k_n\}$ with the central wave vector k_0 and θ_0 is the angular field of view. If we assume that the bandwidth occupied by each transmitted wave is $\Delta k_n = 2\Delta k/M$ (M = number of transmitter elements), then the condition given in the unequality (27) in terms of pulse length (total acquisition time) can be rewritten as

$$T_p > \frac{\pi R \, \lambda^2 \, M}{b^2 \, \alpha \, c} \qquad (28)$$

since $\Delta k_n = 2\pi/T_p \, c$, where $\delta_0 = \lambda R/b\alpha$ ($; b$ is the separation between the receiver elements) and $\theta_0 = \lambda/b$ have been used to obtain equation (28).

3.2 Modulation Coding

The use of different wavelengths for each transmitter appears to be an acceptable method for parallel data acquisition during aperture synthesis. There is, however, an alternative approach where each transmitter output is modulated by a set of codes so that the simultaneously received signals are electronically decomposed into each sub-hologram to perform the aperture synthesis. We now examine coding by a modulation on a single carrier frequency.

It is known[5] that, when the bandwidth of the radiation is much less than the mean frequency ($k_0 c$), the radiation of the source can be expressed in terms of an envelope-function $\bar{B}_n$;

$$S_n(t) = \bar{B}_n(t) e^{-ik_o ct} \tag{29}$$

The wave striking the holographic plane is then given as

$$\tilde{g}_n''(\underset{\sim}{u};t) = \int d^2 x \, f(\underset{\sim}{x}) S_n[t - \tau(\underset{\sim}{x})] \tag{30}$$

where

$$\tau \, [= R'/c; \quad R' = 2R + (\underset{\sim}{u}-\underset{\sim}{x})^2/2R + (\underset{\sim}{x}-\underset{\sim}{\xi}_o-\underset{\sim}{d}_n)^2/2R-\underset{\sim}{x}^2/R]$$

is a distance from the source to the hologram plane.

Therefore, the holographic information corresponding to the above wave is obtained by multiplying reference waves $\bar{A}*(t-t')e^{ik_o c(t-\tau')}$ and integrating it, where $\bar{A}_n(t)$ is an envelope function of the reference wave defined in a similar manner as that of equation (29) and τ' is the time delay of the reference wave. That is

$$g_n''(\underset{\sim}{u}) = \int d^2 x \, I_{nn}(\rho) f(\underset{\sim}{x}) e^{i\frac{k_o}{2R}[(\underset{\sim}{x}-\underset{\sim}{\xi}_o-\underset{\sim}{d}_n)^2+(\underset{\sim}{x}-\underset{\sim}{u})^2]} \tag{31}$$

where

$$I_{nn}(\rho) = \int \bar{A}_n^*(t) \, \bar{B}_n(t-\rho) \, dt$$

and

$$\rho = \tau-\tau'$$

$I_{nn}(\rho)$ is the correlation of $\bar{A}_n$ and $\bar{B}_n$. We are therefore discussing correlation decoding of the transmitter signals, and it is of course necessary to aim at a choice of the set $\bar{A}_n$, $\bar{B}_n$ which provides (a) the required orthogonality (i.e., $I_{nn'}$ is small when $n \neq n'$) and (b) the required drop-off in I_{nn} as ρ increases past the desired range.

The most important aspect of this result is that this approach provides a coherent hologram, even though the transmitted signal is broadband. Apart from the $I_{nn}(\rho)$ factor, equation (31) is the same expression as equation (1) of Section 2 in that the only wave-vector which is present in g''_n is k_0. Hence, the aperture can be synthesized in much the same way as in Section 2.1.

We shall now examine $I_{nn}(\rho)$ in a little more detail. Our study indicates that binary phase moculdation using a Gold code[6] is best suited for the present applications since the Gold code not only satisfied the above mentioned orthogonality requirement, but the electronic processing for both generation and decomposition is also rather simple. However, the problem of using a Gold code is that the code consists of a large number of bits and the length of a bit cannot be less than the minimum pulse length determined (in Section 2.4) for the coherence requirement. That is, the minimum length of one bit has to be longer than T_s to obtain acceptable holographic information. If the code consists of N_C bits, the total length of the pulse is greater than $N_C T_s$ which is larger than MT_s since in general $N_C > M$.

3.3 Comparison of the Methods

From the results obtained in the preceding section, we can compare these three methods (1) the uncoded transmission, (2) the frequency coding approach and (3) the modulation coding approach in terms of range resolution and total acquisition time. The results are summarized in Table I. For range resolution, the last method (3) is best [compared equations (20) and (28)]. On the other hand, the frequency coding method allows the shortest acquisition time, since T_p is likely to be less than MT_s in a practical system.

It should be noted here that, in the above discussion, the presence of cross-talk and other noises which degrade the image quantity have been completely ignored for simplicity of the discussion. Although we do not discuss the image degradation problem in the present paper, our study on the subject indicates that the conclusion obtained above remains much the same in the presence of the noise.

Table I

Approach	Range Resolution	Acquisition Time	Equations to be referred in the text
Uncoded Transmission	CMT_s	MT_s	eq. (20)
Frequency Coding	CT_p	T_p	eq. (28)
Modulation Code	CT_s	$N_c T_s$	eq. (20)

NOTE: $T_s = \lambda \alpha \sqrt{M}/2c$ and $T_p = \pi R \lambda^2 M/b^2 \alpha c$ and λ = wavelength; α = number of resolution element of subhologram; M = number of transmitter elements; R = range; b = element spacing in receiver array; c = velocity of acoustical wave and N_c = number of bits in Gold Code.

Furthermore, it should be pointed out that, in the frequency-coding approach, we have taken the Fourier transform of each sub-hologram using k_n instead of k_0 [see equation (23)]. This approach makes the direct use of Fast Fourier Transform (FFT) difficult so that computer processing times increase significantly. If we limit ourself to a region where the FFT is applicable, the total acquisition time of T in equation (28) becomes identical to that of the sequential acquisition (MT_s).

4.   EXPERIMENT

An experiment was conducted to synthesize an aperture using the Underwater Viewing System (UWVS)[4,7], and to compare the images obtained with this aperture with those obtained from a single large-aperture hologram.

The UWVS consists of a 20 x 20 receiver array (element spacing 9 mm) and a 2 x 2 transmitter array (element spacing 9 cm). In the first experiments, a 10 x 10 portion of the 20 x 20 array was used to produce four 20 x 20 complex sub-holograms with the data being sequentially collected. The result can then be compared with the result from the entire real 20 x 20 receiver array and a single transmitter.

The UWVS is essentially similar to the system used in previous experiments.[4],[7] It was operated at a depth of 5 feet in a pool which is 13.5 feet wide, 45 feet long, and 10 feet deep. The transmitters radiate an acoustic wave which can be pulse modulated CW or biphase-coded by a Barker code of up to 13 bits. The pulse length is 5.2 msec, providing range resolution of about 4 cm for the coded pulse. The code is not essential to this experiment since sequential data acquisition is used. However, for simultaneous acquisition, described in the preceding section, the Barker code is an important element and was therefore used in these experiments.

The acoustic radiation is reflected from a target located 5 feet from the transmitter-receiver plane. The return signal received by the 20 x 20 array is multiplied electrically with a reference wave to obtain the holographic information. Interfering returns (e.g., surfaces within the pool) are rejected by range-gating circuitry. The received signal is processed by a computer using the "image-sum" method to synthesize the image.

The resultant image is displayed on a 40 x 40 array. This[9] is necessary to properly display the image from a 20 x 20 receiver array since, if the spatial frequency bandwidth of the amplitude and phase distribution is W, the intensity distribution is 2W.

Prior to image synthesis, calibration data are obtained from each sub-hologram to compensate for phase and intensity variations between the transmitters. Each hologram is calibrated as follows. First, four sub-holograms of a single "point" target (a 3 cm sphere) are obtained. The intensity at the point where the intensity is a maximum is determined from each sub-hologram. This data is used to compensate the sub-hologram correspond to more general object distributions. After this process, the sub-holograms are reconstructed and then superposed, finally being displayed.

Experimental Results. Figures 3, 4, and 5 are typical examples of the reconstructed imagery. The target was two small glass-shelled spheres (3 cm in diameter) with a separation of 4 cm. Figure 3 shows the reconstructed images

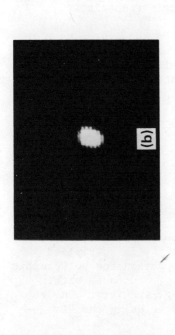

(b)

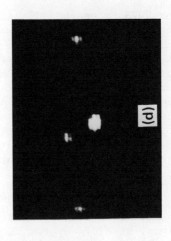

(d)

(a)

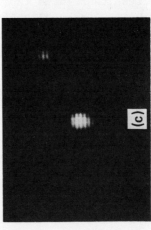

(c)

Figure 3. Reconstructed Images from 10 x 10 Receiver Array Holograms Produced by Radiation from Different Transmitters.

Figure 4. Reconstructed Image from 20 x 20 Receiver Array
Holograms Produced by Radiation from Different
Transmitters.

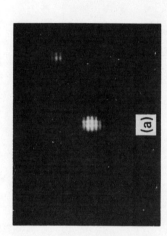

Figure 5. Reconstructed Images by 10 x 10 Receiver Array Hologram (Top), by Synthetic Aperture (20 x 20) Hologram (Left) and by 20 x 20 Receiver Array Hologram (Right) with a Single Transmitter.

obtained from each sub-hologram (10 x 10 array) separately.
The lateral resolution expected with this aperture is 8 cm
in terms of the Rayleigh criterion[10] (for simplicity in the
discussion, the Rayleigh criterion is used as the resolution
although the system uses a coherent or quasi-coherent
radiation).

Figure 3 shows that the two points are unresolved by
this limited aperture. Figure 4 shows the reconstructed
images of the same target using a real 20 x 20 array hologram and each of the 4 transmitters. The resolution of this
system is 4 cm. Photographs (c) and (d) show a well-
resolved image. However, photographs (a) and (b) indicate
that the two points are not separated. This is due to the
fact that, under coherent illumination, the resolvability
of the two adjacent objects depends on the phase difference
between the two points.

Figure 5(b) shows the synthesized image formed by the
coherent superposition [see equation (11)] of the four images
shown in Figure 3. Note that the object distribution is, in
general, a function of the location of the transmitter. A
successfully synthesized image should be equivalent to the
image obtained by a single 20 x 20 array hologram with the
transmitter located at a position which would provide the
result shown in photograph (c) [same as Figure 4(c)].

In order to make a quantitative comparison, intensity
distributions in the images are plotted in Figures 6 and 7.
Figure 6 and Figure 7 are the horizontal and vertical sections of the distributions; (b) and (c) in both figures are
those of the synthesized image and the large real aperture,
respectively. These intensity distributions are normalized
in such a manner that the total intensity over the vertical
section is identical. The two distributions indicate good
agreement. The traces of (a) in Figures 6 and 7 are the
distributions obtained by a single 10 x 10 hologram. The
two points are clearly not resolved with this limited
aperture.

These results indicate that aperture synthesis is a
promising approach to obtain high-resolution images with
smaller receiver apertures and, hopefully, with significant
cost savings.

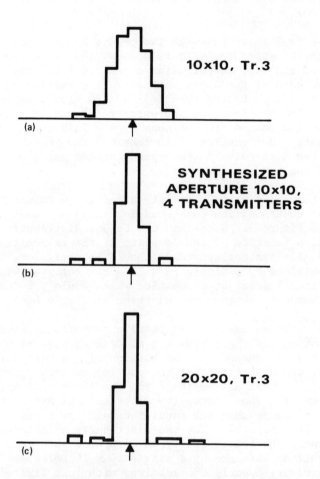

Figure 6. Intensity Distribution of the Reconstructed
Images Along a Horizontal Cross Section; (a) 10
x 10 Array, (b) Synthetic Aperture (20 x 20)
and (c) 20 x 20 Array.

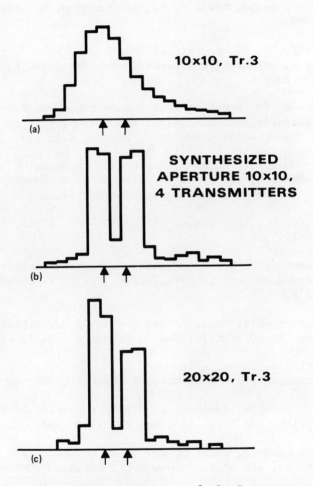

10x10, Tr.3

(a)

SYNTHESIZED
APERTURE 10x10,
4 TRANSMITTERS

(b)

20x20, Tr.3

(c)

Figure 7.   Intensity Distribution of the Reconstructed
            Image Along a Vertical Cross Section; (a) 10
            x 10, (b) Synthetic Aperture (20 x 20) and
            (c) 20 x 20 Array.

REFERENCES

(1) Mueller, R., Proc. IEEE, Vol. 59, 9, 1319, 1971.

(2) For example, Koch, W. E., Electronics, October 12, 1970.

(3) Wells, W. H., Acoustical Holography, Vol. 2, ed. by A. I. Metherell and L. Larimore, New York, Plenum, 87, 1970.

(4) Marom, E., Mueller, R. K., Koppelmann, R. F., and Zilinskas, G., Acoustical Holography, Vol. 3, ed. by A. F. Metherell, New York, Plenum, 191, 1971.

(5) For example, Goldman, S., Optical Processing of Information, ed. by D. K. Pollack, C. J. Koester and J. T. Tippett, Baltimore, Spartan, 31, 1963.

(6) Gold, R., IEEE Trans. Inform. Theory, IT-13, 619, 1967.

(7) Keating, P. N., Koppelmann, R. F., and Mueller, R. K., Acoustical Holography, Vol. 6, New York, Plenum Press, 1975.

(8) For example, Gold, B. and Rader, C. M., Digital Processing of Signals, New York, Mc-Graw Hill, 1969.

(9) Private communication with R. K. Mueller, Eng. Dept., University of Minnesota, Minnesota.

(10) For example, Goodman, J. W., Introduction to Fourier Optics, San Francisco, McGraw-Hill, 1968.

(11) For example, Born, M. and Wolf, E., Principle of Optics, 4th edition, New York, Pergammon Press, 1970.

SYNTHETIC APERTURE APPROACH TO MULTI-BEAM SCANNING

ACOUSTICAL IMAGING

Kazuhiko Nitadori

Oki Electric Industry Co., Ltd.

Hachioji, Japan 193

An acoustical imaging method is described which realizes high resolution imaging with relatively small number of transmitters and receivers on the basis of a synthetic aperture technique, but does not require a severe phase stability. It uses a sparse planar transmitting array for generating narrow transmitting multi-beams and a dense planar receiving array for resolving the transmitted multi-beams into independent pencil beams and scans the transmitting multi-beams electronically to obtain the whole picture elements of an image. It differs from the usual synthetic aperture method only in that the transmitters are used for beamforming instead for sequential scanning. Under ideal circumstances, both methods are almost equivalent. But, under practical circumstances, the present method has many advantages over the usual method as the following: i) unstability of a transmitter/receiver platform and a medium causes less influences on resolution and image dynamic range, ii) no speckle noise occurs, iii) memory capacity and processing time required for signal processing can be saved, and iv) higher acoustic power can be transmitted under the cavitation limited environment.

INTRODUCTION

In a holographic acoustical imaging method, the number of picture elements of a reconstructed image coincides approximately with the number of receivers arranged within a receiving aperture,[1] so that a large number of receivers must be used to obtain high resolution images. In order to

reduce the amount of the receivers without degrading resolution, the synthetic aperture technique is used,[2],[3] in
which the transmitting signals are radiated succesively from
each transmitter in a transmitting array, the reflected
echoes from an object are received with a receiving array,
and the received signals are stored during the whole scanning period, then synthesized to obtain a large effective
receiving aperture. By the technique we can obtain as many
picture elements as the product of the number of transmitters
times the number of receivers. In order for this to be possible, however, a transmitter/receiver platform, a transmission medium, and an object must be stationary during the
whole transmitter scanning period, which may be restricting
the applications of this technique.

    As a method for overcoming the fault of the usual syn-
thetic aperture method, which we refer as a transmitter
scanning method or TS method, for simplicity, we propose
here a multi-beam scanning method or MBS method, which uses
the same transmitting and receiving array as the TS method,
but uses the transmitting array for beamforming instead using the transmitters individually and sequentially. Under
ideal circumstances, the present method is almost equivalent
to the TS method. But, under practical circumstances, it
has many advantages over the TS method.

                    SYNTHETIC APERTURE TECHNIQUE

    We first introduce the theory of synthetic aperture
technique for a fixed transmitting and receiving array pre-
paratory to deriving the MBS imaging method. Suppose transmitters and receivers are arranged on the same plane (T/R
plane) and an object is placed on the object plane that is
parallel to and at a distance z from the T/R plane, as shown
in Fig. 1. Each of the transmitters and the receivers are
assumed to have the identical pupil functions $g_t(x,y)$ and
$g_r(x,y)$,respectively, and to be centered at (u_t,v_t) and
(u_r,v_r) on the T/R plane, respectively. Let $S_t(u_t,v_t)$ denote
the complex amplitude of the driving signal for the transmitter at coordinates (u_t,v_t) and $S_r(u_r,v_r;u_t,v_t)$ denote the
complex amplitude of the corresponding received signal by
the receiver at coordinates (u_r,v_r).

    When the transmitting signal $S_t(u_t,v_t)$ is radiated, the
complex field of the sound at a point (x_o,y_o) on the object

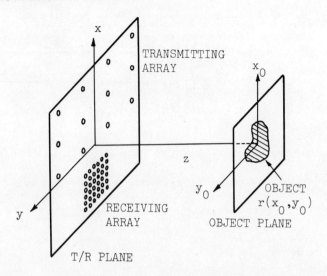

Fig. 1. Geometry of synthetic aperture imaging method.

plane is

$$U_O(x_O,y_O;u_t,v_t) = A(z) \iint g_t(x,y)S_t(u_t,v_t)$$
$$\exp\left\{i\,\frac{k}{2z}\left[(x+u_t-x_O)^2 +(y+v_t-y_O)^2\right]\right\} dxdy \qquad (1)$$
$$A(z) = \frac{1}{i\lambda z}\exp(ikz)$$

where k is a wave number, λ is a wavelength, and the Fresnel approximation is assumed to hold. Let $r(x_O,y_O)$ denote the complex reflectance of the object on the object plane, then the complex field of the sound scattered from the object and returned to the T/R plane is represented by

$$U_r(x,y;u_t,v_t) = A(z) \iint r(x_O,y_O)U_O(x_O,y_O;u_t,v_t)$$
$$\exp\left\{i\,\frac{k}{2z}\left[(x-x_O)^2+(y-y_O)^2\right]\right\}^2 dx_O dy_O \qquad (2)$$

Thus, the received signal is expressed as

$$S_r(u_r,v_r;u_t,v_t) = \iint g_r(x,y)U_r(x+u_r,y+v_r;u_t,v_t)dxdy \qquad (3)$$

The image of the object $r(x_O,y_O)$ can be reconstructed by carrying out the operation

$$f(x_i,y_i) = \sum_{(U_t,V_t)} \sum_{(U_r,V_r)} S_r(u_r,v_r;u_t,v_t)$$

$$\exp\left\{- i\frac{k}{z}\left[\left(x - \frac{u_t+u_r}{2}\right)^2 + \left(\frac{u_t-u_r}{2}\right)^2\right.\right.$$
$$\left.\left. + \left(y_i - \frac{v_t+v_r}{2}\right)^2 + \left(\frac{v_t-v_r}{2}\right)^2\right]\right\} \quad (4)$$

for the received signals $S_r(u_r,v_r;u_t,v_t)$,[3] where (x_i,y_i) is image coordinates.

Substituting Eqs.(1)$\sim$(3) into Eq.(4), we obtain

$$f(x_i,y_i) = A(z)^2 a(x_i,y_i) * \iint r(x_o,y_o)$$

$$h_t(x_i,y_i;x_o,y_o)h_r(x_i,y_i;x_o,y_o)dx_o dy_o \quad (5)$$

where

$$a(x,y) = \exp\left\{i\ \frac{k}{z}\ (x^2+y^2)\right\} \quad (6)$$

$$h_t(x_i,y_i;x_o,y_o) = \iint g_t(x,y)\exp\left\{i\ \frac{k}{2z}\left[(x-x_o)^2 + (y-y_o)^2\right]\right\}$$

$$P_t((x_o-x_i-x)/\lambda z,(y_o-y_i-y)/\lambda z)dxdy \quad (7)$$

$$h_r(x_i,y_i;x_o,y_o) = \iint g_r(x,y)\exp\left\{i\ \frac{k}{2z}\left[(x-x_o)^2 + (y-y_o)^2\right]\right\}$$

$$P_r((x_o-x_i-x)/\lambda z,(y_o-y_i-y)/\lambda z)dxdy \quad (8)$$

$$P_t(f_x,f_y) = \sum_{(U_t,V_t)} S_t(u_t,v_t)\exp\left\{-i2\pi(u_t f_x+v_t f_y)\right\} \quad (9)$$

$$P_r(f_x,f_y) = \sum_{(U_r,V_r)} \exp\left\{-i2\pi(u_r f_x+v_r f_y)\right\} \quad (10)$$

and the asterisk denotes complex conjugate. Equation (5) has the form of a superposition integral with the weighting function $h_t(x_i,y_i;x_o,y_o)h_r(x_i,y_i;x_o,y_o)$. Hence, if the weighting function has a sharp peak at $(x_i=x_o,y_i=y_o)$, $f(x,y)$ becomes a good image of $r(x,y)$.[4] The width of the peak of the weighting function determines the resolution of the imaging method.

Now assume both of the pupils of the transmitter and the receiver are sufficiently small compared with the wavelength λ, thus $g_t(x,y)$ and $g_r(x,y)$ are approximated by a delta function, then Eqs.(7) and (8) are written as

$$h_t(x_i,y_i;x_o,y_o) = a(x_o,y_o)^{1/2}\ P_r((x_o-x_i)/\lambda z,$$
$$(y_o-y_i)/\lambda z) \quad (11)$$

$$h_r(x_i,y_i;x_o,y_o) = a(x_o,y_o)^{\frac{1}{2}} \; P_r((x_o-x_i)/\lambda z,$$

$$(y_o-y_i)/\lambda z) \quad (12)$$

respectively. Thus, Eq.(5) is represented by

$$f(x_i,y_i) = A(z)^2 a(x_i,y_i) * \iint r(x_o,y_o)a(x_o,y_o)$$

$$P_t((x_o-x_i)/\lambda z,(y_o-y_i)/\lambda z)P_r((x_o-x_i)/\lambda z,$$

$$(y_o-y_i)/\lambda z)dx_o dy_o \quad (13)$$

This relation indicates $f(x,y)a(x,y)$ becomes the image of $r(x,y)a(x,y)$ with the weighting function $P_t((x_o-x_i)/\lambda z, (y_o-y_i)/\lambda z)P_r((x_o-x_i)/\lambda z,(y_o-y_i)/\lambda z)$. $P_t(f_x,f_y)$ and $P_r(f_x,f_y)$, defined by Eqs.(9) and (10), are the two-dimensional Fourier transforms of the transmitting and the receiving aperture, respectively, assuming the transmitters and the receivers are sufficiently small, hence $P_t(x/\lambda z,y/\lambda z)$ and $P_r(x/\lambda z,y/\lambda z)$ represent the directional patterns of the transmitting and the receiving array, respectively. The impulse response of the imaging method is the product of the transmitting and the receiving directional pattern. By the convolution theorem in Fourier transformation theory, this composite impulse response agrees with the directional pattern of the synthesized array obtained by spatially convolving the transmitting array with the receiving array. Hence, the aperture synthesis is realized.

Next we examine the effect of the pupils of the transmitters and the receivers. Assume the Fraunhofer approximation is valid in the pupils of every transmitter and receiver, then Eqs.(7) and (8) are written as

$$h_t(x_i,y_i;x_o,y_o) = a(x_o,y_o)^{\frac{1}{2}} \sum_{(u_t,v_t)} \exp\left\{-i\frac{k}{z}\left[(x_o-x_i)u_t\right.\right.$$

$$\left.\left.+(y_o-y_i)v_t\right]\right\}G_t((x_o-u_t)/\lambda z,(y_o-v_t)/\lambda z) \quad (14)$$

$$h_r(x_i,y_i;x_o,y_o) = a(x_o,y_o)^{\frac{1}{2}} \sum_{(u_r,v_r)} \exp\left\{-i\frac{k}{z}\left[(x_o-x_i)u_r\right.\right.$$

$$\left.\left.+(y_o-y_i)v_r\right]\right\}G_r((x_o-u_r)/\lambda z,(y_o-v_r)/\lambda z) \quad (15)$$

respectively, where

$$G_t(f_x,f_y) = \iint g_t(x,y)\exp\left\{-i2\pi(xf_x+yf_y)\right\} dxdy \quad (16)$$

$$G_r(f_x,f_y) = \iint g_r(x,y)\exp\left\{-i2\pi(xf_x+yf_y)\right\}dxdy \qquad (17)$$

are the Fourier transforms of the transmitter and the receiver pupil function, hence represent the directional patterns
of the individual transmitter and the individual receiver,
respectively. Equations (14) and (15) indicate that the
weighting function is shaded by the directional patterns
$G_t(f_x,f_y)$ and $G_r(f_x,f_y)$ of the transmitter and the receiver
according as the angles at the transmitters and the receivers subtended by the object are increasing. This means the
field of view of the imaging system is limited by the directional characteristics of the transmitters and the receivers.
By this effect, it becomes possible to suppress the aliasing
phenomenon, the influence of the sidelobes of the directional patterns of the transmitting and receiving array. This
effect may be indicated more clearly by making approximations
$x_0 \gg u_t, u_r$ and $y_0 \gg v_t, v_r$ for every (u_t, v_t) and (u_r, v_r) in Eqs.
(14) and (15) to obtain

$$h_t(x_i,y_i;x_0,y_0) = a(x_0,y_0)^{\frac{1}{2}} \quad G_t(x_0/\lambda z,y_0/\lambda z)$$

$$P_t((x_0-x_i)/\lambda z,(y_0-y_i)/\lambda z) \qquad (18)$$

$$h_r(x_i,y_i;x_0,y_0) = a(x_0,y_0)^{\frac{1}{2}} \quad G_r(x_0/\lambda z,y_0/\lambda z)$$

$$P_r((x_0-x_i)/\lambda z,(y_0-y_i)/\lambda z) \qquad (19)$$

and making a comparison with Eqs.(11) and (12). When the
pupil functions $g_t(x,y)$ and $g_r(x,y)$ are symmetric with respect to the origin, $G_t(f_x,f_y)$ and $G_r(f_x,f_y)$ become real,
thus the shading in Eqs.(14) and (15) becomes amplitude
shading.

As a typical example of the transmitting and receiving
array, we consider the rectangular planar array consisting
of $M_x \times M_y$ transmitters and $N_x \times N_y$ receivers. Suppose the coordinates of the center of the transmitters and the recei-
vers are

$$u_t = u_{to} +mL_x \qquad (m=0,1,\ldots,M_x-1)$$

$$v_t = v_{to} +nL_y \qquad (n=0,1,\ldots,M_y-1)$$

$$(20)$$

and

$$u_r = u_{ro} + mL_x/N_x (m=0,1,\ldots,N_x-1)$$

$$v_r = v_{ro} + nL_y/N_y (n=0,1,\ldots,N_y-1)$$

(21)

respectively, and no transmitter shading is used, that is, $S_t(u_t,v_t)=1$ for all (u_t,v_t), then the transmitting and the receiving beam pattern are written, respectively, as

$$P_t(f_x,f_y) = \exp\{-i2\pi(u_{to}f_x+v_{to}f_y)\}$$

$$\frac{1-\exp(-i2\pi M_x L_x f_x)}{1-\exp(-i2\pi L_x f_x)} \quad \frac{1-\exp(-i2\pi M_y L_y f_y)}{1-\exp(-i2\pi L_y f_y)}$$

(22)

$$P_r(f_x,f_y) = \exp\{-i2\pi(u_{ro}\ f_x+v_{ro}\ f_y)\}$$

$$\frac{1-\exp(-i2\pi L_x f_x)}{1-\exp(-i2\pi L_x f_x/N_x)} \cdot \frac{1-\exp(-i2\pi L_y f_y)}{1-\exp(-i2\pi L_y f_y/N_y)}$$

(23)

Hence, the composite beam pattern of the transmitting and receiving array is

$$P_t(f_x,f_y)\ P_r(f_x,f_y) = \exp\{-i2\pi[(v_{to}+u_{ro})f_x+(v_{to}$$

$$+v_{ro})f_y]\}$$

$$\frac{1-\exp(-i2\pi M_x L_x f_x)}{1-\exp(-i2\pi L_x f_x/N_x)} \cdot \frac{1-\exp(-i2\pi M_y L_y f_y)}{1-\exp(-i2\pi L_y f_y/N_y)}$$

(24)

which agrees with the beam pattern of the rectangular planar array consiting of $(M_x \times N_x) \times (M_y \times N_y)$ elements arranged with regular internals L_x/N_x and L_y/N_y along x and y direction, respectively.

TRANSMITTER SCANNING IMAGING METHOD (TS METHOD)

A usual procedure for performing the image reconstruction operation Eq.(4) is to scan each transmitter in sequence for transmitting the signal $S_t(u_t,v_t)=1$, to store the received signals corresponding to each transmitted pulse signal, and to carry out the operation Eq.(4) for the stored received signals after the transmitter scanning is completed. Since Eq.(4) is written as

$$f(x_i,y_i) = a(x_i,y_i)^* \sum_{(u_t,v_t)} \sum_{(u_r,v_r)} S_r(u_r,v_r;u_t,v_t)$$

$$\exp\left\{-i\ \frac{k}{2z}\left[u_t^2\ +v_t^2\ +u_r^2\ +v_r^2\ -2x_i\right.\right.$$

$$\left.\left.(u_t\ +u_r)-2y_i(v_t\ +v_r)\right]\right\} \quad (25)$$

the image $f(x_i,y_i)$ is calculated by the simple operation

$$f(x_i,y_i) = a(x_i,y_i)^*\ \sum_{(u,v)} S_{sy}(u,v)\exp\left\{i\frac{k}{z}\right.$$

$$\left.(x_iu+y_iv)\right\} \quad (26)$$

where the summation is taken over all combinations of

$$u = u_t+u_r, \quad v = v_t\ +v_r$$

and $S_{sy}(u,v)$ is synthesized from the received signals by the rule

$$S_{sy}(u_t\ +u_r,v_t\ +v_r) = S_r(u_r,v_r;u_t,v_t)$$

$$\exp\left\{-i\ \frac{k}{2z}\ (u_t^2\ +v_t^2\ +u_r^2\ +v_r^2\ )\right\} \quad (27)$$

The summation in Eq. (26) is the two-dimensional Fourier transform of the synthesized signal $S_{sy}(u,v)$. It becomes the two-dimensional DFT (discrete Fourier transform) when the rectangular planar array with coordinates expressed by Eqs. (20) and (21) is used, hence it is calculated efficiently by the use of the FFT (fast Fourier transform) algorithm.

A flow chart indicating the imaging procedure in the TS method is shown in Fig.2.

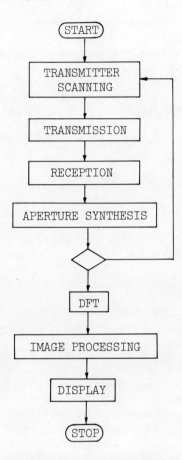

Fig. 2. Image reconstruction procedure in TS method.

MULTI-BEAM SCANNING IMAGING METHOD   (MBS METHOD)

Now we describe a new procedure for obtaining the image $f(x_i,y_i)$, which we call MBS method. We first rewrite the imaging operation Eq.(4) as

$$f(x_i,y_i) = a(x_i,y_i)* \sum_{(u_r,v_r)} S_c(u_r,v_r;x_i,y_i)$$
$$\exp\left\{- i \frac{k}{2z} (u_r^2 +v_r^2 -2x_iu_r-2y_i v_r)\right\} \qquad (28)$$

where

$$S_c(u_r,v_r;x_i,y_i) = \sum_{(u_t,v_t)} S_r(u_r,v_r;u_t,v_t)$$
$$\exp\left\{-i \frac{k}{2z} (u_t^2 +v_t^2-2x_iu_t-2y_iv_t)\right\} \qquad (29)$$

Equation (29) is interpreted as the received signal when the transmitting signals

$$S_t(u_t,v_t;x_i,y_i) = \exp\left\{-i \frac{k}{2z} (u_t^2 +v_t^2-2x_iu_t-2y_iv_t)\right\}(30)$$

are transmitted simultaneously from every transmitter. The transmitted signal

$$\sum_{(u_t,v_t)} S_t(u_t,v_t;x_i,y_i)$$

forms the transmitting beam focused at a point (x_i,y_i) on the object plane. The image $f(x_i,y_i)$ at the point (x_i,y_i) is obtained by carrying out the operation Eq.(28) for the received signals $S_c(u_r,v_r;x_i,y_i)$ corresponding to the transmitted signal $\Sigma S_t(u_t,v_t;x_i,y_i)$. The whole image can be formed by transmitting successively the pulse signals with the transmitting beams focused at each point (x_i,y_i) on the object plane and performing the above signal processing for the corresponding received signals.

In general, as many pulse signals as the whole number of picture elements need to be transmitted for obtaining an image. By properly selecting the transmitting array, however, it becomes possible to generate multi-beams focused at many points on the object plane, thus to resolve many picture elements by a single pulse transmission and to reduce the number of transmission pulses required to form an image.

Whether many picture elements can be resolved by a single transmission depends on the functional form of the received signals $S_c(u_r,v_r;x_i,y_i)$ in Eq.(29) with respect to (x_i,y_i). Substituting Eqs.(1) ~ (3) into Eq.(29), we obtain

$$S_c(u_r,v_r;x_i,y_i) = A(z)^2 \iint r(x_0,y_0)h_t(x_i,y_i;x_0,y_0)$$
$$dx_0dy_0 \left[\iint g_r(x,y)\exp\left\{ i \frac{k}{2z}\left[(x+u_r-x_0)^2+(y+v_r-y_0)^2\right]\right\} dxdy \right] \quad (31)$$

where $h_t(x_i,y_i;x_0,y_0)$ is in Eq.(7) in which $P_t(f_x,f_y)$ is replaced by

$$P_t(f_x,f_y) = \sum_{(u_t,v_t)} \exp\left\{-i2\pi(u_t f_x+v_t f_y)\right\} \quad (32)$$

Thus, the only term $h_t(x_i,y_i;x_0,y_0)$ depends on (x_i,y_i) in the right hand side of Eq.(31). Suppose (x_i,y_i) denote the image coordinates in the imaging operation Eq.(28) and (x_i',y_i') denote the steering coordinates in the transmitting beamforming operation Eq.(30), then the image obtained by the operation Eq.(28) is written as

$$f(x_i,y_i) = A(z)^2 a(x_i,y_i)* \iint r(x_0,y_0)h_t(x_i',y_i';x_0,y_0)$$
$$h_r(x_i,y_i;x_0,y_0)dx_0dy_0 \quad (33)$$

in the same way as in Eq.(5). In order for multiple picture elements to be resolved by a single transmission, it is necessary to hold

$$h_t(x_i,y_i;x_0,y_0) = h_t(x_i',y_i';x_0,y_0) \quad (34)$$

for all (x_0,y_0) and for different (x_i,y_i) and (x_i',y_i'). For this to be possible, it is sufficient that the function

$$P_t((x_0-x_i-x)/\lambda z,(y_0-y_i-y)/\lambda z) \quad (35)$$

in Eq.(7) is periodic with respect to (x_i,y_i), namely $P_t(f_x,f_y)$ in Eq.(32) is periodic over a two-dimensional plane (f_x,f_y). The rectangular planar array mentioned above is a typical example of the transmitting array with such a property.

Suppose $M_x\times M_y$ transmitters are placed at the coordinates (u_t,v_t) expressed by Eq.(20), then the transmitting beam pattern $P_t(f_x,f_y)$ represented by Eq.(22) is periodic with respect to f_x and f_y with periods $1/L_x$ and $1/L_y$, respectively, provided the origin of the coordinates is selected so that u_{to} and v_{to} are multiples of L_x and L_y, respectively. Then, Eq.(35) is periodic with respect to x_i ,y_i with periods

$$\Delta_x = \lambda z/L_x, \quad \Delta_y = \lambda z/L_y \tag{36}$$

respectively, and Eq.(34) holds for all such (x_i,y_i) that are represented by

$$x_i = x_i' + m\Delta_x, \quad y_i = y_i' + n\Delta_y \tag{37}$$

where m and n are arbitrary integers. Hence, by the single transmission steered at (x_i',y_i'), all the picture elements at the points (x_i,y_i) expressed by Eq.(37) can be resolved at the same time. The amplitude response of the transmitting beam pattern $P_t(x/\lambda z,y/\lambda z)$ is then

$$\left| P_t(x/\lambda z,y/\lambda z) \right| = \left| \frac{\sin M_x \pi x/\Delta_x}{\sin \pi x/\Delta_x} \quad \frac{\sin M_y \pi y/\Delta_y}{\sin \pi y/\Delta_y} \right| \tag{38}$$

The cross section along y axis is shown in Fig.3.

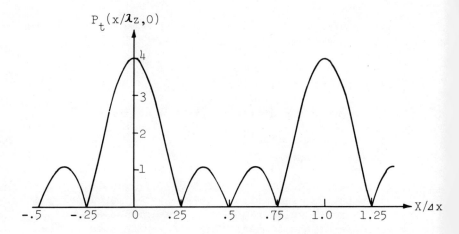

Fig.3. Cross section of the amplitude response of the transmitting beam pattern of a rectangular array $(M_x=4)$.

Also suppose $N_x \times N_y$ receivers are placed at the coordinates (u_r, v_r) expressed by Eq.(21), then the imaging operation Eq.(28) becomes

$$f(x_i + m\Delta_x, y_i + n\Delta_y) = a(x_i + m\Delta_x, y_i + n\Delta_y)^*$$

$$\exp\left\{i2\pi(mu_{ro}/L_x + nv_{ro}/L_y)\right\} \sum_{p=0}^{N_x-1} \sum_{q=0}^{N_y-1} \exp\left\{i2\pi(mp/N_x\right.$$

$$\left. + nq/N_y)\right\} S_{cp}(u_{ro} + pL_x/N_x, v_{ro} + qL_y/N_y; x_i, y_i) \quad (39)$$

where

$$S_{cp}(u_r, v_r; x_i, y_i) = S_c(u_r, v_r; x_i, y_i)$$

$$\exp\left\{-i \frac{k}{2z} (u_r^2 + v_r^2 - 2x_i u_r - 2y_i v_r)\right\} \quad (40)$$

The summation in the right hand side of Eq.(39) is the two-dimensional DFT of $S_{cp}(u_r, v_r; x_i, y_i)$, which is the phase compensated version of the received signal $S_c(u_r, v_r; x_i, y_i)$ in Eq.(29) through the operation Eq.(40). By these operations $N_x \times N_y$ picture elements are resolved for each transmission. In order to obtain more picture elements, the steering (x_i, y_i) of the transmitting beam needs to be changed.

Since the amplitude response of the composite beam pattern in Eq.(24) is written as

$$\left| P_t(f_x, f_y) P_r(f_x, f_y) \right| = \left| \frac{\sin M_x \pi L_x f_x}{\sin \pi L_x f_x/N_x} \quad \frac{\sin M_y \pi L_y f_y}{\sin \pi L_y f_y/N_y} \right| (41)$$

its beamwidths expressed by a Rayleigh distance, namely, a distance between its peak point and the first zeros, are

$$\Delta f_x = 1/M_x L_x, \quad \Delta f_y = 1/M_y L_y$$

along x and y direction, respectively. If these are expressed in a distance on the image plane, they become

$$\delta_x = \lambda z/M_x L_x, \quad \delta_y = \lambda z/M_y L_y \quad (42)$$

respectively, which are $1/M_x$ and $1/M_y$ of the periods of the transmitting beams Δ_x and Δ_y in Eq.(36), respectively. Let the intervals between picture elements be selected to δ_x and δ_y along x and y direction, respectively, then $M_x \times M_y$ transmitting steerings of

$$x_i = m\delta_x \quad (m=0,1,\ldots,M_x-1)$$

$$y_i = n\delta_y \quad (n=0,1,\ldots,M_y-1)$$

are sufficient to obtain the whole picture elements. Therefore, an image can be obtained by $M_x \times M_y$ transmissions as in
the case of the TS method.

A flow chart indicating the imaging procedure in the
MBS method is shown in Fig.4.

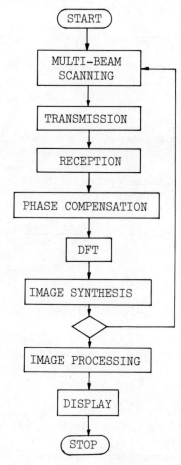

Fig. 4.   Image reconstruction procedure in MBS method.

COMPARISON BETWEEN TS AND MBS METHOD

Under ideal circumstances that a transmitter/receiver platform, a transmission medium, and an object are stationary during the whole scanning period required to form an image, the TS and the MBS imaging method would obtain almost e- quivalent images because both methods have the same weight- ing function $h_t(x_i,y_i;x_o,y_o)h_r(x_i,y_i;x_o,y_o)$. Under practi- cal circumstances, however, the MBS method proposed here has vaious advantages over the TS method.

First, the environmental stability requirement can be relaxed in the MBS method. In the TS method, picture ele- ments are resolved by carrying out the Fourier transform operation for the signals received and stored during the whole scanning period, regarding them as the signals received by the synthesized aperture. Hence, the environmental vari- ations correspond to the alignment error of the synthesized aperture and result in considerable degradations of resolu- tion and dynamic range of an image.[5] Therefore, in order to obtain good images by this method, a very high environ- mental stability is required, which may be limiting its ap- plications for such areas as underwater viewing where the environmental conditions are largely varying.

In the MBS method, on the other hand, picture elements are resolved locally at each pulse transmission. Hence, the environmental variations do not contribute to degrading resolution and image dynamic range, but only cause position- al errors of the picture elements that are resolved by dif- ferent pulse transmissions. Hence, the variations comparable to the dimension of a picture element can be allowed during the whole scanning period, which correspond to the tilt of the transmitter/receiver platform comparable to the wave- length λ and the transverse movements of the platform and the object comparable to the dimension of a picture element, while, in the TS method, the variations comparable to the wavelength λ of any types will not be allowed.

Secondly, the MBS method does not cause the speckle noise that is often troublesome in coherent imaging, be- cause picture elements are resolved mainly by the sharp trans- mitting beams and the picture elements resolved by a single transmission are spatially separated, thus no interference occurs among the adjacent picture elements.

The third point to be mentioned is to be able to simplify signal processing. In the TS method, memory capacity for storing MN complex samples is required, where M and N are the number of transmitters and receivers, respectively, because all the signals received during the transmitter scanning period need to be stored. In the MBS method, however, memory capacity for storing N complex samples is sufficient, because the received signals are processed immediately after detection at each pulse transmission. Also, if we compare processing time by means of the number of complex multiplications required to calculate the DFT's which are the main part of the signal processing operations in both imaging methods, we see the MBS method requires less processing time. Assume n-th order DFT is performed by $(n/2)\log_2 n$ complex multiplications by the use of the FFT algorithm, then the TS method requires $(MN/2)\log_2 MN$ complex multiplications for performing MNxMN-th order two-dimensional DFT, while the MBS method requires $(MN/2)\log_2 N$ ones for performing M times NxN-th order two-dimentsional DFT. Thus, the processing time as well as memory capacity required for signal processing is saved by the use of the MBS method, which results from the fact that a part of the imaging operation is carried out in the transmitting beamforming operation.

The fourth point is concerned with transmission power. In the TS method, only one transmitter is activated at each transmission while, in the MBS method, all the transmitters are activated simultaneously. Thus, N times larger acoustic energy can be transmitted by the MBS method than the TS method when the maximum transmission power of each transmitter is limited. Such a situation occurs, for example, under the cavitation limited environment or when each transmitter has an individual power amplifire.

As disadvantages of the MBS method we point out the following: i) the transmitters are slightly more complex because the transmitting beamforming and the beam scanning are required, ii) the object plane distance z needs to be specified not only in the processing of the received signals but in the transmitting beamforming. But, these are not serious problems and can easily be overcome.

CONCLUSIONS

In a holographic acoustical imaging method, a large number of receivers are required to obtain high resolution

images. In order to reduce the amount of receivers without
degrading resolution, the synthetic aperture technique can
be used. But, in the usual method (TS method) in which
transmitters are scanned sequentially, the stability requirement is very severe. In order to overcome the fault of the
usual method, we propose a new imaging method (MBS method)
on the basis of the synthetic aperture technique in which a
transmitting array is used for multi-beamforming. It is almost equivalent to the TS method under ideal circumstances,
but has many advantages over the TS method under practical
circumstances as the following:

i)   Unstability of a transmitter/receiver platform, a
     medium, and an object causes less influences on
     resolution and image dynamic range.

ii) No speckle noise occurs.

iii) Memory capacity and processing time required for
     signal processing can be saved.

iv) Higher acoustic power can be transmitted under the
     cavitation limited environment.

REFERENCES

1.   H.R. Farran, E. Maron and R.K. Mueller, "An Underwater
     Viewing System Using Sound Holography", Acoustical
     Holography, Vol.2. p.173, Plenum Press, New York (1970).

2.   E. Maron, R.K. Mueller, R.F. Koppelmann and
     G. Zilinskas, "Design and Preliminary Test of an Under-
     water Viewing System Using Sound Holography", ibid.,
     Vol.3. p.191 (1971).

3.   J.K. Kreuzer, "A Synthetic Aperture Coherent Imaging
     Technique", ibid., Vol.3. p.287 (1971).

4.   J.W. Goodman, "Introduction to Fourier Optics",
     McGraw-Hill, New York (1968).

5.   J. Thorn, "Gain and Phase Variations in Holographic
     Acoustic Imaging Systems", Acoustical Holography,
     Vol.4. p.509, Plenum Press, New York (1972).

MAXIMIZATION OF RESOLUTION IN THREE DIMENSIONS

P.N.Keating, R.F.Koppelmann, and R.K.Mueller*

Bendix Research Laboratories, Southfield, Mich.

ABSTRACT

The maximization of resolution in three dimensions (range and two lateral dimensions) is discussed, with emphasis on sampled acoustical holograms. An analysis of holography using wide-bandwidth biphase-coded acoustical signals is presented, and experimental results showing substantial increases in range resolution without significant loss of image quality are presented. It is shown that in order to maximize the lateral resolution it is necessary to have a larger number of picture elements in the image intensity display than in the sampled hologram. Experimental results clearly illustrating this point are presented.

I. INTRODUCTION

Many aspects of acoustical holography and imaging are based on concepts from the fields of either Optics or Underwater Acoustics. For example, the traditional importance of the lateral resolution of optical imaging systems has carried over into acoustical imaging, perhaps even with increased emphasis because of the limited apertures available. On the other hand, while range resolution is rarely a concern in optics, it is of vital importance in underwater acoustics, often more important than lateral resolution. In

---

*Present address: Electrical Engineering Department, University of Minnesota.

acoustical holography and imaging, both range and lateral
resolution are important and we are thus now concerned,
perhaps uniquely, with resolution in three dimensions, i.e.,
range and the two transverse dimensions.

The purpose of the present paper is to examine the maximization of this three-dimensional resolution. In the next
section, we consider the problem of obtaining good range
resolution without seriously degrading the lateral resolution and, in the third section, we explore the requirements
on image display which are necessary to ensure maximization
of the lateral resolution of the system for a given aperture size. Experimental results are also presented in order
to demonstrate the points made.

II.   RANGE RESOLUTION AND COHERENCE REQUIREMENTS

Two different target ranges can be distinguished by
means of the time difference between the returns reaching
the detector, or array of detectors, after traversing the
source-target-detector paths. This time difference is

$$\Delta T = \frac{2}{V} \Delta L, \tag{1}$$

where L is the slant range and V is the acoustic wave velocity. In general, the minimum time difference which can be
measured by the detection system (via correlation process-
ing, for example) is in the order of $(\Delta\omega)^{-1}$, where $\Delta\omega$ is the
bandwidth of the acoustic signal reflected or scattered back
from the target. The range resolution obtained is therefore

$$\Delta L = \frac{V}{2 \, \Delta\omega} \, . \tag{2}$$

Thus, to obtain good range resolution, it is necessary to
use wideband radiation, in the form of a short pulse or an
FM chirp, for example.

II.1   Coherence Considerations

It is well known that holography is in general based
on coherent radiation, and this is not obviously consistent

with the use of wide-band radiation. Most existing holographic systems are minimally concerned with range resolution and use relatively narrow-bandwidth signals. Clearly,
a trade-off between range resolution and coherence will be
necessary for a system with good resolution in three dimensions. It is the purpose of this section to explore this
trade-off and to evaluate the effects of the associated
reductions in coherence.

Another effect of attaining good time resolution is a
mismatch between the times at which a given segment of the
return reaches different parts of the array. Thus, if the
time resolution ΔT is small, the signal may be accepted by
the center channels of the array but not by the outer channels of the array, particularly in the case of off-axis
returns. (See Figure 1.) Hence, the effective size of the
array (i.e., the part which is actually utilized) is reduced, and the lateral resolution will therefore be reduced.
Interestingly, this effect is closely related to the coherence problem described above and leads to the same semi-
quantitative results.

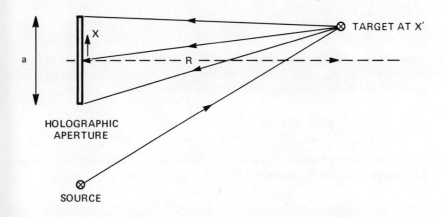

Figure 1.   The imaging geometry considered in Section II.

We begin by considering the coherence problem. It can fairly readily be shown that the reconstructed field due to the geometry shown in Figure 1, consisting of a point target at $\underset{\sim}{x}'$ (two-dimensional vector) which is illumated by a point source by wideband radiation, has the form

$$F(\underset{\sim}{\xi}) = -i \int \frac{dk G(k)}{k-k_o} \exp \left[i(k_o \underset{\sim}{\xi}-k\underset{\sim}{x}')^2/2R(k-k_o) \right] \qquad (3)$$

if the distribution function $G(k)$ represents the wave-vector content in the radiation and if the bandwidth Δk is sufficiently large (or the hologram aperture is sufficiently large) so that

$$\Delta k a^2 \gg R, \qquad (4)$$

where R is the range and a the holographic aperture (see Figure 1). Hence, Equation (3) becomes

$$F(\underset{\sim}{\xi}) = i \int dK \frac{G(k_o+K)}{K} \exp \left[\frac{i[k_o(\underset{\sim}{\xi}-\underset{\sim}{x}')-K\underset{\sim}{x}']^2}{2KR} \right]. \qquad (5)$$

For the finite bandwidth Δk to have a small effect on resolution, we require

$$[k_o(\underset{\sim}{\xi}-\underset{\sim}{x}')-\Delta k\underset{\sim}{x}']^2 < 2\pi\Delta kR \qquad (6)$$

when $|\underset{\sim}{\xi}-x'|$ is replaced by the aperture-limited resolution. In other words, the finite bandwidth Δk_c will have a significant effect on lateral resolution when

$$(\frac{\pi R}{a} - \Delta k_c x')^2 \underset{\sim}{\sim} 2\pi\Delta k_c R. \qquad (7)$$

In the off-axis case ($x' \gg a$), one can readily show that the solution to Equation (7) for Δk_c gives

$$\Delta k_c \underset{\sim}{\sim} \frac{\pi R}{a x'} \text{ (off-axis).} \qquad (8)$$

In other words, the available range resolution (for field targets) before very serious degradation of the lateral resolution occurs is

$$\Delta L_C = V/2 \; \Delta T_C = \frac{1}{2\Delta k_c} \sim \frac{a \tan \theta}{2\pi} \; , \tag{9}$$

where θ is the angular position of the target.

In the on-axis case ($x' \sim 0$)

$$\Delta k_c \sim \frac{\pi R}{2a^2} \; \text{(on-axis)} \tag{10}$$

so that

$$\Delta L_C \sim \frac{a^2}{\pi R} \; .$$

We note that, for on-axis operation in the far-field, it is possible to obtain extremely fine range resolution, as one would expect since the phasefronts are then parallel to the array. If a fine range resolution ΔL is used, then the lateral resolution for far-field on-axis objects will be good but will tend to degrade as one goes away from the axial position according to

$$\Delta\theta \sim \frac{\tan \theta}{2k_o \Delta L} \sim \frac{\Delta k}{k_o} \tan \theta, \tag{11}$$

where k_o is the "center wave-vector". Optical imaging systems in which the resolution is very good on-axis but is poorer for off-axis objects are very common, and include the human eye, of course.

A similar calculation can be carried out for the time mismatch effect mentioned above and the resolution degradation to be expected. Very similar results are obtained via this approach, and differ from those given above only by factors in the order of $2/\pi$, which are within the uncertainties present in the above estimate.

For the 250 kHz Bendix experimental underwater viewing system[1,2] and a short range of 150 cm, the lateral resolution, as estimated above, is shown as a function of target position for different values of range resolution in Figure 2. It will be noted from this figure that at a range of 150 cm a resolution of about 4 cm in all three dimensions is feasible, using an acoustic signal providing a time resolution of ~50 μsec, which is about 13 cycles of the 250 kHz carrier.

In addition to the degraded lateral resolution, some loss of signal will occur with wideband transmissions, because of the destructive interference which takes place. This need not necessarily degrade signal-to-noise performance, however, since there will be appreciably less reverberation content in the returned signal because of the improved range selectivity.

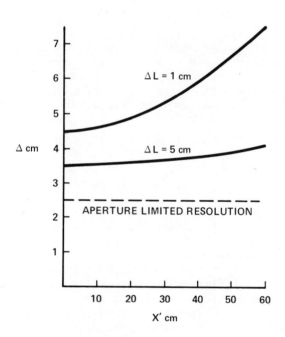

Figure 2.   Lateral resolution as a function of lateral target position at 150 cm range for different values of range resolution.

II.2 Experimental Results

The range resolution described in the previous subsection can be obtained by a short (about 10 cycles) pulse of 250 kHz carrier or by an FM pulse of appropriate bandwidth. However, for reasons which are not of immediate interest, it was decided to use Barker-coded pulses.[3] Underwater experiments have been carried out in which a source placed about 5 ft in front of the Bendix holographic underwater-viewing system[1,2] and emitting a 13-bit Barker-coded 250 kHz pulse was imaged, using the same coded signal for the reference (after an appropriate delay). Since 10 cycles were used for each bit of the biphase code, the time resolution expected is about 40 µsec, similar to that suggested above. The results using such a 13-bit coded pulse 130 µsec long were compared with results obtained with a similar, but uncoded, 130 µsec pulse containing the same energy. The intensity at the peak of the reconstructed image is shown as a function of time delay in Figure 3(a) for a source near the axis and Figure 3(b) for a source not far from the edge of the field of view (at about 50 cm).

It will be noted that the range resolution obtained from either target position is much better in the wideband case than in the narrowband case, and is about 45 µsec at the 3 dB points, corresponding to a range resolution of about 3.5 cm. The effects of the wide bandwidth can also be seen in the reduction in reconstructed image strength in the coded case, especially in Figure 3(b) for the off-axis target. No obvious effects on the lateral resolution were noted in the reconstructed images, however.

III. LATERAL RESOLUTION AND IMAGE DISPLAY REQUIREMENTS

In a recent review paper,[4] Mueller has pointed out the need, in discretely sampled holographic imaging systems, for over-sampling the image intensity display over and above that resulting naturally from the holographic sampling of the wavefront. In other words, the full lateral resolution capability of an N-point holographic system is not maximized unless the image intensity is displayed at 2N points. It is the purpose of this section to explore and demonstrate this point further and to present results from an acoustical underwater viewing system which illustrates it experimentally.

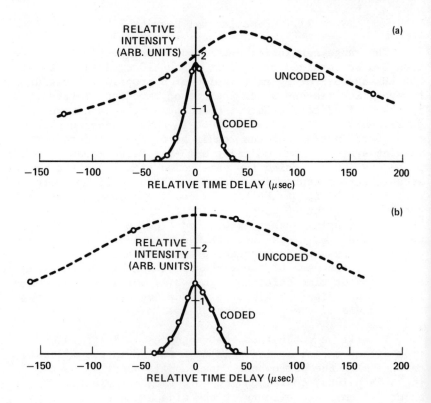

Figure 3.   Peak image intensity in the reconstructed image
of a single small source as a function of system
range setting (relative values in μsec) for both
coded and uncoded transmissions: (a) nearly on-
axis source, (b) off-axis source.

III.1   Resolution and Image Sample Spacing

The discussion here will be directed primarily toward
systems in which the linear field is sampled to form a complex hologram. This type of hologram is quite different
from the usual optical hologram where, at best, only the
real part of the optical field is recorded. However, in
optical holography, limited sampling is not normally a problem; it is in acoustical and microwave holography where
sampling becomes expensive and also where linear fields are

measureable. We shall begin by considering one-dimensional
objects and one-dimensional holograms for simplicity of presentation and generalize to two dimensions later.

There is a maximum spatial frequency (wave vector k_m)
in the holographic field and, to avoid aliasing problems,
it is necessary to sample at least at the Nyquist rate.
Thus, the hologram is sampled using a sampling interval

$$\Delta x < \pi / k_m . \tag{12}$$

In both scanned systems and in systems using digital FFT
reconstruction it is natural to produce an image display
with the same number of sample points as the hologram. However, as pointed out by Mueller[4] and demonstrated below,
this sampling rate is insufficient for the image intensity
display.

After a digital FFT reconstruction, for example, the
complex reconstructed field is determined at intervals of

$$\Delta r = \frac{\lambda R}{N \Delta x} , \tag{13}$$

where R is the range and N the number of sampling points.
The incoherent resolution[5] of the system with a square aperture is given by the Rayleigh criterion[6] as

$$\Delta \theta_r = \lambda / L = \frac{\lambda}{N \Delta x} , \tag{14}$$

where L is the total distance over which the hologram is
sampled. Near the center of the image, the sampling point
separation is, from Equation (13),

$$\Delta \theta_s \overset{\sim}{=} \frac{\Delta r}{R} = \frac{\lambda}{N \Delta x} , \tag{15}$$

which is the same as $\Delta \theta_r$ of Equation (14).

Now, the idea of resolvability rests on the presence of
an observable dip in the image intensity due to two point-
object sources which are separated by $\Delta \theta_r$. However, in the
present case, this dip is not observed because it is never
sampled, as shown diagrammatically in Figure 4. As can be

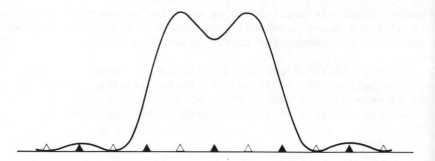

Figure 4. Theoretical intensity distribution for two sources
separated by the Rayleigh resolution and either
incoherent or, if coherent, in quadrature. The
white triangles represent the natural image display sampling rate (i.e., N samples if N is the
number of holographic samples taken). The black
triangles represent the additional samples which
are necessary.

seen, the natural sampling rate (represented by the white
triangles) does not allow the dip to be contrasted with the
two peaks and it is necessary to double the image sampling
rate to obtain the Rayleigh resolution.

So far, we have discussed only the incoherent resolution of the system, primarily because the resolution with
coherent illumination is dependent upon the relative phases
of the signals from the two points in the object and is
therefore less readily characterized. For example, if the
two signals are in phase at the center of the hologram, then
the coherent resolution is poorer than the incoherent resolution. If they are in antiphase, however, the coherent
resolution is higher. For a 90 degree phase-difference,
the incoherent resolution is again obtained. The Rayleigh
criterion thus remains a useful criterion, even for coherent
systems.

In the more general case of two-dimensional holograms,
it would be necessary to quadruple the number of samples
since the line joining the two points to be resolved could
be oriented in any direction. In this case, it might be

more efficient, perhaps, to display the complex image field since it is then only necessary to double the number of values displayed. However, the eye is not familiar with such a display, and efficiency in this area is not normally an important requirement.

III.2 Experimental Results

The acoustical holographic underwater viewing system used for these experiments has been described elsewhere[1,2] and the method of processing the holographic data is the same as in Reference 2, with one exception which is to be described here. The experiment consisted of placing two small acoustic sources about 120 cm away from the hologram sampling array of 20 x 20 hydrophone transducers. At 120 cm and 250 kHz, the Rayleigh resolution given by Equation (4) is approximately 4 cm (although it should be pointed out that the range used is well into the Fresnel zone). The sources were therefore placed in the order of 4 cm apart (4.4 cm). With uncoded, coherent signals at 250 kHz applied to the two sources so that the received signals were in anti-phase at the center of the array, the 20 x 20 complex hologram obtained in the manner described in Reference 2 was processed in two ways. In the first case, a 20 x 20 image display was obtained by minicomputer reconstruction in the manner described in Reference 2. In the second case, a 40 x 40 image display was obtained by minicomputer reconstruction of a 40 x 40 complex hologram array formed from the 400 complex elements of the 20 x 20 complex array and 1200 zero elements arranged in such a way that the FFT operation itself interpolates[7] to give the complex image field at the desired additional points. The 40 x 40 image intensity display is thus obtained merely by forming the square modulus of each of the 40 x 40 complex field values.

As mentioned in the previous subsection in connection with Figure 4, we expect the increased sampling of the image intensity distribution to be necessary whenever the object field is such that the Rayleigh resolution or better is required. The experimental results shown in Figure 5 provide experimental demonstration of this point. In Figure 5(a), it can be noted that the two sources cannot be resolved in the 20 x 20 intensity display, even though their

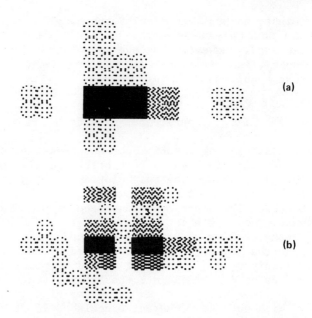

Figure 5. Experimental imagery obtained for two coherent
point sources in antiphase via (a) 20 x 20 intensity display and (b) 40 x 40 interpolated
intensity display. The two points which are
not resolved in (a) are clearly resolved in (b).

separation is slightly greater than the Rayleigh resolution.
The limitation lies in the display, as Figure 5(b) shows,
and not in the system. The image display shown in Figure
5(b) was obtained from exactly the same holographic data
as that in Figure 5(a) and the only difference in processing
was the interpolation, as described above, to obtain the
40 x 40 display. The two sources are not clearly resolved
in the image display. This experimentally confirms that
the resolution capability of a holographic system with N-
point uniform linear sampling requires a linear 2N-element
image display if intensity is being displayed. For a two-
dimensional system with M x N sampling, an interpolated
2M x 2N image intensity sampling is required. If the complex field is displayed, however, only an N-point, or M x N

display (of two quantities) is required because the phase
provides the extra information necessary to resolve the two
points whenever the system is capable of resolving them.

The above considerations also apply to older systems
similar to holographic systems, of course, including phased-
array systems.

IV. SUMMARY

The question of maximizing system resolution in three
dimensions (range and two lateral dimensions) has been
addressed. The effect on lateral resolution of the large
bandwidths which are needed to obtain good range resolution
has been examined and it has been shown that an existing
system can provide about 4 cm resolution in all three dimensions at a range of 150 cm. If only on-axis resolution
is of importance, better range resolution is obtainable.
An experimental demonstration of the three-dimensional range
resolution obtainable with Barker-coded transmissions has
been provided.

The need to provide at least twice as many picture
elements in the image intensity display as the number of
sampled values in the hologram has been explored. Experimental results which show the loss of lateral resolution
which can occur if this is not done have been given.

REFERENCES

(1) H. R. Farrah, E. Marom, and R. K. Mueller, "An Under-
     water Viewing System Using Sound Holography",
     Acoustical Holography, Vol. 2 (1970); E. Marom,
     R. K. Mueller, R. F. Koppelmann, and G. Zilinskas
     "Design and Preliminary Test of an Underwater Viewing
     System Using Sound Holography", Acoustical Holography,
     Vol. 3 (1971).

(2) P. N. Keating, R. F. Koppelmann, R. K. Mueller, and
     R. F. Steinberg, "Complex On-Axis Holograms and Recon-
     struction Without Conjugate Image", Acoustical
     Holography, Vol. 5, p. 515 (Plenum Press, New York,
     1974).

(3) See, for example, K. E. Toerper, "Biphase Barker –
     Coded Data Transmission", I.E.E.E. Trans. Aerospace
     Electronic Systems, AES–4, 278 (1968).

(4) R. K. Mueller, "Acoustical Holography – A Survey", to
     be published.

(5) While the system uses coherent radiation, the inco-
     herent resolution remains a useful parameter, as dis-
     cussed later in the section.

(6) This criterion is, of course, that the first zero in
     the point-spread-function (PSF) for the first point
     fall on the maximum of the PSF for the second point.

(7) B. Gold, C. M. Rader, Digital Processing of Signals,
     pp. 199–201, (McGraw-Hill, New York, 1969).

EXPERIMENTAL IMPLEMENTATION OF ADVANCED PROCESSING IN ACOUSTIC HOLOGRAPHY*

R.F.Steinberg, P.N.Keating, and R.F.Koppelmann

Bendix Research Laboratories

Southfield, Michigan

ABSTRACT

Two advanced spatial-processing techniques, one of which was analyzed at an earlier symposium, have been implemented experimentally, with interesting results. The previously described weak-signal enhancement technique (WSET) has been shown experimentally to be insensitive to phase-errors, as predicted, and to work well in certain circumstances. In other circumstances, however, the WSET is appreciably less effective and the reasons for this are presented. A second and more recent processing technique is adaptive in nature and removes strong highlights, together with the energy in the associated subsidiary maxima due to the pointspread function of the aperture. This approach, in which the reconstruction minicomputer automatically optimizes the highlight removal, has also been tested experimentally, with excellent results.

1. INTRODUCTION

The use of active and passive sonar and acoustic imaging techniques for surveillance, classification, and tracking problems has been limited by the difficulty of detecting weakly scattering targets in the presence of strong noise sources. These strong "noise" sources can include environmental noise and interference (including deliberate jamming),

---
*Work supported in part by the Office of Naval Research under Contract N00014-73-C-0258.

reflections from interfaces, and often in active classification problems, strong specular reflections from the target
itself. Similar "noise" problems are also encountered in
other applications where sampled holograms are used, including seismic and microwave holography and medical applications of acoustical holography.

Two spatial processing methods, the "weak signal enhancement technique (WSET)"[1,2] and a "null-processing technique (NPT)"[3] have been developed at Bendix Research Laboratories to help alleviate these problems. The first of
these methods, WSET, can be used to enhance the image of
weak targets when accompanied by a strong reflection, if
some *a priori* knowledge of the scene exists. It is applicable to problems where the following two conditions are
met: (1) the signal is coherent with the desired target
signal and is spatially localized, and (2) some spatial
separation of the weak target and the strong signal virtual
source is obtainable. It has been predicted[2] that WSET can
be especially useful if the system is phase-error limited.

The second method, NPT, is a software-oriented, null-
processing technique. It involves spatial homodyning, is
holographic in nature, and possess similarities to both
DICANNE[4] and WSET.[1,2] NPT removes the sidelobes from the
strong source image as well as the main lobe. This technique, which can be made adaptive, is intended for the
removal of a strong source in a particular frequency band
which, even after filtering, masks a possible weak source
in a nearby direction because of the sidelobes in the beam
pattern of the array. Experimental tests which demonstrate
the removal of sidelobes as well as the main lobe have been
reported[3] for the non-adaptive technique.

In this article, we describe experimental tests of
WSET and an adaptive version of NPT, using the underwater
acoustic imaging system developed previously.[5] It should
be noted that complex hologram acquisition and reconstruction techniques[6] are used with the acoustic imaging system
and, in addition, interpolation of image points[7] is used to
obtain a final 40x40 pixel image display from the 20x20 element receive array. Experimental results are presented and
discussed. For completeness, the main features of the two
techniques are also reviewed briefly.

2. WEAK SIGNAL ENHANCEMENT TECHNIQUE

In its original form,[1] the weak signal enhancement technique was based on (1) the use of the signal field for reconstruction, and (2) high-pass filtering of the desired information. This reconstruction method is most conveniently carried out by means of digital computer reconstruction techniques.

In a conventional hologram the image information is returned as the product R*S, where $S = S_1 + S_2$ is the total signal field at the hologram plane, S_1 is the strong "noise" signal, S_2 the weak desired signal, and R the reference field. Conventional reconstruction involves an additional multiplication by R', a reconstruction field at the hologram plane, followed by a transformation (e.g., a Fourier transform in the Fraunhofer case, a Fresnel transform in the Fresnel approximation) to the image plane. With a suitable R', this transformation gives, at the image plane, the field $S = S_1 + S_2$ which is the image corresponding to the holographic field.

The original WSET then involved the following steps: (a) recording of S, as is conventional holography; (b) the formation of $|S|^2 = |S_1 + S_2|^2$ from S, or directly; (c) high pass filtering of $|S|^2$, giving $|\tilde{S}|^2$; (d) multiplication of $|\tilde{S}|^2$ by S and reconstruction. If $|S_1|^2$ is slowly varying, and S_1 and S_2 are spatially separated, high pass filtering can be used [step (c)] to obtain the reduction

$$|S_1|^2 + |S_2|^2 + 2\,\text{Re}\,(S_1\,S_2^{\;*}) \rightarrow |S_2|^2 + 2\,\text{Re}\,(S_1\,S_2^{\;*})$$

Then, step (d) gives

$$I = (S_1 + S_2)\,(|S_2|^2 + S_1\,S_2^{\;*} + S_1^{\;*}\,S_2)$$

$$= (|S_1|^2 + |S_2|^2)\,S_2 + 2\,|S_2|^2\,S_1 + S_1^{\;*}\,S_2^{\;2} + S_1^{\;2}\,S_2^{\;*}$$

The last two terms provide signals in the image plane which are spatially separated from the images of interest and are not considered further. The third term yields the image S_1 but reduced by the small factor $2\,|S_2|^2$. The first term is the desired image term S_2 multiplied by the large factor $|S_1|^2$ (since $|S_1|^2 \gg |S_2|^2$). The weak image has thus been enhanced relative to the strong one by a factor $1/2 \; |S_1|^2/ |S_2|^2$.

The condition that $|S_1|^2$ is a slowly varying intensity in the holographic plane is required so that rejection of $|S_1|^2$ by high pass filtering is effective, and the final enhanced image is not degraded. The condition that S_1 and S_2 are spatially separated is necessary so that $S_1 S_2^*$ contains sufficiently high spatial frequencies that can be readily filtered from $|S_1|^2$.

It has been shown[2] that the WSET is insensitive to phase error limitations on dynamic range. This is because the most important part of the processing procedure is the filtering of the intensity, and since this latter is insensitive to phase, the weak signal is already enhanced by the time phase errors have any effect. This feature has been experimentally verified, and this will be described below.

In the implementation of the WSET, not all of the above processing steps are really necessary if one does not require the enhanced weak image to appear in the same place as it does in the original reconstruction. Therefore, an algorithm was used in which certain somewhat superfluous steps were removed. The simplified version, which allows the critical portions of the enhancement technique to be emphasized, consists of the following:

(a) The holographic field is used to calculate a holographic intensity. In our case, using the 20x20 element receive array of the acoustic imaging system,[5] if $\{F_{IJ}\}$ is the 20x20 complex array obtained after the focusing routine, we form the 20x20 array $\{G_{IJ}\}$, which is given by

$$\text{Re} \; (G_{IJ}) = [\text{Re} \; (F_{IJ})]^2 + [\text{Im} \; (F_{IJ})]^2$$

$$\text{Im} \; (G_{IJ}) = 0$$

(b) The Fourier transform $\{G_{KL}\}$ of $\{G_{IJ}\}$ is formed via the FFT routine, and the low spatial-frequency portions of $\{G_{IJ}\}$ are filtered out by multiplying each element $\{G_{KL}\}$ by the element H_{KL} of a filter array. The values of H_{KL} go smoothly to zero in the middle of the array $(K = L = 0)$ and tend smoothly to unity toward the sides.

(c) The result is Fourier-transformed back to a modified array $\{\tilde{G}_{IJ}\}$ which is then reconstructed to give a 40x40 image and displayed in the normal manner.

With the above processing, the whole image is shifted relative to the conventionally reconstructed image in such a way that the strong highlight position lies at the center of the image. A conjugate image of the weak target is also produced.

Two kinds of filtering were used. In one case,

$$H_{KL}^{A} = \left(\frac{K^2}{K_o^2 + K^2} \right) \left(\frac{L^2}{K_o^2 + L^2} \right)$$

where K_o is an adjustable filter parameter. In the second case,

$$H_{KL}^{B} = \frac{1}{2} \left(\frac{K^2}{K_o^2 + K^2} + \frac{L^2}{K_o^2 + L^2} \right)$$

The first tends to filter out information along the vertical and horizontal axes of the final image, while the second is closer to being circularly symmetric and is a preferable approach.

3. NULL PROCESSING TECHNIQUE

The development and analysis of the null-processing technique (NPT) has been previously described.[3] The technique was developed for use with underwater acoustic arrays and was designed to remove a strong noise signal which, even after appreciable temporal filtering, masks weak targets in nearby directions because of the side-lobes in the beam-pattern of the array.

The basis of the technique is the use of three steps:

(a)   Spatial homodyning to bring a given "spatial frequency" in the complex holographic data array $\{F_I\}$ down to "dc".

(b)   Removal of this "spatial frequency" by subtracting from each F_I' the mean value $<F>$, where the prime denotes the homodyned set.

(c)   The return of the set $\{F_I'\}$ to its original form, apart from the removed segment.

The above operations, for a one-dimensional N-element array in the far field, are mathematically represented by:

(a) $F_I' = F_I \exp \dfrac{2\pi i I\mu}{N}$ $(I = 0,1,2,\cdots:N-1)$

(b) $B_I = F_I' - \dfrac{1}{N} \displaystyle\sum_{I'=0}^{N-1} F_{I'}'$

(c) $\tilde{F}_I = B_I \exp\left(-\dfrac{2\pi i I\mu}{N}\right)$

where μ is, in effect, the null direction, i.e., $\mu = \dfrac{Na}{\lambda} \sin\theta$, θ is the direction of the far field null, λ is the acoustic wavelength, a is the array element separation, N is the number of array elements, and the label, I, is either integral or half-integral.

In the case of two-dimensional arrays, as was used here, the scalar label I is replaced by a two-dimensional vector label $\underline{I} = (I,J)$ and the scalar variable μ is replaced by $\underline{\mu} = (\mu_i,\mu_j)$. Products such as $I\mu$ become $I\cdot\mu = I\mu_i + J\mu_j$. Further, when working in the Fresnel zone (as was the case in our experimental work), it is necessary to multiply the set $\{F_I\}$ by the proper Fresnel phase factor before applying the algorithm.

The analysis for the NPT given in reference 3 showed that, for the case of a single strong point source, the application of the above algorithm removes the strong source including its side-lobe contributions without any loss in resolution. Further, the technique can be cascaded if more than one strong localized source is present, as is shown experimentally below.

NPT processing could be made adaptive by using an algorithm to locate the approximate position of the strongest source and, using that as a starting value of μ, require the software processor to determine a nearby value of μ for which the strong source image intensity is a minimum. This is essentially the approach that was taken, but with one important difference described below.

The approximate direction in which the null is to be placed is determined by first conventionally reconstructing the original complex hologram[6,7] and by determining from the image field data the location of the most intense image point. The position so determined will be one of the 1600 pixels in the 40x40 image display. This is used as the starting value of μ which is an input to the adaptive routine. Other nearby values of μ between adjacent sample points might then be examined to obtain a minimum in the strong source intensity. In actual fact, the criterion, or index of performance, used by the adaptive routine is the minimization of the total intensity in the hologram. This criterion can be used instead of the more obvious criterion because, by Parseval's Theorem,

$$\int_{-\infty}^{\infty} |F(k)|^2 \frac{dk}{2\pi} = \int_{-\infty}^{\infty} |f(x)|^2 dx,$$

where F and f are Fourier transforms of each other. Hence, a minimization of the strong-source light intensity in the image plane is nearly equivalent to a minimization of the total hologram intensity, since the strong image intensity is assumed much greater than other target image intensities. Making use of this theorem saves considerable processing time since the search for the optimum μ then takes place without making any Fourier transforms, which is a time-consuming step in the overall processing.

The adaptive NPT routine, as used here, fits into the entire holographic processing and reconstruction process as shown in Figure 1. The analog complex hologram data is digitized and loaded into the computer. The data is then focused (multiplied by Fresnel phase factors) and Fourier-transformed to obtain the image field. The intensity values are formed and the data is normalized and displayed on a CRT. The location of the maximum intensity image point is determined within the computer routine at the time the intensity values are formed.

The location of the maximum intensity point (approximate null direction) is then input to the adaptive null processing algorithm. The optimum value of μ is determined in the manner described above and, using this optimum μ value, the null-processing algorithm is applied to the holographic data. The homodyned and filtered data is then reconstructed in the same manner as before.

4. EXPERIMENTAL RESULTS

The experimental data used to evaluate the WSET and NPT was gathered using the experimental arrangements shown in

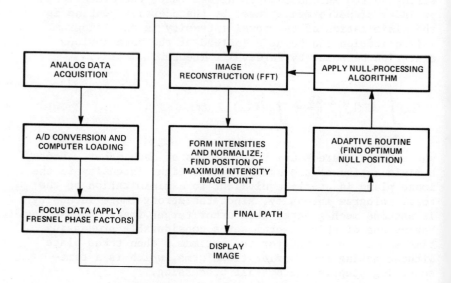

Figure 1. Flow diagram of holographic data processing using adaptive null-processing technique.

Figure 2. The arrangement in Figure 2(a) simulates a weak desired target spatially separated from a stronger localized source by a small angle. In Figure 2(b), reflection from the wall of the experimental test tank provided a constant background reflector (which might simulate a bottom reflection). A small active source near the wall provided the weak signal. In Figure 2(c), the strong specular reflections from the extended target provided the strong source; weaker signals scattered from the target edges or discontinuities were those to be enhanced.

The results obtained using the enhancement techniques were also sometimes compared to those from a simple "clipping" method. This process simply consisted of setting each image intensity value which was greater than a preset value, I_c, equal to I_c, allowing the limited dynamic range available in the CRT display to be more effectively used.

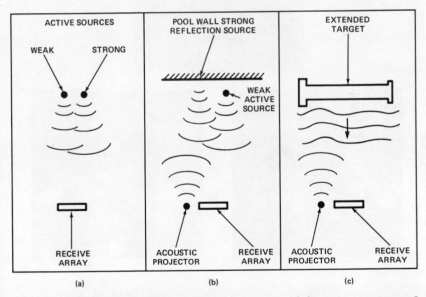

Figure 2. Experimental arrangements for (a) suppression of strong point source; (b) suppression of strong background reflection; (c) suppression of strong specular reflection from extended targets.

4.1　WSET Results

The image enhancement of a weak signal using WSET processing is shown in Figure 3. Here the data was obtained using localized strong and weak sources [Figure 2(a)]. The weak signal was 18 dB below the strong one and spatially separated by about 12 degrees. Figure 3(a) is a conventional image reconstruction; the weak signal is barely visible where indicated (its intensity is actually less than other noise present). Figures 3(b) and 3(c) are enhanced images of the same data for two different values of the filter parameter, K_O, using WSET processing. In both

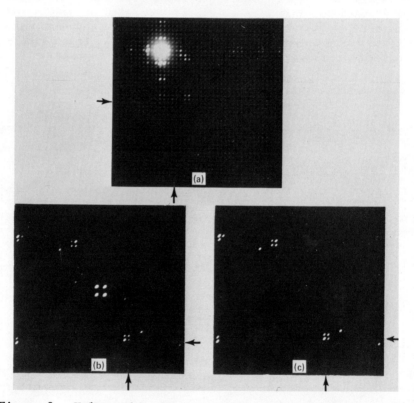

Figure 3.　Holographic image reconstructions of strong and weak point sources:　(a) conventional reconstruction with weak (-18 dB) signal barely visible; (b) using WSET and a low filter parameter; (c) same as (b) but with higher filter parameter.

cases, the weak image, along with its conjugate, is clearly
visible. (Note that as discussed in Section 2, the position
of the image has been shifted such that the strong signal
position lies at the center of the display). Figure 3(b)
was processed using the smaller filter parameter (i.e., the
spatial frequency cutoff is low) and some residual strong
target signal can be seen. Increasing the filter parameter
(raising the spatial frequency cut-off) eliminates this
residue as shown in Figure 3(c). The filter parameter value
is essentially the number of picture elements in one direction which are suppressed by filtering. Consequently, when
the angular separation between the weak and strong signals
is significant, a large K_0 can be used. A smaller value is
required for targets with smaller separations.

The relative magnitude of the weak signal with respect
to the strong one which can be imaged and resolved using
WSET is limited by the imaging system's background noise
and by the angular separation of the two sources. For
small spatial separation angles the side-lobe structure of
the strong source interferes with and tends to obscure the
weak target. The maximum magnitude that can be imaged is
limited by the array gain, 26 dB in our case, but this can
only be achieved for relatively large target separation
angles, greater than about 10 degrees. Experimentally
images of weak targets 21 dB below the strong one (S/N $\approx$ 1)
have been identified. Even weaker targets have been imaged
but their intensities are less than background noise so they
could not be identified without some other *a priori* knowledge. In a real situation, this extra knowledge can often
be the fact that the desired weak target is moving (in successive image displays) while the noise background is
stationary.

Images of data obtained using a weak target in the
presence of a strong background reflection, using the experimental setup of Figure 2(b), are illustrated in Figure 4.
Here the weak target is separated angularly by 12 degrees
and is 15 dB less intense than the strong reflector. The
conventional reconstruction of both the weak and strong
signals is shown in Figure 4(a). Figure 4(b) is the reconstruction of the same data using WSET; the real and conjugate images are clearly visible and the strong source is
completely suppressed. The pseudo-enhancement obtained by
simply intensifying and clipping the conventionally reconstructed image is shown in Figure 4(c); the weak source is

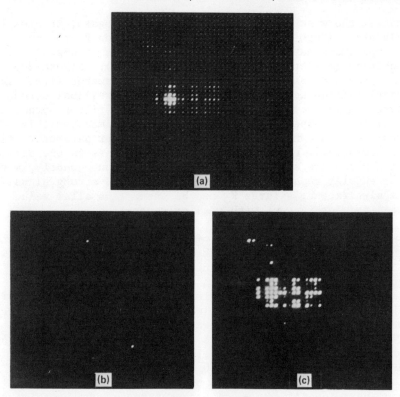

Figure 4.   Holographic image reconstructions of strong wall
reflection and weak (−15 dB) point source:
(a) conventional reconstruction; (b) using WSET;
(c) pseudo−enhancement by clipping conventional
reconstruction.

visible although noise in the vicinity of the strong source
image is also intensified and is somewhat distracting.

The value of WSET processing for cases when the dynamic
range of the system is limited by phase errors is illustrate
in Figure 5. This data was also obtained using the experimental setup of Figure 2(b), but random phase errors were
purposely introduced into the measured array data. The
intensity of the weak source is 13.5 dB below the strong

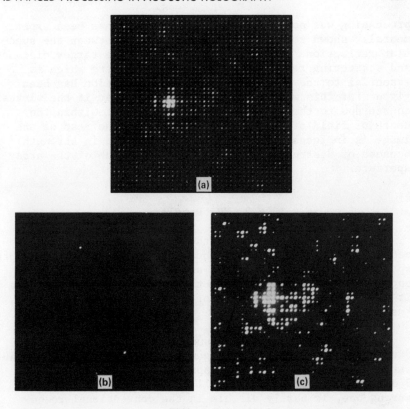

Figure 5. Holographic image reconstructions of strong wall
           reflection and weak (-13.5 dB) point source with
           random phase errors introduced in hologram:
           (a) conventional reconstruction; (b) using WSET;
           (c) pseudo-enhancement using clipping.

one. The WSET processed image, Figure 5(b), shows no evidence of noise due to the phase errors; however, the intensified and clipped conventional reconstruction in Figure 5(c)
is extremely noisy and the desired weak image cannot be
easily identified.

     In the above results, the sources of both the weak and
strong targets have been well-localized. When an extended
target was used [experimental setup in Figure 4(c)], WSET

processing was not effective. The problem has been experimentally shown to be lack of a cross-term between the specular reflection (virtual source at twice the target distance) and scattering returns. It is this cross-term which is essential to the WSET. A plausible explanation has been given[8] for this which shows, in essence, that if the virtual source due to the specular reflection is not within the depth of field of the imaging system when the rest of the image is in focus, the cross-term is reduced in strength because of destructive interference over much of the array aperture.

4.2 NPT Results

It was noted above that the effectiveness of WSET processing diminished (a) when the weak and strong targets were spatially separated by only a small angle and (b) in the suppression of specular reflections from extended targets. It is shown here that NPT is often effective in both of these cases.

Figure 6 shows image reconstructions of experimental data obtained using the arrangement of Figure 2(b) when the strong and weak sources were spatially separated by only 5 degrees. The weak target, 11 dB less intense than the strong one, is barely visible in the conventional reconstruction of Figure 6(a). When NPT processing is used, however, the weak image is clear and the strong localized reflection essentially completely suppressed as shown in Figure 6(b). Pseudo-enhancement obtained by clipping, Figure 6(c), also intensifies the weak image but it is not clearly identifiable in the scene.

The effect of NPT processing in suppressing strong specular reflections from extended targets is illustrated in Figure 7. The target here was in the form of a cylinder with several smaller structures mounted on it. The strongest reflection, shown in the conventional reconstruction of Figure 7(a), originated from the body of the cylinder. Other less intense reflections are also observed from some of the mounted structures. When a null is placed in the approximate null direction, using NPT but without the adaptive routine, the image in Figure 7(b) results. The strong source image intensity has been reduced by 5.8 dB, and although some residue remains in the vicinity of the null,

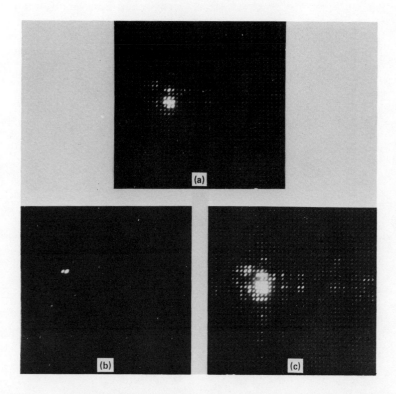

Figure 6. Holographic image reconstructions of strong wall
reflection and weak (-11 dB) point source:
(a) conventional reconstruction; (b) using null-
processing; (c) pseudo-enhancement of (a) by
clipping.

the secondary reflection to the left of the original highlight is now dominant. When the adaptive routine was used
with NPT, Figure 7(c), the original strong source was reduced an additional 1.2 dB. As can be seen from Figures 7(b)
and 7(c), scattering from other parts of the cylinder and
the smaller mounted structures are more obvious as a result
of NPT processing.

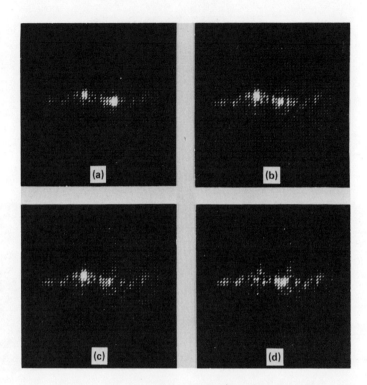

Figure 7.   Holographic images of cylinder with extra struc-
            tures: (a) conventional reconstruction; (b) using
            null-processing; (c) using adaptive null process-
            ing; (d) repetitive (2 nulls) adaptive null-
            processing.

     Figure 7(d) illustrates the use of successive nulls in
suppressing multiple highlights. Here a second null was
placed on the strongest remaining image point of Figure 7(c)
resulting in relatively higher intensities of the weaker
scattering signals.

5. CONCLUSIONS

Two signal enhancement techniques previously described, the weak signal enhancement technique (WSET)[1,2] and a null-processing technique (NPT)[3] have been reduced to practice in conjunction with an underwater acoustic imaging system and tested with real data. An adaptive routine has also been developed and implemented for use with NPT.

The experiment results obtained verify that both techniques work in practice under certain conditions. WSET was shown to be effective when the strong signal is spatially separated from the weak one, e.g., a weak localized target and a bottom reflection or strong jamming signal. In these cases, if the weak target is sufficiently spatially separated from the strong one, WSET can pull the weak one "out of the mud" even when its intensity is 20 dB or more below the strong one. For small spatial separations, the value of WSET diminishes because of the side-lobe limitations on the dynamic range and because the spatial frequencies of the weak target approach those of the strong target and the filtering inherent in WSET is less effective. It was shown, however, that NPT is effective for small separations between weak and strong targets. The experimental results also confirmed the earlier theoretical prediction[2] that WSET markedly extends the effective dynamic range when the latter is limited by phase errors. WSET was generally not effective, however, in suppressing highlights during the acoustic imaging of extended targets. The problem here is thought to be that the cross-term between the weak and strong signals which is important in WSET processing is not properly formed.

NPT was shown to be effective not only in suppressing strong signals resulting from reverberations and jamming-like noises, but in the suppression of strong specular reflections from extended targets. In this manner, more detailed target information could be obtained. It was also shown that adaptive techniques are desirable when using NPT in order that improved images can be obtained in a cost-effective manner.

Finally, although WSET and NPT were tested using only acoustical data from underwater arrays, the techniques should be useful in processing other types of sampled hologram data, including acoustical holograms for medical applications, seismic holograms, and microwave holograms.

REFERENCES

1. R. K. Mueller, R. R. Gupta, and P. N. Keating, "Holographic Weak Signal Enhancement Technique", J. Appl. Phys., 43, 457 (1972).

2. P. N. Keating, R. K. Mueller, and R. R. Gupta, "Conventional and Weak Signal Enhancement Holography in the Presence of Measurement Errors", Acoustical Holography, Vol. 4, 251, Plenum Press, New York (1972).

3. P. N. Keating, R. F. Koppelmann, R. K. Mueller, R. F. Steinberg, and G. Zilinskas, "Adaptive Null Processing – A Holographic Approach and Experimental Results", J. Acoust. Soc. Am., in press.

4. V. C. Anderson, "DICANNE, a Realizable Adaptive Process", J. Acoust. Soc. Am., 45, 398 (1969).

5. H. R. Farrah, E. Marom, and R. K. Mueller, Acoustical Holography, Vol. 2, p. 173, Plenum Press, New York (1970); E. Marom, R. K. Mueller, R. F. Koppelmann, and G. Zilinskas, Acoustical Holography, Vol. 3, p. 191, Plenum Press, New York (1971).

6. P. N. Keating, R. F. Koppelmann, R. K. Mueller, and R. F. Steinberg, "Complex On-Axis Holograms and Reconstruction Without Conjugate Images", Acoustical Holography, Vol. 5, p. 515, Plenum Press, New York (1973).

7. P. N. Keating, "On the Holographic Approach to High Resolution Sensors", JUA(USN), January 1974; P. N. Keating, R. F. Koppelmann, R. K. Mueller, "Maximization of Resolution in Three Dimensions", Acoustical Holography, Vol. 6.

8. "An Experimental Study of Highlight Suppression in Acoustical Imaging: Final Report", Bendix Research Laboratories, Report No. 7056, April 1974, ONR Contract No. N00014-73-6-0258.

A NEW PROCESSING TECHNIQUE FOR SCANNED ULTRASONIC HOLOGRAPHY

Michel J-M. Clément[†]

Department of Electronic and Electrical
Engineering, University of Technology,
Loughborough, Leicestershire, England

SUMMARY

The weak signal enhancement technique (W.S.E.T.) of
Mueller et al has been extended to scanned ultrasonic
holography via the use of electronic holographic reconstruction and recording.

The method obviates the need for critical a posteriori
optical alignments.   Furthermore, extra processing time is
avoided and the recording time is merely as required for a
normal scanned hologram.   The W.S.E. reconstruction is
achieved from the W.S.E. hologram, say with coherent light
illumination as from normal holograms.

Using an electronic implementation, it becomes possible
to introduce convenient non-linearities in the processing.
In this way, phase-only W.S.E. holograms can be recorded.
Experimental results which were obtained with circular
scanning are given, showing weak-scattering targets enhanced
from strong coherent backgrounds.   The insonifying wave
frequency was typically 2 MHz and the recording performed
in the source-receiver scanning mode.

INTRODUCTION

The holographic weak-signal enhancement technique
first developed by Mueller et al[1] and Keating et al[2], can

[†] New address: Laboratoire de Biophysique, C.H.U. Cochin,
        Université de Paris - V, France.

be extended to scanned ultrasonic holography with the introduction of two extra steps.[3,4]

Their original method is an extention of the normal holographic reconstruction method and is basically achieved through two successive reconstructions. It involves the following steps:

(1) recording of the object field (complex amplitude $\underline{S}(x,y)$), as in conventional holography (i.e. recording of a normal acoustic hologram).

(2) formation of its intensity $|\underline{S}(x,y)|^2 = |\underline{S}_1 + \underline{S}_2|^2$, where $\underline{S}_1(x,y)$ is the complex amplitude of the strong unwanted field and $\underline{S}_2(x,y)$ the complex amplitude of the weak field of interest.

(3) high-pass filtering of $|\underline{S}|^2$ (denoted as $|\underline{\hat{S}}|^2$).

(4) multiplication of $\underline{S}$ by $|\underline{\hat{S}}|^2$ and reconstruction of $\underline{S}_2$ enhanced with respect to $\underline{S}_1$.

Such an approach relies on the assumption that the intensity of the strong field $|\underline{S}_1|^2$ is spatially varying slower than the intensity of the weak field $|\underline{S}_2|^2$. This must be so in order that step 3 can be performed and the final enhanced image of S_2 will not be degraded. Step 3 also requires the strong field S_1 and the weak field S_2 to be spatially or angularly separated, so that $\underline{S}_1 \underline{S}_2{}^*$ and $\underline{S}_1{}^* \underline{S}_2$ contain high spatial frequencies, which can be readil filtered out from $|\underline{S}_1|^2$.

The enhancement of the method can be seen from the following analysis. Consider the intensity of the object field which can be written[1](step 2):

$$|\underline{S}|^2 = |\underline{S}_1 + \underline{S}_2|^2 = |\underline{S}_1|^2 + |\underline{S}_2|^2 + \underline{S}_1{}^*\underline{S}_2 + \underline{S}_1\underline{S}_2{}^* \quad (1)$$

After high pass filtering and under the assumption previous stated, the intensity of the object field becomes:

$$|\underline{\hat{S}}|^2 = |\underline{S}|_2{}^2 + \underline{S}_1{}^*\underline{S}_2 + \underline{S}_1\underline{S}_2{}^* \quad (2)$$

If the original object field is multiplied by equation (2), we obtain the holographically reconstructed complex amplitude:

$$\underline{S}\,|\underline{\overset{\vee}{S}}|^2 = (\underline{S}_1 + \underline{S}_2)|\underline{\overset{\vee}{S}}|^2 = (\underline{S}_1 + \underline{S}_2)(|\underline{S}_2|^2 + \underline{S}_1{}^*\underline{S}_2 + \underline{S}_1\underline{S}_2^*)$$

(3)

Rearranging the terms of equation (3), the space dependent complex amplitude of the reconstruction can be written:

$$\underline{S}|\underline{\overset{\vee}{S}}|^2 = \left[|\underline{S}_1|^2 + |\underline{S}_2|^2\right]\underline{S}_2 + 2|\underline{S}_2|^2\underline{S}_1 + \underline{S}_1{}^*\underline{S}_2^2 + S_1{}^2S_2^*$$

(4)

The first term of equation (4) corresponds to the wanted image term S_2 multiplied by a factor $\left[|\underline{S}_1|^2 + |\underline{S}_2|^2\right]$. The second term $2\,|\underline{\overset{\vee}{S}}_2|^2$. $\underline{S}_1$ corresponds to the image S_1 multiplied by a small factor $2|\underline{S}_2|^2$. The remaining last two terms will give images separated from the images of interest, as they are propagating at greater angles off-axis. Therefore, the weak image has been enhanced with respect to the strong one by the large factor $|\underline{S}_1|^2/2|\underline{S}_2|^2$ (since $|\underline{S}_1|^2 \gg |\underline{S}_2|^2$).

WEAK SIGNAL ENHANCEMENT WITH ELECTRONIC HOLOGRAPHIC RECONSTRUCTION

When applying the weak-signal enhancement method to an ultrasonic holographic system, using optical reconstruction, two types of difficulty are encountered. First, it is not very easy to perform the optical alignment of the transparency $\overset{\vee}{S}|^2$ in relation to the reconstructing wave $\underline{S}$. Secondly, the reconstructing wave $\underline{S}$ is not immediately available in optical form for reconstruction, but rather as an ultrasonic wave or an electronic signal.

In this paper the technique is extended via the use of holographic electronic reconstruction. The method consists essentially in performing Mueller's method onto the detected electronic signal (including the holographic reconstruction in step 4) thus obtaining the complex amplitude (4) electronically. The corresponding electronic signal is then recorded holographically (step 5) to allow optical reconstruction. The technique involves therefore two more steps than that of Mueller et al.:

(1) linear detection of the object field (e.g. by scanning) leading to an electronic signal:

$$s(x,y,t) = S(x,y)\left[\cos\omega_s t + \phi(x,y)\right] = \text{Real}\{\underline{S}(x,y)e^{i\omega_s t}\}$$

where S is its amplitude, ϕ its phase, $\underline{S}$ its corresponding complex amplitude and $\omega_s/2\pi$ the ultrasound frequency.

(2) point by point formation of the intensity
$$|\underline{S}(x,y)|^2 = |\underline{S}_1 + \underline{S}_2|^2$$

(3) temporal high pass filterings of $|\underline{S}|^2$ (denoted $|\underline{\tilde{S}}|^2$).

(4) multiplication of s by $|\underline{\tilde{S}}|^2$.

(5) record of s.$|\underline{\tilde{S}}|^2$, as in conventional holography, multiplied by a reference signal R.

(6) reconstruction (e.g. optically) of the "weak-signal" hologram obtained in step (5).

The multiplication in step (4) being performed onto electronic signals, it effectively represents   electronic holographic resonstruction, where the reconstructed field of representation $|\underline{\tilde{S}}(x,y)|^2.s(x,y,t)$ is obtained point by point as a time varying electronic signal.

The analysis remains similar, although the space dependant complex amplitudes (1) - (4) become implicitly dependent on the scanning time τ.   For simplicity, the calculations are done with complex amplitudes (i.e. exponenti form) rather than with the electronic signal representation (i.e. cosine form).   This gives a reconstructed complex amplitude (eq.(4)) (step 5):

$$\underline{S}_p(\tau) = S_p(\tau)e^{i\phi_p(\tau)} = (|\underline{S}_1|^2 + |\underline{S}_2|^2)\underline{S}_2 + 2|\underline{S}_2|^2\underline{S}_1 +$$
$$S_1^*S_2^2 + S_1^2 S_2^* \qquad (5)$$

As the steps 1 to 4 are performed electronically, the wanted reconstructed wave of complex amplitude $\underline{S}_p$ (eq.(5)) is not available physically. Instead, a signal $s_p(t)$ analog to it is obtained, whose amplitude and phase correspond to that of the complex amplitude (5), i.e.:

$$s_p(t) = |\underline{\hat{S}}(\tau)|^2 \cdot s(t,\tau) = S_p(\tau)\cos\left[\omega_s t + \phi_p(\tau)\right] \qquad (6)$$

This signal is then mixed (e.g. multiplied) with a reference signal R as in conventional ultrasonic holography, yielding the weak-signal enhanced (W.S.E.) hologram:

$$H_{wse}(x,y) = C + (\underline{S}_p\underline{R}^* + \underline{S}_p{}^*\underline{R}) \qquad (7)$$

where C is a constant and the implicit time dependent complex amplitude $\underline{S}_p(\tau)$ has been converted into a space varying one by a suitable hologram recorder. When illuminating the W.S.E. hologram (7) with a reconstructing wave $\underline{U} e^{i\omega t}$, and for simplicity assuming that the hologram has not been demagnified, the complex amplitude of the reconstructed wave of interest would be:

$$C\underline{U}e^{i\omega t} + \underline{S}_p(\underline{R}^*\underline{U})e^{i\omega t} + \underline{S}_p^*(\underline{R}\ \underline{U})e^{i\omega t} \qquad (8)$$

Assuming $\underline{R}^*\underline{U}$ to be real (e.g. by taking the reference and reconstructing waves as identical plane waves), then it can be seen that the second term of the complex amplitude (8) is identical to that of (5) and (4), in which we find the enhanced image $(|\underline{S}_1|^2 + |\underline{S}_2|^2)\underline{S}_2$. A synoptic comparison of the basic method and of its electronic implementation are shown in table 1.

SPATIAL FREQUENCY SPECTRUM

There is quite a number of components present in the W.S.E. reconstructed waves (8). To see if the enhanced weak-field can be properly extracted from them, the Fourier spectrum of the reconstruction (8) must be considered.

It is assumed that the reference signal simulates an offset plane wave and that the reconstructing beam is an on-axis plane wave. Also, for the sake of simplicity, the hologram is assumed not demagnified. In these circumstances, it follows that the three components of equation (8) can be fully written as:

Table 1. Comparison between various
implementations of W.S.E. techniques

STEPS	W.S.E.T. IMPLEMENTATION WITH OPTICAL COMPONENTS OR WITH COMPUTER CALCULATIONS (after R.K. Mueller et al.[1])	W.S.E.T. IMPLEMENTATION WITH ELECTRONIC PROCESSING
1	HOLOGRAPHIC RECORD (H) OF $S = S_1 + S_2$	LINEAR DETECTION OF $S = S_1 + S_2$ GIVING AN ELECTRONIC SIGNAL $s=s_1+s_2$
2	OBTAINING $\lvert S \rvert^2$ FROM RECONSTRUCTION OF (H) OR DIRECTLY FROM S	OBTAINING $\lvert S \rvert^2$ DIRECTLY FROM SIGNAL s
3	HIGH PASS FILTERING OF $\lvert S \rvert^2$, GIVING $\lvert \tilde{S} \rvert^2$	HIGH PASS FILTERING OF $\lvert S \rvert^2$, GIVING $\lvert \tilde{S} \rvert^2$
4	MULTIPLICATION OF $\lvert \tilde{S} \rvert^2$ BY S, RECONSTRUCTING ENHANCED IMAGES OF S_2	MULTIPLICATION OF $\lvert \tilde{S} \rvert^2$ BY s, RECONSTRUCTING ENHANCED SIGNAL s_2
5	——	HOLOGRAPHIC RECORD (H') OF $s.\lvert \tilde{S} \rvert^2$ BY MIXING WITH A REFERENCE SIGNAL
6	——	RECONSTRUCTION OF THE HOLOGRAM(H') GIVING ENHANCED IMAGES OF S_2

$$Ce^{i\omega t}$$

$$+ e^{i\omega t}e^{-ix\frac{2\pi}{\lambda}\sin\theta}\left\{\left[|\underline{S}_1|^2 + |\underline{S}_2|^2\right]\underline{S}_2 + 2|\underline{S}_2|^2\underline{S}_1 + \underline{S}_1^*\underline{S}_2^2 + \underline{S}_1^2\underline{S}_2^*\right\}$$

$$+ e^{i\omega t}e^{ix\frac{2\pi}{\lambda}\sin\theta}\left\{\left[|\underline{S}_1|^2 + |\underline{S}_2|^2\right]\underline{S}_2 + 2|\underline{S}_2|^2\underline{S}_1 + \underline{S}_1^*\underline{S}_2^2 + \underline{S}_1^2\underline{S}_2^*\right\}$$

$$(9)$$

where λ is the reconstructing light wavelength and θ the offset angle of the simulated reference.

One-Dimensional W.S.E. Holograms

As mentioned in the introduction, the spatial frequency spectrum of the object wave is assumed to be composed of two distinct parts: a strong low-frequency component and a weak high-frequency component.

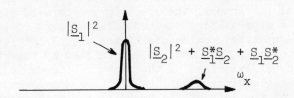

igure 1. Spatial frequency spectrum of the intensity $|\underline{S}|^2$

'urthermore, the weak component is assumed to lie in the 'ositive frequencies only (or negative only) so that the 'nhanced image S_2 will not be deteriorated by its out of 'ocus conjugate image S_2^* (fig.1).

When recording by scanning a one-dimensional (1-D) 'ologram of the object along the x direction, the spatial 'requency spectrum of the object signal intensity $|\underline{S}|^2$ fig. 1) would be obtainable from its temporal spectrum. his indeed follows directly from the scanning equation $= v_x t$ and from the scaling formula of Fourier transforms:

$$f(x) \longleftrightarrow F(\omega_x)$$

$$F(t) = f(\frac{x}{v_x}) \longleftrightarrow v_x F(v_x \omega_x) = v_x F(\omega_t) \qquad (10)$$

where v_x is the constant scanning speed along x . For a constant scanning speed v_x, the temporal spectrum of $|\underline{S}|^2$ is proportional to the ω_x cross section of its full 2-D spatial spectrum. This is shown in fig. 2a.

The spatial filtering invoked earlier (step 3) can now be performed while scanning, as an equivalent result is obtained by time-filtering the intensity $|\underline{S}(\tau)|^2$ instead. The resulting high-pass filtered spectrum is seen in figure 2b. The filtered intensity $|\overset{\sim}{\underline{S}}(\tau)|^2$ can be multiplied by the original object signal s, leading to the electronically reconstructed W.S.E. signal $|\overset{\sim}{\underline{S}}|$ .s.

The temporal frequency spectrum of $|\overset{\sim}{\underline{S}}|$ .s is shown in figure 2c and can also be found in ref. 2. The space frequency spectrum of the 1-D W.S.E. hologram and its corresponding reconstructed waves is shown in fig. 2d. As in normal holography, the spectrum is composed of a spike at ω_x = o representing the undiffracted zero order wave. Also the spectral components of the signal $|\overset{\sim}{\underline{S}}|^2$ .s are found in the positive frequencies, but shifted by the offset frequency $2\pi/\lambda \sin\theta$ (see on the right of figure 2d). There ar as well the conjugate replica of these components, but located on the left of figure 2d (i.e. in the negative frequencies).

From the spatial spectrum of the W.S.E. hologram it can be seen that the reconstructed enhanced wave S_2 can be separated from the other ones which propagate at different angles. The offset frequency should however, be sufficient so as to separate the real and conjugate waves (2nd and 3rd term in equation (9)). If not so, the wave of complex amplitude $\underline{R}\underline{S_1}^{*2}\underline{S_2}$ would propagate at almost the same angle as the wave of complex amplitude $\underline{R}^* \left[|\underline{S_1}|^2 + |\underline{S_2}|^2 \right] \underline{S_2}$ and thus deteriorates the enhanced weak field S_2.

Two-Dimensional W.S.E. Holograms

The determination of the spatial spectrum of a 2-D W.S.E. hologram is somewhat more complicated than with the

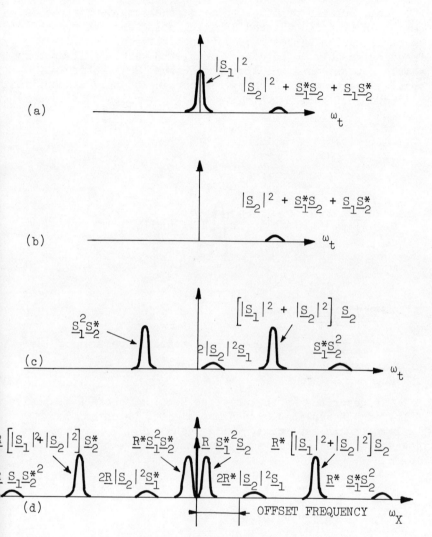

Figure 2. Various spectra occurring in W.S.E. scanned
holography (one-dimensional scanning):
(a) temporal spectrum of the scanned object-signal intensity
(b) high-pass filtered spectrum of (a)
(c) W.S.E. reconstructed signal spectrum
(d) Spectra of W.S.E. hologram and reconstructed waves

1-D case, but the principle remains the same.   For the same
reason as that encountered with 1-D holograms, the weak-field
spectrum has to be restricted to positive (or negative)
frequencies only.   In other words, the weak signal spectrum
should not surround the strong spectrum $|\underline{S}_1|^2$ entirely.   A
spatial spectrum of $|\underline{S}(x,y)|^2$ according to this assumption
is shown in figure 3 (which is the 2-D version of figure 1).

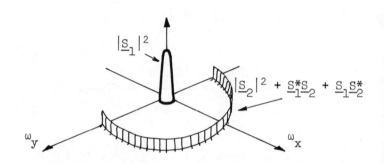

$$|\underline{S}_1|^2$$

$$|\underline{S}_2|^2 + \underline{S}_1^*\underline{S}_2 + \underline{S}_1\underline{S}_2^*$$

Figure 3. Two-dimensional spatial spectrum of $|\underline{S}|^2$

As described in the previous section, the 2-D spatial
filtering is replaced by a temporal one, which can be
implemented in real-time, while scanning.   For 2-D scans,
however, the kinematic equations now involve time varying
speeds:

$$x = v_x(t).t$$

$$y = v_y(t).t \tag{11}$$

For example, with a cartesian scan, such as that of figure 4,
$v_x(t)$ is a saw-tooth function of time, while $v_y(t)$ is a
stair-case function of time.

A 2-D spatial spectrum $F(\omega_x,\omega_y)$ of a function $f(x,y)$
is converted into a temporal spectrum $G[\omega_t(t)]$, when a 2-D
scanning, such as described by equation (11), is used.
Note that this time F and G are different functions, while

they were identical for the rectilinear 1-D scan.   There is, however, no need to work out explicitly the non-linear analytic relation between F and G.   The function G is indeed merely introduced for a qualitative understanding of the effect in the space frequency domain of a temporal filtering.

   If a raster scan, such as that in figure 4 is performed to record a hologram, the spatial spectrum of $|\underline{S}(x,y)|^2$ shown in figure 3 would be converted

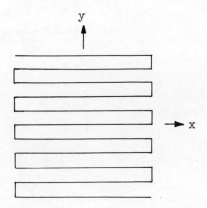

Figure 4. Two-dimensional cartesian scan

into a temporal spectrum such as in figure 5a.   A high-pass filtering of the scanned intensity $|\underline{S}(t)|^2$ would lead to a new temporal spectrum such as in figure 5b.   The 2-D spatial spectrum corresponding to this filtered temporal spectrum is shown in figure 5c, where the high-pass filtering (in the temporal frequency domain) has effectively removed the ω_x-frequencies below a given threshold.   The spectrum 5c is thus a good approximation of the spectrum of $|\underline{S}_2|^2 + \underline{S}_1^*\underline{S}_2 + \underline{S}_1\underline{S}_2^*$, the discrepancy being only in the low ω_x-frequency region.

   A cross section along $\omega_y = 0$ of the 2-D spectrum 5c is shown in figure 5d.   The 2-D spectrum of the W.S.E. hologram

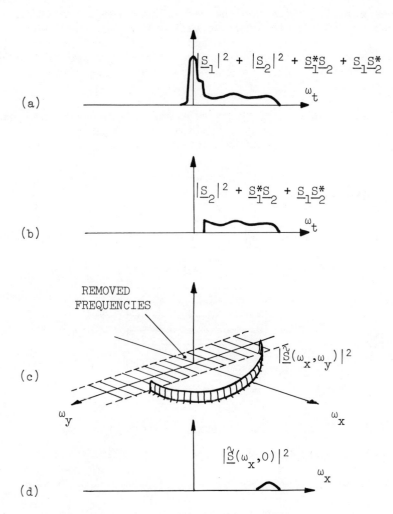

Figure 5. Various spectra occurring in W.S.E. scanned
holography (two-dimensional scanning):
(a) temporal spectrum of the scanned object-signal
intensity $|\underline{S}(t)|^2$
(b) high-pass filtered spectrum (a)
(c) 2-D spatial spectrum corresponding to the temporal
spectrum in (b) (for cartesian scan along x)
(d) x-frequency content of spectrum (c) (i.e. its ω_x
cross section)

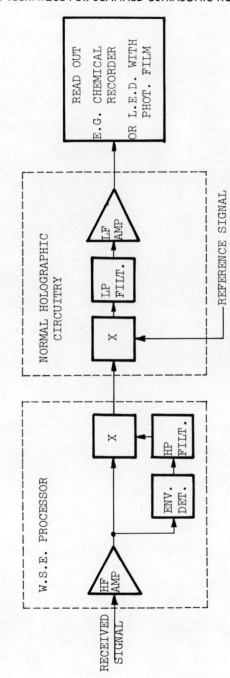

Figure 6. Ultrasonic holographic detection circuit incorporating a W.S.E. processor

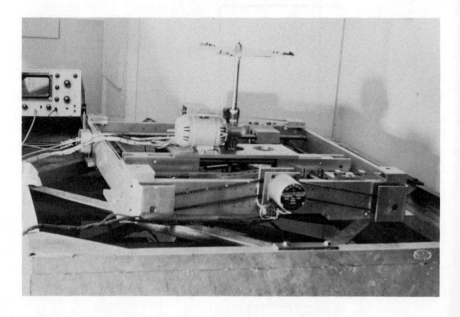

Figure 7. The translated-circular scanner

would be somewhat complicated to draw, but its x-frequency content is analogous to figure 2d, due to the similarity between the spectra in figures 2b and 5d.

EXPERIMENTAL RESULTS

Apparatus

A block diagram shown in figure 6 describes the electronic circuitry which has been used for the experiment. It is essentially composed of a W.S.E. processor followed by a normal holographic circuitry.

In the W.S.E. processor the object signal is firstly amplified then its intensity obtained by an envelope detection. The intensity is high-pass filtered, nondispersively , to remove the $|S_1|^2$ term and then fed to a mixer which multiplies it by the original R.F. signal.

The W.S.E. R.F. signal delivered by the processor is fed to a normal holographic circuitry (e.g. ref 5), which multiplies it by a coherent reference signal and filters out the unwanted doubled frequency. The resulting slowly varying signal is used to modulate the brightness of a scanning light-emitting-diode (L.E.D.). A photographic camera then records the 2-D light intensity pattern of the L.E.D. to give the W.S.E. hologram.

A rotational scanner (translated-circular type) was used for sampling the hologram, and is shown in figure 7. Its radius of scan can be as large as 0.25 m, while its linear translation can reach 0.55 m. The sampling properties of such scanners have been studied in references 6-7 and it has been shown that their scanning arc length must be restricted[7]. Under this assumption, the kinematic equations of the scan can be approximated to that of a cartesian one (i.e. $v_x(t)$ is close to a saw-tooth function of time).

Experiments

The experimental arrangement used to test the feasibility of the method is shown in figure 8. A toy gun is laid on a plate which is a strong reflector. When insonifying these two objects with a collimated beam, such as with the synthetic aperture method[8], the reflected wave is composed of a collimated part (from the plate reflector) and a diverging part (from the gun). At the hologram plane the divergent ultrasonic waves contributing to the received signal have a much smaller intensity than the collimated ones. Using the W.S.E. method just described, it should be possible to enhance the reflection from the gun relative to the strong background, provided there is sufficient angular separation between the two waves.

Ultrasonic holograms of the gun laid on rubber mats and on the mirror, respectively, are shown in figures 9 and 10. A normal and a W.S.E. hologram of the arrangement of figure 8 can be seen in the figure 10. The normal hologram of the unprocessed object field (figure 10a) shows strong fringes related to the supporting plate; inclination of these fringes indicates that the plate was slightly tilted. The

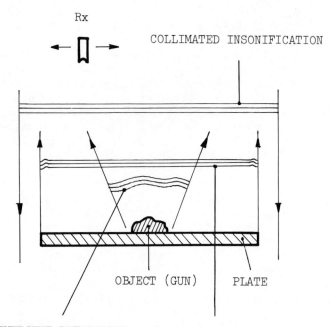

Rx

COLLIMATED INSONIFICATION

OBJECT (GUN)   PLATE

DIVERGENT REFLECTION

COLLIMATED REFLECTION
(OR SLIGHTLY DIVERGENT)

Figure 8. Experimental arrangement for testing the
feasibility of the W.S.E. processing for
scanned holography

hologram did not register any fringes from the gun, as its
signal was much too small relative to that of the plate
and reference signals.   Figure 10b shows the W.S.E.
hologram recorded with the hardware described in figures
6-7.   The fringes originated from the plate reflector
are now very much attenuated, while the fringes of the gun
have clearly been registered as the result of enhancement.

Optical reconstructions of the gun from its ultrasonic
holograms are shown in figure 11.   In figure 11a, the

Figure 9. Normal ultrasonic hologram of a
toy gun laid on an anechoic support
(insonification : 2 MHz - CW - object
size : 15 cm - object distance 30 cm)

reconstruction is obtained from the ultrasonic hologram
recorded with continuous waves and with the gun lying on
an anechoic rubber mat (figure 9).   In figure 11b, however,
the reconstruction is from the W.S.E. hologram recorded with
the gun laid on the plate (figure 10b).   Although the
reconstruction is of a modest quality, it nevertheless shows
that the image of the gun has been considerably enhanced with
respect to the image of the plate.

CONCLUSIONS

The W.S.E. method of Mueller et al has been extended to
scanned ultrasonic holography, via the introduction of two
extra steps.   No alignments are required and the processing
involved does not require any extra time.   The experimental
results obtained at 2 MHz with a mechanical scanner indicate
that the method is feasible and that the expected enhancement
takes place.   Improved quality would be achieved by using

(a)

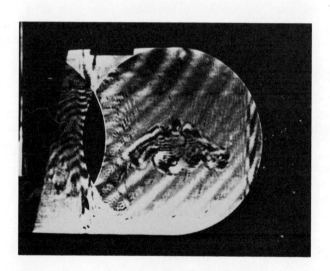

(b)

Figure 10. Ultrasonic holograms of a toy gun laid on a plate
(insonification : 2MHz-CW object size : 15 cm –
object distance : 30 cm)
(a) normal hologram
(b) W.S.E. hologram

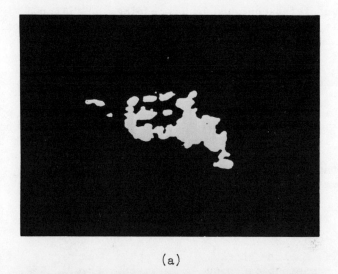

(a)

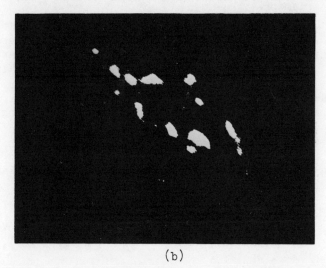

(b)

Figure 11. Optical reconstructions from ultrasonic holograms
of a toy gun (frequency of insonification :
2 MHz-CW)
(a) from a normal hologram (fig. 9) with object
on anechoic support
(b) from a W.S.E. hologram (fig. 10b) with object
on a plate

an offset reference, and by processing the filtered
intensity non-linearly, effectively recording phase-only
W.S.E. holograms.

ACKNOWLEDGEMENT

This work has been carried out with the support of
Procurement Executive, Ministry of Defence (U.K). The
author is very grateful for the permission to publish this
paper. He also would like to thank Professor J.W.R.
Griffiths and Dr. J. Szilard for their interest and
encouragement.

REFERENCES

1.   R.K. Mueller et al., "Holographic weak-signal
      enhancement technique", Jnl. Appl. Phys., 1972, Vol.43,
      No 2, pp. 457-462.

2.   P.N. Keating et al., "Effects of phase errors in
      conventional and weak-signal enhancement holography",
      Jnl. Appl. Phys., 1972, Vol. 43, No 3, pp. 1198-1203.

3.   M.J-M. Clément, "Extension of the weak-signal
      enhancement technique to scanned acoustical holography",
      Proc. 8th Int. Cong. of Acous., London, 1974, p 716
      (Abstract).

4.   M.J-M. Clément, Ph.D. Thesis, Loughborough (England),
      1973.

5.   E.E. Aldridge et al. "Ultrasonic holography in
      nondestructive testing", Acoustical Holography, Vol.3
      Ed: A.F. Metherell, Plenum Press, New York, 1971,
      Chapter 7, pp 129-145.

6.   B.P. Hildrebrand and B.B. Brenden "Introduction to
      acoustical holography", Plenum Press, New York, 1972.

7.   M.J-M. Clément and J.W.R. Griffiths "A study of
      'translated-circular' scanning for acoustical holography
      Nondestructive Testing , Aug.73, Vol 6, No 4, pp 195-199

8.   E.E. Aldridge "Acoustical holography", (Merrow
      Publishing Co, Watford, 1971).

HOLOGRAPHIC INTERFEROMETRY WITH ACOUSTIC WAVES

Katsumichi Suzuki* and B. P. Hildebrand**

*Atomic Energy Research Laboratory, Hitachi Ltd.,
Ozenji, Tomaku, Kawasaki, Kanagawa, Japan

**Battelle, Pacific Northwest Laboratories
Richland, WA 99352

ABSTRACT

Holographic interferometry has become one of the most important applications for holography. Except for a few preliminary experiments very little has been done to transfer the optical technology to acoustics. In this paper we report an extensive series of experiments demonstrating the practicality of acoustical holographic interferometry. Specifically, we have demonstrated double exposure single hologram interferometry, double exposure two-hologram interferometry, two-hologram aberration correction, two-frequency contouring and multiple-frequency contouring. Applications envisaged for this technique include large deformation analysis, periodic inspection of critical structures, imaging and mapping of low contrast objects such as the sea bottom, mapping of inaccessible inner surfaces of metal structures, and aberration free imaging through curved barriers. The results of these experiments will be shown and extrapolated to potential applications.

INTRODUCTION

Ultrasonic holography holds promise for imaging flaws in solid materials,[1] imaging objects through metal barriers,[2,3,4] and for medical diagnosis.[5,6] All of these

577

applications rely on the imaging property of holograms.
This property is important, but as in the case of optical
holography, applications beyond imaging are possible. We
refer to interferometry, especially those forms useful for
deformation analysis,[7,8] contour generation[9,10] and
aberration correction.[11,12] In fact interferometry has
become the most useful application for optical holography
as evidenced by the many articles and books published on
the subject.[13,14]

Acoustical holography has followed in the footsteps of
optical holography although the relative importance of the
imaging property is much greater. Only recently has the
possibility of interferometry been suggested for acoustical
holography.[15,16] In this paper we present the results
of an extensive experimental program designed to show the
capability that holography brings to acoustical interferometry and suggest some possible applications.

EXPERIMENTAL RESULTS

All of the experiments were performed on the Battelle-
Northwest digital scanner developed originally for the
Holotron Corp. and now commercially available from Holosonics
Corp. The system is shown in block diagram form in Figure 1.
This is a versatile system. The simulated reference beam
angle can be chosen at will by driving the phase shifter
at the desired repetition rate. The scan line density can
be chosen to be as low as 4/cm and as high as 66/cm. The
accuracy of the scan lines is sufficient to allow multiple
holograms to be recorded on a single piece of film by
interlacing.

The general procedure is quite simple. A focused F/2
transducer is used in the pulse-echo mode to scan the ultrasonic field reflected from the object. The electronic
signal is phase detected and used to modulate the intensity
of a light source scanned in synchronism with the transducer. A camera photographs the moving light source in a
time exposure to form the hologram. The hologram is inserted
in an optical system for image formation as shown in Figure 2. The various types of interferometry are then performed as described.

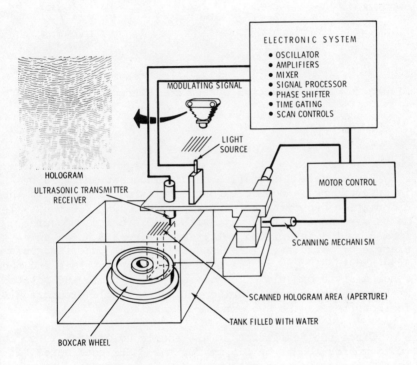

Figure 1. Block Diagram of the Battelle-Northwest Holographic Scanner.

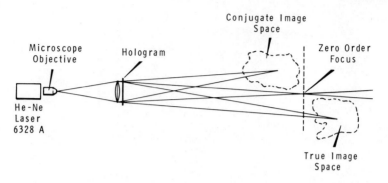

Figure 2.   Reconstruction System for Ultrasonic Holograms.

Deformation Analysis

For surface deformation or displacement analysis by
interferometry, a double exposure hologram is made with the
deformation occurring between exposures. Our first experiments were done in this way, with the separate exposures
formed by two interlaced holograms. Figure 3 shows an
example of targets (Figure 3a) used and the interferograms
3(b) obtained. One of the targets is a brass plate and the
other a styrofoam plate mounted on a tilt table and rotated
2° between exposures. For comparison a single exposure 3(c)
is shown beside the two-exposure image.

Since the resolution in ultrasonic holography is so low
(0.15 mm for 10 MHz or 0.5 mm for the 3 MHz used here), it
is not necessary to double expose. It is possible to make
two completely separate holograms with the object in the
two positions and then register them on the optical bench
with sufficient accuracy to form interference fringes.
Figure 4 shows examples of this technique. Five holograms
were made of the same object (an acrylic disk with a slot
and holes) at different tilt angles with respect to the
scan plane. The angles chosen were 0°, 1°, 2°, 3° and 4°.
They were then placed in the liquid gate two at a time (the
0° image and one of the others). Figure 4(a) shows the
fringes obtained by interfering the 0° image with the 1°
image. Figures 4(b), (c), (d) show interference of the

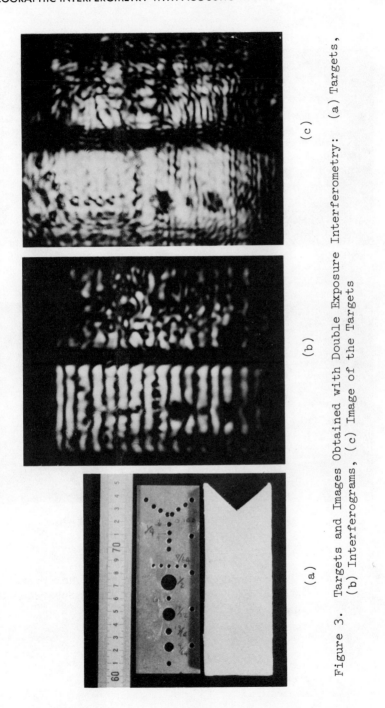

(c)

(b)

(a)

Figure 3. Targets and Images Obtained with Double Exposure Interferometry: (a) Targets,
(b) Interferograms, (c) Image of the Targets

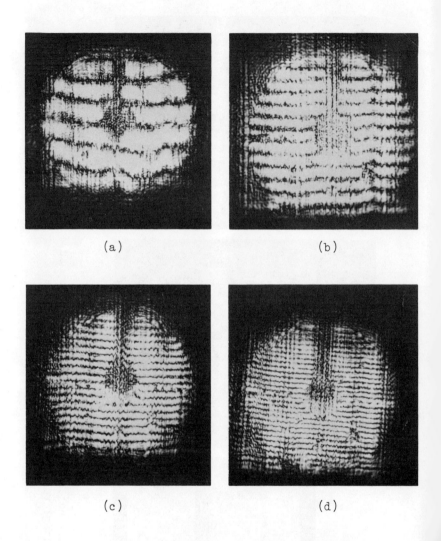

(a)   (b)

(c)   (d)

Figure 4.   Interferograms Obtained from Two Holograms Taken
            with the Object in Two Positions Separated in
            Angle by (a) 1°, (b) 2°, (c) 3°, (d) 4°.

0° image with 2°, 3°, and 4° images respectively. A final
example is shown in Figure 5 where a thin brass plate is
deformed between exposures. Figure 5(a) shows a sketch
of the deformed plate, 5(b) a double exposure fringe pattern
and 5(c) a two-hologram fringe pattern. The first position
of the plate was flat and the second deformed as shown. The
fringes therefore give an indication of the shape at the
second position. The first hologram can therefore always
be used to find the shape of any other specular object
occupying a similar position.

Contour Holography

Another form of holographic interferometry involves
multiple exposures with a change in frequency between exposures or a single exposure with a multiple frequency
source.[17] The resulting image contains fringes representing contours of constant height with the interval equal
to $C/2\Delta f$ where Δf is the difference in frequency between
exposures and C is the sound velocity. The sharpness of
the fringes, and hence the accuracy of the contours, is
dependent upon the number of frequencies used. The general
fringe visibility is given by

$$\left|\frac{\sin(N\pi\Delta fz/c)}{\sin(\pi\Delta fz/c)}\right|^2$$

where N is the number of frequencies spaced by Δf and
z is the distance from the hologram to the object point
at which the fringe is located.

Figure 6 shows some results for a styrofoam slab
inclined by 2°, 4°, 6°, 8° and 10° from the scanning plane.
For Figures 6(a), (b), (c) and (d) the frequency interval
was 585 KHz centered about 3 MHz yielding a contour spacing
of 1.28 mm. In Figure 6(e) the frequency interval was
320 KHz for a contour spacing of 2.34 mm. These images
were obtained from double exposure holograms although two
single exposure holograms can also be used.

Figure 7 shows some contour maps of the surface of a
cylindrical aluminum slab 7(a) 5.08 cm thick and 60.96 cm
radius. Figure 7(b) shows an image of the convex top surface with the axis of the cylinder nearly parallel to the

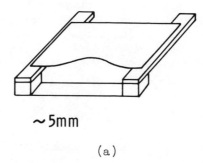

~5mm

(a)

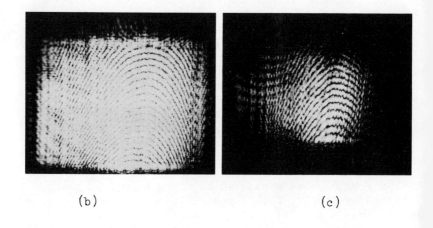

(b)   (c)

Figure 5.   Interferograms of an Object in Two Positions,
             One Flat and the Other Buckled: (a) Sketch of
             Buckled Object, (b) Interferogram by Double
             Exposure Hologram, (c) Interferogram by Two-
             Holograms.

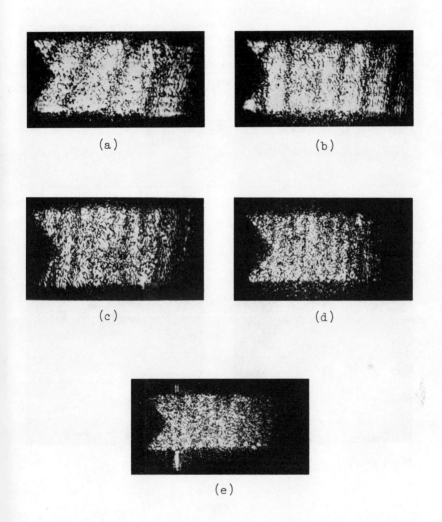

Figure 6.   Images of a Slanted Styrofoam Target Obtained
            by Two-Frequency Contour Holography: (a) 2°
            Slant, (b) 4° Slant, (c) 6° Slant, (d) 8° Slant,
            (e) 10° Slant.

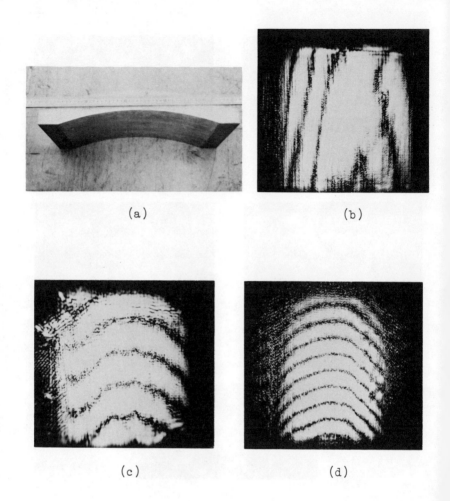

Figure 7.   Images of a Curved Cylindrical Surface Obtained
            by Two-Frequency Contour Holography: (a) Target
            (b) Axis of Cylinder Parallel to the Scan Plane,
            (c) Axis of Cylinder Slanted by 5°, (d) Same
            Orientation but Image of Inside Surface of
            Cylinder.

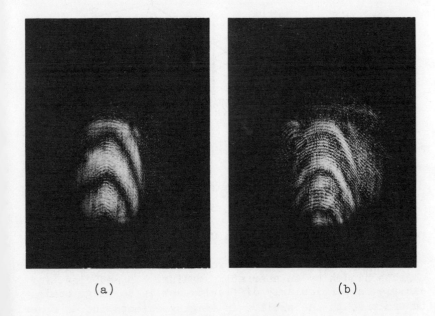

(a)   (b)

Figure 8.   Images of Curved Cylindrical Surface Obtained
            by Contour Holography: (a) Two-Frequency,
            (b) Four-Frequency.

scan plane with a contour interval of 1.28 mm. Figure 7(c)
shows the same surface tilted with respect to the scan plane
and with a 2.56 mm interval. Figure 7(d) shows the inner
or bottom surface of the slab, isolated by time gating,
with 1.28 mm contours.

A final experiment designed to demonstrate fringe
sharpening by multiple-frequency holography is shown in
Figure 8. Figure 8(a) shows the contour map of a tilted
cylinder derived from a two-frequency, and 8(b) from a
four-frequency hologram. In both cases, the frequency
interval was 300 KHz. Note the same contour interval but
sharpened fringes in the latter image.

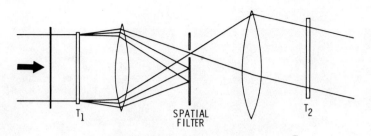

Figure 9.   Reconstruction System for Aberration Correction
Experiments.

Aberration Correction

In many applications we would like to be able to image
objects hidden by an aberrating medium. In medicine, for
example, a particularly difficult task is to image brain
tumors through the skull. In imaging objects inside spherical or cylindrical containers distortions due to the curved
surfaces also occur. It is possible, using interferometric
techniques, to correct such distortions.

The basic procedure is to divide object space into two
subspaces; one comprising the aberrating medium and the
other the object. In ultrasonics this is easily done by
using the appropriate time gate. One hologram of each sub-
space is made and the two combined on the reconstruction
bench. Figure 9 shows the optical arrangement used to
effect the distortion correction.

The hologram of the aberrating structure, T_1, is imaged
onto the hologram of the object as seen through the aberrating medium, T_2. The precision required in this step is
directly proportional to the spatial bandwidth of the distorting function. For simple structures such as a cylindric
shell, the alignment procedure can be quite casual. In the
transform plane for the first hologram, an aperture is used
to select the conjugate wavefront and allow it to propagate.
Thus, the second hologram is illuminated with a wavefront
carrying the distortion with opposing phase. The wavefront
reconstructed by the second hologram when thus illuminated
yields a distortion free image of the object.

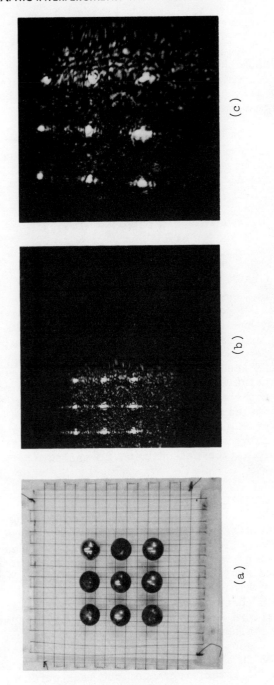

Figure 10. Images of Point Targets Seen Through a Cylindrical Barrier: (a) Target, (b) Image, (c) Corrected Image.

Figure 10 shows an experiment using a square matrix of ball bearings as the target to be imaged through a 5.08 cm thick aluminum cylinder with 60.96 cm radius. Figure 10(a) shows the object, 10(b) the uncorrected image, and 10(c) the corrected image. Figure 11 shows a similar sequence using a circular acrylic plate and F. In both cases the astigmatism introduced by the cylindrical plate has been removed.

Attempts were made to perform corrections through severely distorting media without success. This appeared to be because the distortion was severe in both amplitude and phase. Since the technique demonstrated above works perfectly only for phase distortions, we surmise that the amplitude was too severely distorted to provide a recognizable image.

APPLICATIONS

The capability to perform ultrasonic interferometry, as demonstrated, has many potential applications. An obvious application is to measure movement or deformation of structures over long periods of time. In this application a portable scanner would be positioned with respect to a fiduciary point in the foundation of a nuclear reactor, for example. An ultrasonic hologram of a particular component would be made and filed. After some period, say a year, the scanner would be set up again and a second hologram made. The two holograms would then be used to determine if motion or deformation had occurred in the intervening time interval.

A second possible application is in imaging of flaws at or very near a reflecting surface. Normally, time-gating is used to reject large reflections from front and back surfaces. Since gating is not perfect, a small volume near these surfaces is not imaged. If we wish to image rough areas in a smooth surface, or voids very near the surface, we can do so by making two exposures of the surface itself rotated by 2 or 3° between them. This will result in a linear fringe pattern except where a shadow of the flaw exists, thus outlining the flaw by interrupted fringes. A simple example of such a case is shown in Figure 12 where a milled letter F in a 12.32 cm thick aluminum block is imaged by making an interferogram of the back surface.

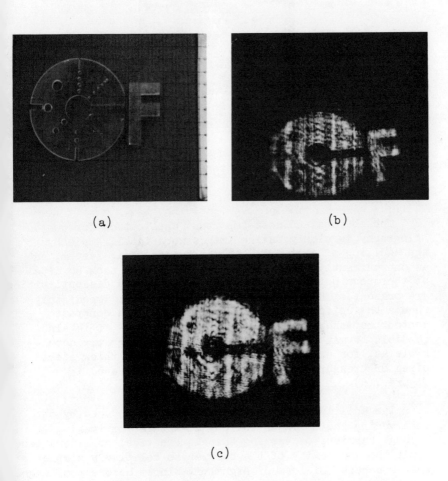

(a)

(b)

(c)

Figure 11. Images of Extended Target Seen Through a
Cylindrical Barrier: (a) Target, (b) Image,
(c) Corrected Image.

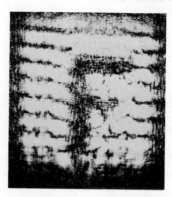

Figure 12.   Image of a Void in a Flat Surface as seen
             by Holographic Interferometry.

The obvious and most important application for contour
holography is simply making images more intelligible by
the addition of contours. An underwater sonar system, such
as one currently being constructed,[18] will show no detail
of the ocean bottom unless sharp features and discontinuities
are present. The addition of contouring will immediately
change the uniformly bright image into a high contrast
contour map showing terrain shape and slope. It is also
possible to map the shape of surfaces not directly acces-
sible as, for example, the inner surfaces of thick steel
pipes or pressure vessels. This was demonstrated in
Figure 7(d).

The ability to correct distortions introduced by the
intervening medium is an obvious advantage for imaging
through barriers of varying shape. Practical considerations
limit the complexity of the barrier to relatively simple
shapes because of loss of signal if slopes become too large,
and to homogeneous materials. The former problem can be
partially rectified by curve-following techniques which
keep the transducer nearly parallel to the surface at all
times. The latter restriction results because only phase
corrections are made in the described technique and excessive scattering causes too much amplitude distortion.

CONCLUSION

An extensive experimental program was carried out to prove feasibility of performing holographic interferometry with acoustical waves. We have demonstrated double-exposure and two-hologram interferometry for deformation measurement, two-and multiple-frequency holography for contour imaging, and two-hologram aberration correction of images distorted by an intervening medium. These experiments are easily performed on existing equipment and require no special modifications or training. We have touched on some useful applications and believe there are many more.

REFERENCES

1. B. P. Hildebrand and H. D. Collins, "Evaluation of Acoustical Holography for the Inspection of Pressure Vessel Sections," MTRSA vol. 12(12), pp. 23-31, 1972.

2. B. P. Hildebrand, "Acoustical Holography for Nuclear Instrumentation," Trans. Am. Nuc. Soc. 11, pp. 533-534, 1968.

3. H. Toffer, R. W. Albrecht and B. P. Hildebrand, "Applications of Ultrasonic Holography in the Nuclear Industry," Proc. of ANS 19th Conference on Remote Systems Technology, Idaho Falls, 1972.

4. H. Toffer, B. P. Hildebrand and R. W. Albrecht, "Ultrasonic Holography Through Metal Barriers," Acoustical Holography, vol. 5, edited by P. S. Green, Plenum Press, New York, pp. 133-157, 1974.

5. M. R. Sikov, F. R. Reich and J. L. Deichman, "Biomedical Studies Using Ultrasonic Holography," Acoustical Holography, vol. 4, edited by Glen Wade, Plenum Press, New York, pp. 147-157, 1972.

6. R. E. Anderson, "Potential Medical Applications for Ultrasonic Holography," Acoustical Holography, vol. 5, edited by P. S. Green, Plenum Press, New York, pp. 505-514, 1974.

7. K. A. Haines and B. P. Hildebrand, "Interferometric
    Measurements on Diffuse Surfaces by Wavefront
    Reconstruction," IEEE Transactions IM-15, pp. 149-161,
    1966.

8. J. E. Sollid, "Holographic Interferometry Applied to
    Measurements of Small Static Displacements in
    Diffusely Reflecting Surfaces," Appl. Opt., vol. 8,
    pp. 258-265, 1968.

9. B. P. Hildebrand and K. A. Haines, "Multiple-Wavelength
    and Multiple-Source Holography Applied to Contour
    Generation," J. Opt. Soc. Am., vol. 57, pp. 155-162,
    1967.

10. B. P. Hildebrand, "The Role of Coherence Theory in
    Holography with Application to Measurement,"
    The Engineering Uses of Holography, edited by
    E. R. Robertson and J. M. Harvey, Cambridge Univer-
    sity Press, Cambridge, UK, pp. 401-435, 1970.

11. R. J. Collier, C. B. Burkhardt and L. H. Lin,
    Optical Holography, Academic Press, New York,
    pp. 368-375, 1971.

12. B. P. Hildebrand and B. B. Brenden, An Introduction
    to Acoustical Holography, Plenum Press, New York,
    pp. 197-201, 1972.

13. Applications of Holography in Mechanics, edited by
    W. G. Gottenberg, Am. Soc. of Mech. Eng., New York,
    1971.

14. Holographic Non-Destructive Testing, edited by
    R. K. Erf, Academic Press, New York, 1974.

15. B. P. Hildebrand and B. B. Brenden, An Introduction
    to Acoustical Holography, Plenum Press, New York,
    p. 217, 1972.

16. M. D. Fox, W. F. Ranson, J. R. Griffin and
    R. H. Pettey, "Acoustic Holographic Interferometry,"
    Acoustical Holography, vol. 5, edited by P. S. Green,
    Plenum Press New York, pp. 103-121, 1974.

17. B. P. Hildebrand, "The Role of Coherence Theory in Holography with Application to Measurement," The Engineering Uses of Holography, edited by E. R. Robertson and J. M. Harvey, Cambridge University Press, Cambridge, UK, pp. 401-435, 1970.

18. N. O. Booth and B. A. Saltzer, "An Experimental Holographic Acoustic Imaging System," Acoustical Holography, vol. 4, edited by Glen Wade, Plenum Press, New York, pp. 371-381, 1972.

ACOUSTICAL INTERFEROMETRY USING ELECTRONICALLY SIMULATED

VARIABLE REFERENCE AND MULTIPLE PATH TECHNIQUES

Dr. H. Dale Collins

HOLOSONICS, INC.
2950 George Washington Way
Richland, WA 99352

ABSTRACT

This paper describes a method for imaging various external and internal objects or surfaces by the use of acoustical interferometry techniques. The acoustic reference beam is simulated electronically with a selectable inclination angle. This capability provides a variable control on the fringe density of the object. Using this technique, various objects have been imaged with depth resolution exceeding one-half wavelength. If the object is imaged near the focal plane of the focused transducer, (i.e., focused hologram), the image consists of a series of fringes with their density directly proportional to the slope of the object and the simulated inclination angle of the reference.

Multiple path or echo techniques have been employed with the described methods to dramatically increase the depth resolution at any specified frequency. A resolution of 0.0125mm (i.e., $\lambda/24$) has been achieved at 5MHz using the twelfth echo. These techniques have been used extensively in the nondestructive testing of various pressure vessels for corrosion measurements on the inner surfaces.

The theory developed in the paper accounts for variable angle reference and the multiple path techniques for simple and complex surfaces. Experimental results are shown which graphically illustrate the unique capabilities of these techniques to provide acoustical images with extremely high resolution.

INTRODUCTION

This paper discusses the theory and application of a scanned acoustical interferometry technique for imaging internal and external surface defects in thin walled vessels, etc. The technique parallels the conventional single frequency multi-beam interferometer with some major modifications. The scanned acoustical imaging system is capable of performing several rather unique kinds of interferometry: multiple beam, multiple frequency, multiple exposure, etc. The rather unique feature of this system is its ability to differentiate between negative and positive slopes. The fringe spacing decreases from the reference grating spacing for positive slopes and increases for negative slopes. Previous efforts in acoustical interferometry have been reported by Hildebrand and Brenden[1] and Fox, Parson, Griffen and Pettey[2]. Hildebrand is also presenting a paper this year at the symposium on multiple frequency interferometry[3].

The interferometry technique used by the author is a scanned holographic system operating in the focused hologram mode employing a electronic simulated acoustic reference beam. A definition of this class of hologram is simply the interference pattern between object and reference beams with the condition that the object is in focus at the hologram plane. Aldridge[4] was probably the first to construct a scanned focused acoustical hologram using an electronic simulated on-axis acoustic reference. The contour lines or fringes across the object's surface represent half-wavelength deviations in depth from the scanning plane. The fringe density in this system is not selectable and planar surfaces parallel to the scanning plane exhibit very few fringes, thus extremely small defects are completely obscure in the coarse fringe structure of the interferogram.

The major disadvantage, non-selectable fringe density, is easily circumvented by employing an electronic simulated acoustic reference with a selectable inclination angle (i.e., angle between the object and reference beams). The variable inclination angle provides the capability of controlling the fringe density without the use of different frequencies, as with the more conventional techniques.

Internal or external planar surface defects such as corrosion, deformations, etc., can be imaged with the deviations in the fringe density representing the thickness

defects. One typical industrial application is the inspection of various pressure vessels for corrosion on the
bottom surface. The extremely small corrosion defects
represent deviations from the vessel wall thickness. The
usual NDT pulse echo techniques are inadequate in resolving
small thickness differences using the typical operating
frequencies. Using interferometry techniques (i.e., multiple
path) $\lambda/24$ depth defects have been detected and imaged. At
10MHz this represents thickness defects in the order of
0.025mm in steel or aluminum.

The use of multiple paths (or echoes) dramatically increases the depth resolution by changing the fringe spacing
criterion. The spacing between fringes represents $\lambda/2k$
deviations in depth where k is the number of double paths or
pulse echo returns. Thus simply by going from the first to
the sixth echo the depth deviation between fringes decreases
from $\lambda/2$ to $\lambda/12$.

ACOUSTICAL INTERFEROMETRY

Acoustical interferometry techniques usually parallel
those previously done in the field of optics. Naturally
there are special cases in acoustics which are impossible to
duplicate in optics because the detectors, sources, etc.,
are physically different. The method described in this
paper employs a simultaneous focused source-receiver scanning
technique with a simulated acoustic reference. This system
coupled with the acoustic axicon and multiple path or echo
method represents a very unique interferometry imaging
device. The fringe density or spacing criterion can be
controlled without changing the insonification frequency.

Figure 1 shows the geometry for constructing an acoustical interferogram using a focused or axicon source-receiver transducer. The phase at the receiver point (x,y,z) is

$$\phi_1(x,y,z) = \phi_0(x,y,z) - \phi_r(x,y,z) \tag{1}$$

and

$$\phi_1(x,y,z) = \frac{2\pi}{\lambda s}(Z_0 + Z_1) - \phi_r(x,y,z) \tag{2}$$

where λs = acoustic wavelength in the medium

Z_1 = object to receiver distance

Z_0 = object to source distance.

The equation of the phase at the receiver can be rewritten in terms of the focal length and depth of focus of the transducer

$$\phi_1(x,y,z) = \frac{2\pi}{\lambda s} \left[2(f \pm \delta)\right] - \phi_r(x,y,z) \tag{3}$$

where

$$\delta = 2\lambda s \left(\frac{f}{a}\right)^2 = \text{depth of focus}$$

and $Z_1 = Z_0$.

The use of a lens transducer severely limits the allowable depth deviation of the object if it is to remain in focus.

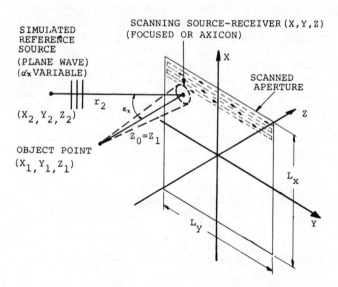

FIGURE 1. GEOMETRY FOR SCANNED ACOUSTICAL INTERFEROMETRY

The object distance (r_1) is essentially bounded by the following relationship

$$f-\delta \leq Z_1 \leq f+\delta. \tag{4}$$

Using Equation (4) and a typical focused transducer with an "F" number of 4, the maximum allowable deviation is $8\lambda s$. If we replace the focused transducer with an acoustical axicon[5,6,7] the depth of the line of focus can be extremely large. The lateral image resolution also remains constant with depth (i.e., f/a = constant). Figure 2 shows the ray diagram of the axicon.

The substitution of the axicon for the focused transducer circumvents the stringent depth of focus requirement and adds tremendous flexibility to the system. We can now rewrite the general phase relationship as before

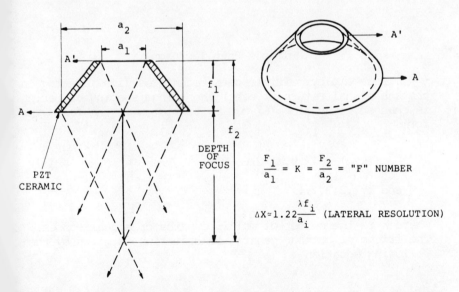

$$\frac{F_1}{a_1} = K = \frac{F_2}{a_2} = \text{"F" NUMBER}$$

$$\Delta X \approx 1.22\frac{\lambda f_i}{a_i} \quad \text{(LATERAL RESOLUTION)}$$

FIGURE 2. ACOUSTICAL AXICON EXHIBITING A CONSTANT "F" NUMBER

$$\phi_1(x,y,z) = \frac{2\pi}{\lambda s}(2Z_1) - \phi_r(x,y,z) \tag{5}$$

where

$$f_1 \pm \delta_1 \leq Z_1 \leq f_2 \pm \delta_2$$

and $\delta_i \simeq 2\lambda s \left(\frac{f_i}{a_i}\right)^2$.

The resultant phase (ϕ_1) is a function of both the object and the reference phases. The reference phase is simulated electronically and thus the contour or fringe density in the interferogram is selectable. The operator has complete control over the fringe density using only one illumination frequency. Fringes are produced when the resultant phase equals $2\pi n$ radians. Equation (6) defines the condition mathematically

$$\frac{2\pi}{\lambda s}(2Z_1) - \phi_r(x,y,z) = \pm 2\pi n \tag{6}$$

where n=0,1,2......

Multiple Path Technique

If we employ multiple path techniques (i.e., gating on the kth signal return or echo) the fringe spacing criterion is changed. Equation (6) can be rewritten in the following form

$$\frac{2\pi}{\lambda s}(2kZ_1) - \phi_r(x,y,z) \simeq 2\pi n \tag{7}$$

and k=1,2,3......

where k is the number of multiple paths or signal echoes. The fringe or contour generation criterion is given by the following equation

$$\Delta Z_1 = \frac{\lambda s}{2k} \tag{8}$$

where the reference phase is assumed constant (i.e., on-axis). Equation (8) shows the fringe density is also selectable depending on the number of multiple paths or echoes. Surface deviations of less than 0.0125mm have been imaged at 5MHz gating on the twelfth echo. This technique is extremely useful in contouring internal surfaces for corrosion, etc.

Multiple Path and Simulated Acoustic Reference

If we introduce the simulated variable angle reference beam into the phase equation, the fringe spacing criterion is altered. The depth deviation is no longer the simple $\lambda s/2k$, but varies as a function of the phase shift grating (or simulation angle*). The depth deviation (Δz_1) can now be expressed in the following form

$$\Delta Z_1 = \frac{\lambda s}{2k} \left(1 - \frac{\Delta X}{d_x}\right) \tag{9}$$

where k = number of multiple paths

$d_x = V_x T_p$ (reference spacing)

Δ_x = fringe spacing on the interferogram.

If we scan in the y direction, the equation is the same except Δy and d_y are substituted for Δx and d_x respectively.

The change in height or depth between fringes is a function of the fringe separation, wavelength of sound, phase shift grating and the number of multiple paths. Δz can be either positive or negative depending on the slope of the object with respect to the equivalent plane of the simulated reference beam. The separation between fringes from the reference spacing (Δx) decreases for positive slopes and increases for negative slopes. The reverse is true if we program the simulated reference angle to be negative.

See selection on simulated acoustic reference

Figure 3 shows the relationship between the depth deviations and the fringe spacing. The simulated reference beam is assumed to have a positive slope. If it is programmed to have a negative slope, the graph would be reversed.

The one objection to the system is the critical slope anomaly. If the object has a negative slope in the x direction ($m_c = -\lambda s/dx$) the fringe separation equals the grating spacing (dx) and two possible depth deviations would exist. This anomaly only occurs in the system at negative x or y slopes depending on the scan direction.

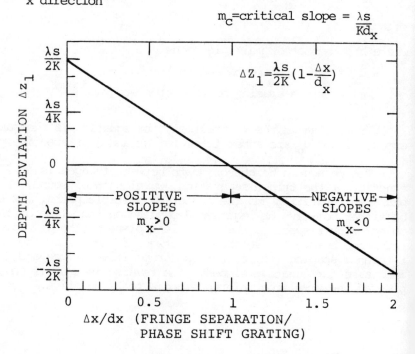

$d_x = V_x T_p$ (phase shift grating) λs=acoustic wavelength
                                         K=number of multiple echoes
Δx=fringe separation in the   or paths
    x direction

m_c=critical slope = $\dfrac{\lambda s}{\overline{K} d_x}$

$$\Delta Z_1 = \frac{\lambda s}{2K}\left(1 - \frac{\Delta x}{d_x}\right)$$

DEPTH DEVIATION Δz_1

$\dfrac{\lambda s}{2K}$

$\dfrac{\lambda s}{4K}$

0

POSITIVE SLOPES $m_x \geq 0$

NEGATIVE SLOPES $m_x \leq 0$

$-\dfrac{\lambda s}{4K}$

$-\dfrac{\lambda s}{2K}$

0   0.5   1   1.5   2

$\Delta x/dx$ (FRINGE SEPARATION/ PHASE SHIFT GRATING)

FIGURE 3. DEPTH DEVIATION IS A FUNCTION OF FRINGE SEPARATION

ELECTRONIC SIMULATED ACOUSTIC REFERENCE

If we assume an acoustic plane wave reference beam, then the reference beam can be expressed as

$$S_r(x,y) = P_r(x,y) \cos \left(wt - \frac{2\pi}{\lambda s} X \sin \alpha_x \right) \qquad (10)$$

where $\phi_r(x,y) = \frac{2\pi}{\lambda s} X \sin \alpha_x$.

No loss of generality results in only considering two dimensions with the object located on the (x-z) plane. The signal contribution to the acoustic receiver by the reference beam is

$$Pr(x,y) \cos \left(wt - \frac{2\pi}{\lambda s} \sin \alpha_x V_x t \right) \qquad (11)$$

where V_x is the scanning velocity of the acoustic receiver and

$$\omega_x = \frac{2\pi}{\lambda s} V_x \sin \alpha_x \qquad (12)$$

Equation (11) represents a sinusoidal wave whose phase (ϕr) is a function of the scanning velocity (Vx) and the inclination angle (αx). Thus, the acoustic plane wave reference beam can be simulated with an electrical signal of this form and combined with the object signal in a balanced mixer or multiplier. If the direction of propagation of the plane wave reference is perpendicular to the x axis then $\omega x = 0$ and the reference signal is simply pR(x,y)cos wt. This electrical signal, when combined with the object signal in a mixer, would simulate an on-axis plane wave acoustic reference beam. Now, if the inclination angle is not zero, then the electrical reference signal must be phase shifted to simulate an off-axis acoustic reference beam. The simulated inclination angle is a function of the scanning velocity, wavelength of sound, and the phase shifter control voltage frequency. Naturally, in three dimensions skewed beams are possible with phase shifting and time delay circuits. The phase of the electronic reference can be shifted with respect to time by the following voltage waveform shown in Figure 4. For a more detailed analysis of the phase shifter and control

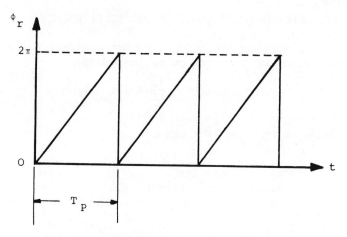

FIGURE 4. PHASE SHIFTER VOLTAGE WAVEFORM

circuits, see Appendix A.

The control voltage radian frequency (ω_p) can be expressed in terms of ω_x:

$$\omega_p = 2\pi f_p = \frac{2\pi}{\lambda s} V_x \sin \alpha_x \tag{13}$$

If the simulated plane wave is inclined with respect to a planar object (i.e., $\alpha x > 0$), then a linear fringe grating will be imposed on the interferogram. The fringe spacing is a function of the scanning velocity (Vx) phase shifter frequency (fp) and the image magnification (m). The fringe spacing on the interferogram is

$$d_x = \frac{V_x T_p}{m} = \frac{V_x}{m f_p} = \frac{\lambda s}{\sin \alpha_x} \tag{14}$$

where $\alpha_x = \sin^{-1} \left(\frac{\lambda s \, m f p}{V_x} \right)$

Figure 5 is a graph illustrating the fringe spacing as a
function of the scanning velocity and the phase shifter
voltage. It should be obvious that any grating spacing or
fringe density can be imposed on the interferogram by proper
adjustment of the phase shifter control voltage frequency or
scanning velocity. The grating spacing can be less than a
wavelength which indicates that electronic simulation is
more versatile than using an acoustic reference beam. The
contour lines are usually constructed perpendicular to the
scanning lines in rectilinear scanning. Figure 6 is a
typical reference grating imposed on the interferogram
(rectilinear scanning).

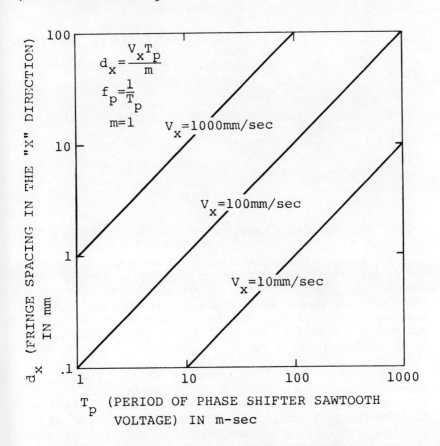

FIGURE 5. FRINGE SPACING AS A FUNCTION OF THE SCANNING
VELOCITY AND THE PHASE SHIFTER VOLTAGE

ACOUSTIC FREQUENCY: 1MHz
SCAN RATE: 76.2mm/sec
PHASE SHIFTER PERIOD: Tp=20m/sec
MAGNIFICATION: 2
FRINGE SPACING: 3mm
REFERENCE ANGLE: $\alpha x=90°$

FIGURE 6. GRATING CONSTRUCTED WITH THE SIMULATED ACOUSTIC
REFERENCE

ACOUSTICAL INTERFEROMETRY IMAGING SYSTEM

Figure 7 is the block diagram of the scanning acoustical interferometry imaging system showing the various components. A brief description of the salient features of the
interferometry system will be discussed. Probably the most
expedient method of describing the system is to simply
follow the signal from the start (i.e., oscillator) to the
final interferogram.

The output of the high frequency oscillator which
represents the coherent source is divided or split into two
signals, the object and reference, which parallels optical
interferometry or holography. The oscillator signal is
gated to produce short pulses which are amplified and drive
the source transducer after impedance matching networks.
The acoustic energy is focused by the axicon on the object
or defect as shown in Figure 8. Plate thickness deviations,

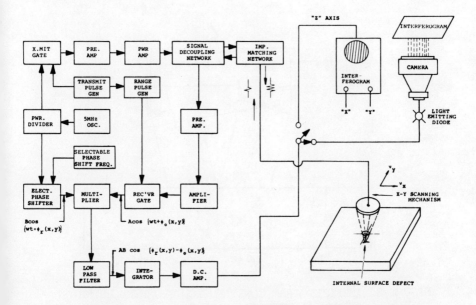

FIGURE 7. SIMPLIFIED BLOCK DIAGRAM OF THE ACOUSTIC INTERFERO-
METRIC IMAGING SYSTEM

as a result of corrosion, represent the internal defects and
are imaged in the interferogram. Deviations in the linear
fringe pattern of the plate identify the internal defects
and their depth. The object signal at the receiver is described by the following equation

$$S_o(x,y) = P_o(x,y) \cos [\omega t + \phi_o (x,y)] \qquad (15)$$

where $P_o(x,y)$ is the amplitude and ϕ_0 the phase. The acoustic
reference signal is simulated electronically and can be
expressed as

$$S_r(x,y) = P_r(x,y) \cos [\omega t + \phi_r (x,y)] \qquad (16)$$

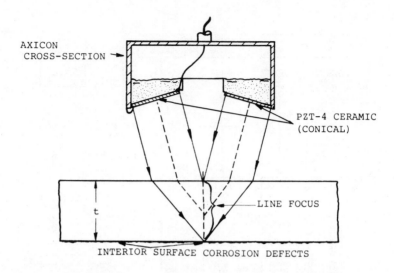

AXICON
CROSS-SECTION →

PZT-4 CERAMIC
(CONICAL)

LINE FOCUS

t

INTERIOR SURFACE CORROSION DEFECTS

FIGURE 8. INTERFEROGRAM CONSTRUCTION GEOMETRY USING AN
AXICON TRANSDUCER

when $Pr(x,y)$ and $\phi r(x,y)$ are selectable. The two signals
are multiplied and then time averaged as shown in Figure 7.
The resultant interferometry signal is given by the following equation

$$S_i(x,y) = \frac{1}{T} \int_0^T S_o(x,y) \, S_r(x,y) \, dt =$$

$$\tfrac{1}{2}P_r(x,y) \, P_o(x,y) \, \cos[\phi_o(x,y-\phi_o(x,y)]. \qquad (17)$$

Fringes or contour lines occur when the resultant phase
equals $\pm 2\pi n$ radians. This signal drives the "Z" axis of the
storage oscilloscope or light emitting diode. The focused
image interferogram is then constructed on either the storage tube or polaroid film.

EXPERIMENTAL RESULTS

Figure 9a shows the interferogram construction geometry of a flat plate parallel to the scanning plane. The fringe spacing is equal to the grating spacing dx and the deviation Δz_1 is zero. Equation (18) represents the fringe spacing in the x direction as a function of the system parameters

$$\Delta X = \frac{m_f \lambda\ s}{2km_x + \dfrac{\lambda s}{d_x}} \tag{18}$$

where mf is the film magnification and mx the object's slope in the x direction. The interferogram fringe spacing is $m_f d_x$ (i.e., 3mm) when the slope is zero (see Figure 9b).

Figure 10a shows the two flat plates exhibiting both positive and negative slopes. The first section (y_1) of the interferogram was constructed with the reference angle at 18°. The fringe spacing is the grating spacing (dx) for the flat plate section. The inclined plates represent section y_2 and the fringe spacing is dramatically different for equal negative and positive slopes (see Figure 10b). The third section y_3 represents the interferogram of the inclined plates with the reference angle at zero.

Figure 11a shows the convex spherical section located on the flat plate approximately paralled to the scanning plane. The interferogram (Figure 11b) shows the comparison between the usual interferogram (lower image) and the one constructed with inclined electronic reference. The flat plate area exhibits very few fringes in the usual interferogram, but many with the inclined reference. Figure 11c illustrates the inclined reference interferogram and dramatically exposes the areas of negative and positive slopes in the x direction. The identical situation would result in the y direction if the interferogram was constructed with a y scan. The ability to detect and image extremely small depth variations such as corrosion defects is graphically verified in these interferograms.

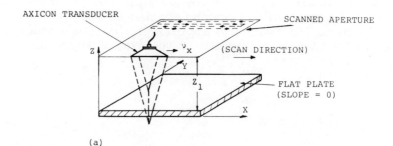

(a)

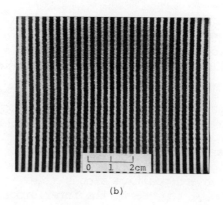

(b)

ACOUSTIC
    FREQUENCY :   1MHz

SCAN RATE(V_x) :   7.62cm/sec.
PHASE SHIFTER
        PERIOD(T_p) :   20m/sec.
MAGNIFICATION(m_f) :   2
FRINGE SPACING :   3mm
REFERENCE ANGLE(α_x) : 90°

FIGURE 9.   (a) INTERFEROGRAM CONSTRUCTION GEOMETRY AND
            (b) INTERFEROGRAM (SIMULATED REFERENCE
            INCLINATION ANGLE $\alpha=90^\circ$)

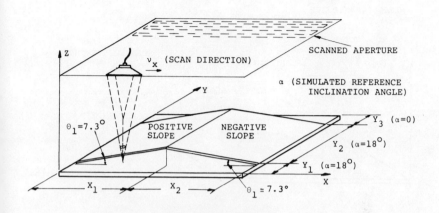

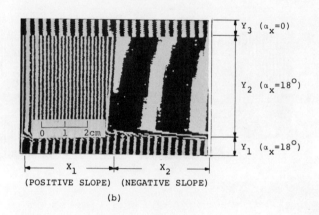

(b)

ACOUSTIC
      FREQUENCY : 3.23MHz
SCAN RATE(V_x): 7.62cm/sec.
PHASE SHIFTER
      PERIOD(T_p): 20m/sec.
MAGNIFICATION(m_f): 2
REFERENCE ANGLE (α_x) : 18°

FIGURE 10 (a) INTERFEROGRAM CONSTRUCTION GEOMETRY AND
      (b) ACOUSTIC INTERFEROGRAM

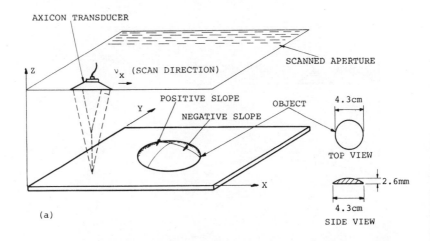

(a)

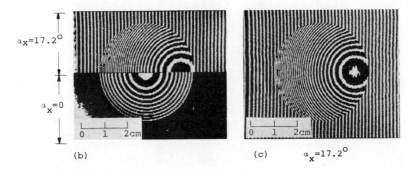

(b)   (c) $\alpha_x=17.2^o$

ACOUSTICAL
   FREQUENCY: 3.23MHz

SCAN RATE(V_x):   7.62cm/sec

PHASE SHIFTER
       PERIOD(T_p):   20m/sec.

MAGNIFICATION(m_f):   1

REFERENCE ANGLE (α_x)   18°, 0°

FIGURE 11.   (a) INTERFEROGRAM CONSTRUCTION GEOMETRY,
           (b) INTERFEROGRAM ($\alpha x=17.2^o$, 0°) AND
           (c) INTERFEROGRAM ($\alpha x-17.2^o$)

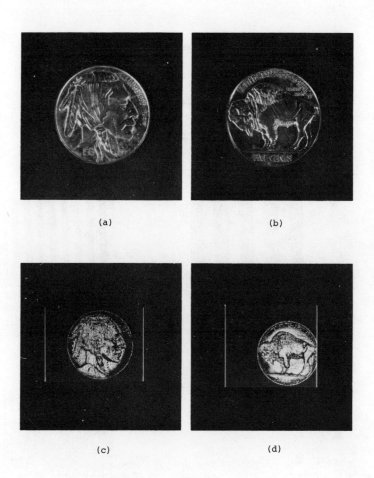

(a)   (b)

(c)   (d)

ACOUSTIC FREQUENCY: 11MHz
SCAN RATE:(V_x): 7.62cm/sec
PHASE SHIFTER
        PERIOD(T_p): 5msec
MAGNIFICATION (m_f): 0.25
REFERENCE ANGLE(αx): =21.3°

FIGURE 12. (a & b) ARE THE OPTICAL IMAGES AND (c & d) THE
           FOCUSED ACOUSTICAL INTERFEROGRAMS

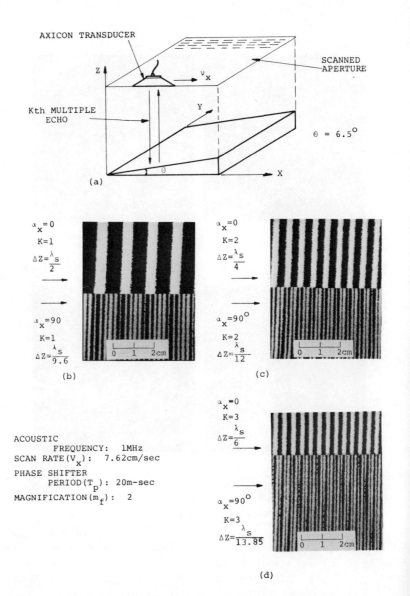

FIGURE 13. (a) INTERFEROGRAM CONSTRUCTION GEOMETRY, (b) INTE
FEROGRAM CONSTRUCTED WITH THE 1st SIGNAL ECHO (k=1), (c)
INTERFEROGRAM CONSTRUCTED WITH THE 2nd SIGNAL ECHO (k=2),
AND (d) INTERFEROGRAM CONSTRUCTED WITH THE 3rd ECHO (k=3)

Figures 12a & b are photographs of an enlarged nickel. This object was imaged using the interferometry system operating at 11MHz with the simulated reference. The fringe lines are extremely close on the head of the Indian and have merged together on the bison (see Figures 12c & d). The fringe density across either the Indian or bison is selectable by varying the inclination angle of the simulated reference electronically. Thus the operator has complete control of the fringe density without changing the frequency.

Figure 13a represents the interferogram construction geometry in conjunction with signal gating on the kth echo (i.e., number of multiple paths). The use of multiple echoes dramatically increases the system's depth deviation sensitivity. The depth deviation between fringes varies inversely as the number of echoes (i.e., $\Delta z_1 = \lambda s/2k$). Figure 13b shows the interferogram constructed by gating on the first echo (i.e., k=1). The reference angle was changed from 0° to 90° for each half of the interferogram. The 0° section exhibits fringes every half wavelength in depth. Interferogram (Figure 13c) was constructed by gating on the second echo and the fringes occur every one-sixth wavelength. The sections where the reference is not 0° exhibits different criteria for the occurrence of fringes and the depth deviations are easily calculated using Equation (9).

APPENDIX A

The Electronic Reference Phase Shifter

The phase shifter consists of six basic units each capable of phase shifts exceeding 170°, and linear to approximately 65°. The phase of the electronic reference signal is shifted by the FET-capacitance network as shown in Figure A.1. The FET's drain to source resistance is varied by the voltage applied between the gate and the source.

The equivalent circuit of the phase shifter is shown in Figure A.2. Writing the equations for the equivalent circuit shows

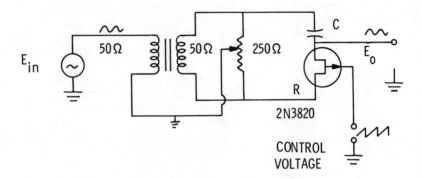

FIGURE A.1. SCHEMATIC OF THE PHASE SHIFTER UNIT

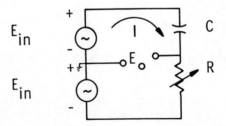

FIGURE A.2. EQUIVALENT CIRCUIT OF THE PHASE SHIFTER UNIT

$$I = \frac{2E_{in}}{R-jX_c}$$

$$\frac{E_o}{E_{in}} = 1 - \frac{2R}{R-jx_c} = \frac{(R + jX_c)}{R-jX_c} = - \frac{e^{j\phi}}{e^{-j\phi}} = e^{j} (\pi + 2\theta)$$

where $\theta = \tan^{-1} \frac{1}{\omega RC}$

The gain of the phase shifter (i.e., E0/Ein) is unity and the output amplitude (E0) is independent of either R or C. If R or C → 0 the phase is shifted 180°. Thus, each section theoretically is capable of shifting the phase 180°. In practice, with 2N3820 FET's the linear phase shift range was approximately 65° per unit and the complete system consisted of six units to obtain 360° phase shift.

The six units are connected in series and controlled by a variable frequency sawtooth generator. The system provides essentially constant output amplitude of the signal over the entire phase shift range.

REFERENCES

1)   B. P. Hildebrand, B. B. Brenden, Introduction to Acous-
     tical Holography, Plenum Press, New York (1972).

2)   M. D. Fox, W. F. Ronson, J. R. Griffin, R. H. Pettey,
     Acoustical Holography, Vol. 5, Editor P. S. Green,
     Plenum Press, New York (1973).

3)   B. P. Hildebrand, A General Analysis of Contour Holo-
     graphy (PhD Thesis), University of Michigan (1967).

4)   E. E. Aldridge, Acoustical Holography, Vol. 3, Editor
     A. F. Metherell, Plenum Press, New York (1971).

5)   J. H. McLeod, "The Axicon: A New Type of Optical
     Element", J. Opt. Soc. Am. 44, 592-597, 1954.

6)   C. B. Burckhart, P. A. Grandchamp, and H. Hoffman,
     Methods for Increasing the Lateral Resolution of B-
     Scan, Acoustical Holography, Vol. 5, Editor, P. S.
     Green, Plenum Press, 1973.

7)   C. B. Burckhart, "Ultrasound Axicon", J. Acoustical
     Soc. Am. 54, P. 1628, 1973.

FRINGE LOCALIZATION IN ACOUSTICAL HOLOGRAPHIC INTERFEROMETRY

W. S. Gan

Department of Physics, Nanyang University
Jurong Road, Singapore 22, Republic of
Singapore

ABSTRACT

We study the localization of fringes in acoustical double exposure, multiple-beams, time-averaged
and linearized subfringe holographic interferometry.
Both scanning and liquid surface methods are used.
The distance of the localization fringes to the object
surface is derived. It can be used to study the surface microstructure. Results from scanning method differ from the optical cases by showing the property of
sound wave as long wavelength and can be detected linearly. We also apply the correlation function method
to derive the intensity of acoustical holographic interference fringes for the scanning method. Results
from liquid surface method are the same as that from
optical holography because we do not consider the time
varying properties of the acoustical fields. We derive the acoustical transfer function for the liquid
surface method which is equivalent to the characteris-
fringe function in optical holography. We finally
discuss the applications of acoustical holographic interferometry.

INTRODUCTION

Acoustical holographic interferometry (or AHI)
was first initiated by M. D. Fox et al[1]. So far
there is no theoretical works on fringe localization

in AHI. In view of the intensive theoretical works on
fringe localization[2] in optical holographic interferometry and recent experimental works on AHI[3], it is useful to develop a theory of fringe localization for
this new field. For the optical case, fringe localization is due to the scattering of light from diffused
object surface. Since acoustical waves obey also
Snell's laws for specular reflections like light waves,
fringe localization will also occur for acoustical
waves scattering from diffused surface. A surface on
which fringes are said to be localized is one for
which fringe contrast or visibility is maximum. In
this paper we will consider this problem for double
exposure, multiple-beams, time-averaged and linearized
subfringe AHI using both liquid surface and scanning
methods.

LIQUID SURFACE DOUBLE EXPOSURE CASE

We first extend the method of R. K. Mueller et
al[4] to double exposure AHI as follows. Let the reference beam = $U_r(y) = P_r \exp(i\zeta_1 y)$, the first object

beam = $P_{o1}(X_i, Y_i) \exp\left[-i(\zeta_1 y + \phi_1(X_i, Y_i)\right]$, and second

object beam = $P_{o2}(X_i, Y_i) \exp\left[-i(\zeta_2 y + \phi_2(X_i, Y_i)\right]$.

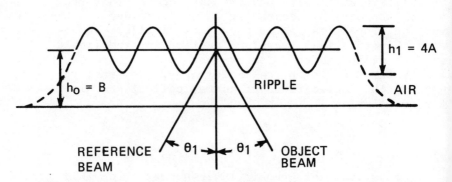

Fig. 1. Acoustical Holographic Interference Fringes
         shown as Ripple Pattern on Liquid Surface.

The meanings of the symbols are: ϕ_1, ϕ_2 = phase

change in the acoustic waves due to the object, $\zeta_1 = \frac{2\pi}{\lambda} \sin\theta_1$ and $\zeta_2 = \frac{2\pi}{\lambda} \sin\theta_2$ where λ = wave-

length of the acoustic wave. The directions of the acoustic waves are shown in Fig. 1. Assuming dealing with only "small" regions in the object, then $\phi_1(X_i,Y_i) = \phi_2(X_i,Y_i) = \phi(X_i,Y_i)$. So the intensity

of the acoustical holographic interference fringes

$$= (U_{o1} + U_r)(U_{o1}^* + U_r^*)/2\rho c + (U_{o2} + U_r)(U_{o2}^* + U_r^*)/2\rho c$$

$$= \frac{P_r^2}{\rho c} + \frac{[P_{o1}(X_i,Y_i)]^2}{2\rho c} + \frac{[P_{o2}(X_i,Y_i)]^2}{2\rho c} + \frac{P(X_i,Y_i)P_r}{\rho c}$$

$$\cos[2\zeta y + \phi(X_i,Y_i)] \tag{1}$$

where $P(X_i,Y_i) = \left\{[P_{o1}(X_i,Y_i)]^2 + [P_{o2}(X_i,Y_i)]^2 + 2P_{o1}(X_i,\right.$

$\left. Y_i)P_{o2}(X_i,Y_i)\cos[2y(\zeta_1 - \zeta_2)]\right\}^{1/2}$ and

$$\tan 2\zeta y = \frac{P_{o1}(X_i,Y_i)\sin 2\zeta_1 y + P_{o2}(X_i,Y_i)\sin 2\zeta_2 y}{P_{o1}(X_i,Y_i)\cos 2\zeta_1 y + P_{o2}(X_i,Y_i)\cos 2\zeta_2 y}$$

with $\zeta = \frac{2\pi}{\lambda} \sin\theta$ and U_{o1}, U_{o2} = object beams.

Now the localization condition is given by $\dfrac{\partial \phi}{\partial \phi_s} = 0$

Using the geometry of Fig. 2, we have

$$\phi = -(2\pi/\lambda)(x\alpha)(\sin\phi_i + \sin\phi_s) \tag{2}$$

and $d\phi/d\phi_s = 0$ gives

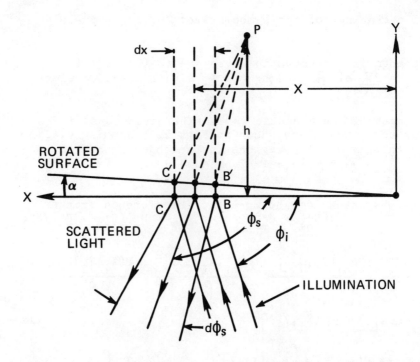

Fig. 2. Geometry for the Localization of fringes.

h = distance from the object surface to a point of
fringe localization = $- \dfrac{x \cos\phi_s \sin^2\phi_s}{\sin\phi_i + \sin\phi_s}$ (3)

The result is the same as that for the optical case
because we do not consider the time varying property
of the acoustical fields.

LIQUID SURFACE MULTIPLE-BEAMS CASE

Similarly for multiple-beams, the intensity of the
acoustical holographic interference fringes

$$= (U_{o1} + U_r)(U^*_{o1} + U^*_r)/2\rho c + (U_{o2} + U_r)(U^*_{o2} + U^*_r)$$

$$/2\rho c + (U_{o3} + U_r)(U_{o3}^* + U_r)/2\rho c + \cdots\cdots\cdots$$

$$+ (U_{on}^* + U_r)(U_{on}^* + U_r)/2\rho c$$

$$= \frac{nP_r^2}{2\rho c} + \frac{\left[P_{o1}(X_i,Y_i)\right]^2}{2\rho c} + \frac{\left[P_{o2}(X_i,Y_i)\right]^2}{2\rho c} + \cdots\cdots\cdots$$

$$+ \frac{\left[P_{on}(X_i,Y_i)\right]^2}{2\rho c} + \frac{P(X_i,Y_i)P_r}{\rho c}\cos\left[2\zeta y + \phi(X_i,Y_i)\right]$$

where
$$P(X_i,Y_i) = \left[P_{o1}(X_i,Y_i)\right]^2 + \left[P_{o2}(X_i,Y_i)\right]^2 + \cdots\cdots\cdots$$

$$+ \left[P_{on}(X_i,Y_i)\right]^2 + 2P_{o1}(X_i,Y_i)P_{o2}(X_i,Y_i)\cos\left[2y(\zeta_1 - \zeta_2)\right] + 2P_{o1}(X_i,Y_i)P_{o3}(X_i,Y_i)\cos\left[2y(\zeta_1 - \zeta_3)\right] +$$

$$2P_{o1}(X_i,Y_i)P_{o4}(X_i,Y_i)\cos\left[2y(\zeta_1 - \zeta_4)\right] + \cdots\cdots\cdots$$

$$+ 2P_{o1}(X_i,Y_i)P_{on}(X_i,Y_i)\cos\left[2y(\zeta_1 - \zeta_n)\right] + 2P_{o2}(X_i,$$

$$Y_i)P_{o3}(X_i,Y_i)\cos\left[2y(\zeta_2 - \zeta_3)\right] + 2P_{o2}(X_i,Y_i)P_{o4}(X_i,$$

$$Y_i)\cos\left[2y(\zeta_2 - \zeta_4)\right] + \cdots\cdots\cdots + 2P_{o2}(X_i,Y_i)$$

$$P_{on}(X_i,Y_i)\cos\left[2y(\zeta_2 - \zeta_n)\right] + 2P_{o3}(X_i,Y_i)P_{o4}(X_i,$$

$$Y_i)\cos\left[2y(\zeta_3 - \zeta_4)\right] + 2P_{o3}(X_i,Y_i)P_{o5}(X_i,Y_i)\cos\left[2y\right.$$

$$(\zeta_3 - \zeta_5)\right] + \cdots\cdots\cdots + 2P_{o3}(X_i,Y_i)P_{on}(X_i,Y_i)$$

$$\cos\left[2y(\zeta_3 - \zeta_n)\right] + \cdots\cdots\cdots + 2P_{o(n-1)}(X_i,Y_i)$$

$$P_{on}(X_i, Y_i) \cos\left[2y(\zeta_{n-1} - \zeta_n)\right]^{1/2} \tag{5}$$

and

$$\tan 2\zeta y = \frac{P_{ol}(X_i, Y_i)\sin 2\zeta_1 y + \ldots + P_{on}(X_i, Y_i)\sin 2\zeta_n y}{P_{ol}(X_i, Y_i)\cos 2\zeta_1 y + \ldots + P_{on}(X_i, Y_i)\cos 2\zeta_n y} \tag{6}$$

with $\zeta = \dfrac{2\pi}{\lambda} \sin\theta$. Now the localization condition

is given by $^2 \dfrac{d\psi}{d\phi}\Big|_s = 0$ where $\phi_1 = \phi_2 = \ldots = \phi$. This

yields the same expression for h as for the double exposure case.

LIQUID SURFACE TIME-AVERAGED CASE

Here we extend the results of K. A. Stetson[5] from the optical case to the acoustical case. Putting the acoustical object field

$$= P_o(X_i, Y_i)\exp\left\{-i\left[\zeta y + \phi(X_i, Y_i)\right]\right\}, \text{ we have the}$$

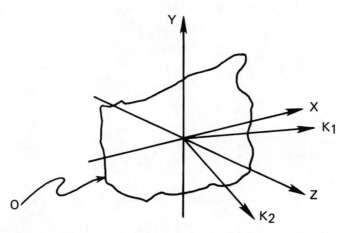

Fig. 3. Coordinate System for the Diffracted Acoustical Waves.

object field modified by the object motion

$$= S_t(X_i,Y_i,t)$$

$$= \iint_{-\infty}^{\infty} P_o(X_i,Y_i)\exp i\phi_t(X_i,Y_i,t)\exp i(-\zeta y)\, d\theta dy \tag{7}$$

The meanings of the symbols in eqn. (7) are shown in Fig. 3. The phase change ϕ_t now describes the modifications in the object field as a function of time. A time average hologram will reconstruct a field equal to the time average of equation (7).

$$\left\langle S_t(X_i,Y_i,t) \right\rangle = \frac{1}{T} \int_0^T \iint_{-\infty}^{\infty} P_o(X_i,Y_i)\exp\!\left[i\phi_t(X_i,Y_i,t) \right]$$

$$\exp\!\left[i(-\zeta y \right]\, d\theta dy dt \tag{8}$$

$$= \iint_{-\infty}^{\infty} P_o(X_i,Y_i)M_t(X_i,Y_i)\exp i(-\zeta y)d\theta dy$$

where M_t is defined by

$$M_t(X_i,Y_i) = \frac{1}{T}\int_0^T \exp\!\left[i\phi_t(X_i,Y_i,t) \right]dt \tag{9}$$

The localization conditions are given by:[5]

$$\frac{\partial}{\partial \theta} M_t(\phi) = \frac{d}{d\phi}\, M_t(\phi)\frac{\partial \phi}{\partial \theta} = 0 \tag{10}$$

and $$\frac{\partial}{\partial y} M_t(\phi) = \frac{d}{d\phi}\, M_t(\phi)\frac{\partial \phi}{\partial y} = 0 \tag{11}$$

According to Ref. (2) and Fig. 2,

$$\phi_t(X_i,Y_i,t) = \frac{-2\pi}{\lambda}\, D(x)\cos\omega t(\sin\phi_i + \sin\phi_s) \tag{12}$$

where $D(x)\cos t$ = displacement of the object. Hence the results are the same as that for the optical case.

LIQUID SURFACE LINEARIZED SUBFRINGE CASE

This is done by introducing an extra phase term
in the   expression of the ordinary AHI(double exposure,
multiple-beams or time-averaged) so as to increase the
sensitivity of the interference fringes to the object
motion.[6]   The results will be the same as those of
A. F. Metherell[6] and they will not be repeated here.

DERIVATION OF THE ACOUSTICAL TRANSFER
FUNCTION FOR THE LIQUID SURFACE METHOD

In liquid surface method acoustical holography,
the acoustical transfer function is defined as the ratio of the hologram ripple amplitude to the object
beam pressure amplitude. This has the equivalent mea-
ning as the characteristic fringe function[5] in optical
holography which is the function commonly observed as
hologram interference.

DOUBLE EXPOSURE CASE

The radiation pressure due to reflection of sound
at liquid surface $= \Pi(X_i,Y_i,y) = 2I(X_i,Y_i,y)/c$

$$= \frac{2}{c}\left\{ \frac{\left[P_{o1}(X_i,Y_i)\right]^2}{2\rho c} + \frac{P_r^2}{\rho c} + P_r P(X_i,Y_i)\cos\left[2\zeta y + \phi(X_i,Y_i)\right]\right.$$

(13)

where $P(X_i,Y_i) = \left\{\left[P_{o1}(X_i,Y_i)\right]^2 + \left[P_{o2}(X_i,Y_i)\right]^2 + 2P_{o1}(X_i,\right.$

$\left. Y_i)P_{o2}(X_i,Y_i)\cos\left[2y(\zeta_1 - \zeta_2)\right]\right\}^{1/2}$

and $\quad \tan 2\zeta y = \dfrac{P_{o1}\sin 2\zeta_1 y + P_{o1}\sin 2\zeta_2 y}{P_{o1}\cos 2\zeta_1 y + P_{o2}\cos 2\zeta_2 y}$

With $\Pi(X_i,Y_i,y) = \rho gz - \gamma\dfrac{\partial^2 z}{\partial y^2}$ (see Ref. 4 and Fig.1)

where γ = surface tension in force per unit length and
substituting the solution of

$$z(X_i,Y_i,y) = 2A(X_i,Y_i)\cos(2\zeta y + \phi(X_i,Y_i) + B(X_i,$$

$Y_i)$ in (13), we have

$$2\rho gA(X_i,Y_i) + 8\zeta^2 A(X_i,Y_i) = \frac{2}{\rho c^2} P_r P(X_i,Y_i)$$

by equating coefficients, giving $A(X_i,Y_i)=\dfrac{P_r P(X_i,Y_i)}{\rho c^2(\rho g+4\zeta^2)}$

and the acoustical transfer function

$$= \frac{A(X_i,Y_i)}{P(X_i,Y_i)} = \frac{P_r}{\rho c^2(\rho g + 4\zeta^2)} \qquad (14)$$

which is same as that for single exposure case.

MULTIPLE-BEAMS CASE

Similarly for the multiple-beams case, the acoustical transfer function is also the same as that for the single exposure case since it is independent of the number of object beams involved.

TIME-AVERAGE CASE

The acoustical transfer function in acoustical holography is equivalent to the characteristic fringe function in optical holography. So the required expression is

$$M_t(X_i,Y_i)=\frac{1}{T}\int_0^T \exp\left[i\phi_t(X_i,Y_i,t)\right]dt$$

which is the same as that for the optical case.

LINEARIZED SUBFRINGE CASE

The result is the same as that for the optical

case because we are not considering the time-varying
property of the acoustical fields.

SCANNING METHOD DOUBLE EXPOSURE CASE

For the scanning method we take account of the
time varying property of the acoustical fields. This
method cannot be treated by the ordinary way of simply
adding the object beam and the reference beam to obtain
the holographic interference fringes. This is because
for time varying functions we have to convert them into exponential functions in order to be multiplied by
their complex conjugates to obtain the intensity of the
holographic interference fringes. In so doing the
time varying terms of the acoustical fields will cancel out and we could not obtain the required solutions.
So here we will use the correlation function method.

Let $a_1(x,y)\cos \omega_1 t + \phi_1(x,y)$, $a_3(x,y)\cos \omega_1 t + \phi_3(x,y)$

and $a_2(x,y)\cos \omega_1 t + \phi_2(x,y)$ represent the two object

beams and the reference beam in the scanned plane respectively. For the first hologram, the cross-
correlation function, $\Gamma_{12}(\tau)$ is

$$\int_0^{2\pi/\omega_1} a_1(x,y)\cos \omega_1 t + \phi_1(x,y)\ a_2(x,y)\cos[\omega_1(t+\tau)$$

$$+ \phi_2(x,y)]\ dt$$

$$= a_1(x,y)a_2(x,y)\frac{\pi}{\omega_1}\left\{\cos \omega_1\tau\cos(\phi_1 - \phi_2) + \sin\omega_1\tau\right.$$

$$\left.\sin(\phi_1 - \phi_2)\right\} \tag{15}$$

Using Wiener-Kintchine theorem, the spectral intensity

$$=M_{12}(\omega)=\frac{1}{2\pi}\int_0^{2\pi/\omega_1} a_1(x,y)a_2(x,y)\frac{\pi}{\omega_1}\left\{\cos \omega_1\tau\cos(\phi_1 - \phi_2)\right.$$

$$+ \sin\omega_1 \sin(\phi_1 - \phi_2) \exp(-i\omega_1\tau)d\tau$$

$$= \frac{1}{2\pi} a_1(x,y)a_2(x,y)\left(\frac{\pi}{\omega_1}\right)^2 \exp\left[-i(\phi_1 - \phi_2)\right] \qquad (16)$$

Similarly for the second hologram,

$$M_{32}(\omega_1) = \frac{1}{2\pi} a_3(x,y)a_2(x,y)\left(\frac{\pi}{\omega_1}\right)^2 \exp\left[-i(\phi_3 - \phi_2)\right] \qquad (17)$$

So the total(holographic interference fringes) spectral intensity $= M(\omega_1)$

$$= \frac{1}{2\pi} a_1(x,y)a_2(x,y)\left(\frac{\pi}{\omega_1}\right)^2 \cos(\phi_1 - \phi_2) + \frac{1}{2\pi}$$

$$a_3(x,y)a_2(x,y)\left(\frac{\pi}{\omega_1}\right)^2 \cos(\phi_3 - \phi_2)$$

$$= \frac{1}{2\pi} a(x,y)a_2(x,y)\left(\frac{\pi}{\omega_1}\right)^2 \cos(\phi - \phi_2) \qquad (18)$$

by taking only the real parts of (16) and (17) and with

$$a(x,y) = \left[a_1^2(x,y) + a_3^2(x,y) + 2a_1(x,y)a_3(x,y)\right.$$

$$\left. \cos(\phi_3 - \phi_1)\right]^{1/2} \text{ and } \tan\phi = \frac{a_1\sin\phi_1 + a_3\sin\phi_3}{a_1\cos\phi_1 + a_3\cos\phi_3}$$

Now the required fringe localization conditions are given by

$$\left(\frac{\partial M}{\partial\phi}\right)_{\omega_1} = 0 \text{ and } \left(\frac{\partial M}{\partial\omega_1}\right)_{\phi} = 0 \qquad (19)$$

$$\left(\frac{\partial M}{\partial\omega_1}\right)_{\phi} = \frac{1}{2\pi} a(x,y)a_2(x,y)\pi^2(-2\omega_1^{-3})\cos(\phi - \phi_2) = 0 \qquad (20)$$

$$\left(\frac{\partial M}{\partial \phi}\right)_{\omega_1} = \frac{1}{2\pi} \, a(x,y)a_2(x,y)\left(\frac{\pi}{\omega_1}\right)^2\left[-\sin(\phi - \phi_2)\right] = 0 \tag{21}$$

By equating (20) and (21), we have

$$\phi = \phi_2 + \tan^{-1}\left(\frac{1}{\omega_1}\right) \tag{22}$$

which gives the required relationship between the phase change due to the object and the sound frequency for localization.

SCANNING METHOD MULTIPLE-BEAMS CASE

Following the same procedure as for the double exposure case, we have the total spectral intensity due to the holographic interference fringes

$$= \frac{1}{2\pi} \, a_1(x,y)a_2(x,y)\left(\frac{\pi}{\omega_1}\right)^2 \cos(\phi_1 - \phi_2) + \frac{1}{2\pi} \, a_3(x,y)$$

$$a_2(x,y)\left(\frac{\pi}{\omega_1}\right)^2 \cos(\phi_3 - \phi_2) + \frac{1}{2\pi} \, a_4(x,y)a_2(x,y)\left(\frac{\pi}{\omega_1}\right)^2$$

$$\cos(\phi_4 - \phi_2) + \cdots\cdots\cdots + \frac{1}{2\pi} \, a_n(x,y)a_2(x,y)$$

$$\left(\frac{\pi}{\omega_1}\right)^2 \cos(\phi_n - \phi_2) \tag{23}$$

$$= \frac{1}{2\pi} \, a(x,y)a_2(x,y)\left(\frac{\pi}{\omega_1}\right)^2 \cos(\phi - \phi_2) \text{ where}$$

$$a(x,y) = a_1^2(x,y) + a_3^2(x,y) + \cdots\cdots\cdots + a_n^2(x,y)$$

$$+ 2a_1(x,y)a_3(x,y)\cos(\phi_3 - \phi_1) + 2a_1(x,y)a_4(x,y)$$

$$\cos(\phi_4 - \phi_1) + \cdots\cdots + 2a_1(x,y)a_n(x,y)$$

$$\cos(\phi_n - \phi_1) + 2a_3(x,y)a_4(x,y)\cos(\phi_4 - \phi_3) +$$

$$2a_3(x,y)a_5(x,y)\cos(\phi_5 - \phi_3) + \ldots\ldots + 2a_3(x,y)$$

$$a_n(x,y)\cos(\phi_n - \phi_3) + \ldots\ldots\ldots + 2a_{n-1}(x,y)$$

$$a_n(x,y)\cos(\phi_n - \phi_{n-1})^{1/2} \text{ and}$$

$$\tan\phi = \frac{a_1\sin\phi_1 + a_3\sin\phi_3 + a_4\sin\phi_4 + \ldots\ldots + a_n\sin\phi_n}{a_1\cos\phi_1 + a_3\cos\phi_3 + a_4\cos\phi_4 + \ldots\ldots + a_n\cos\phi_n}$$

$$(24)$$

Following the same procedure as for double exposure
case, we obtain the localization condition as

$$\phi = \phi_2 + \tan^{-1}\left(\frac{1}{\omega_1}\right) \quad \text{where } \phi \text{ is now given by eqn.}$$

(24).

SCANNING METHOD TIME-AVERAGE CASE

It is not possible to apply the correlation function method to the time-average case because it cannot deal with more than two different time-varying
functions; and for the time-average case we have
whole series of time varying functions. So here we
extend K. A. Stetson's work to the acoustical case.
Putting in the time varying term, we have the object
field modified by the object motion

$$= S_t(X_i, Y_i, \omega_1)$$

$$= \iiint_{-\infty}^{\infty} \mathscr{C}(X_i, Y_i)\exp\left[i\omega_1 t + \phi_t(X_i, Y_i, \omega_1)\right]$$

$$\exp\left[i(\omega_1 t + \underline{K}_2 \cdot \underline{R})\right]dX_i dY_i d\omega_1 \tag{25}$$

$$\therefore \left\langle S_t(X_i,Y_i,\omega_1) \right\rangle = \frac{1}{T} \int_0^T \iiint_{-\infty}^{\infty} \mathscr{C}(X_i,Y_i) \exp\left[i(\omega_1 t + \right.$$

$$\left. \phi_t(X_i,Y_i,\omega_1) \right] \exp(i\underline{K}_2 \cdot \underline{R}) dX_i dY_i d\omega_1 dt$$

$$= \iint_{-\infty}^{\infty} \mathscr{C}(X_i,Y_i) M_t(X_i,Y_i,\omega_1) \exp\left[i(\omega_1 t + \underline{K}_2 \cdot \underline{R})\right]$$

$$dX_i dY_i d\omega_1 \tag{26}$$

where $M_t(X_i,Y_i,\omega_1) = \frac{1}{T} \int_0^T \exp\left[i(\omega_1 t + \phi_t(X_i,Y_i,\omega_1)\right] dt$

with T = exposure time, $\mathscr{C}(X_i,Y_i)$ = acoustic wave spec-

trum and the rest of the symbols have their meanings
shown in Fig. 3. So for the acoustical case, ϕ_t

of the optical case is now modified to $\phi_t' = \omega_1 t + \phi_t(X_i,$

$Y_i,\omega_1)$ and so the required fringe localization con-

ditions are: $\dfrac{\partial}{\partial X_i} M_t(\phi') = \dfrac{d}{d\phi'} M_t(\phi') \dfrac{\partial \phi'}{\partial X_i} = 0 \tag{27}$

$$\dfrac{\partial}{\partial Y_i} M_t(\phi') = \dfrac{d}{d\phi'} M_t(\phi') \dfrac{\partial \phi'}{\partial Y_i} = 0 \tag{28}$$

and $\dfrac{\partial}{\partial \omega_1} M_t(\phi') = \dfrac{d}{d\phi'} M_t(\phi') \dfrac{\partial \phi'}{\partial \omega_1} = 0 \tag{29}$

If we know the expression for the phase change due to
the object motion, we can evaluate eqns. (27) to (29)
and obtain the relationship between phase change due
to object and the sound wave frequency during localization.

SCANNING METHOD LINEARIZED SUBFRINGE CASE

Like the liquid surface case, this is done by adding an extra phase term to the phase change due to object of the ordinary AHI. For the double exposure and the multi-beams cases, the localization equation eqn. (22) is now modified to

$$\phi + \frac{\lambda}{4} = \phi_2 + \tan^{-1}\left(\frac{1}{\omega_1}\right) \qquad (30)$$

where $\omega_{\mp}$ sound wave frequency.

For the time average case, this is done by adding the extra time varying phase modulation $b \cos\omega_1 t$ to $\phi_t'(X_i, Y_i, \omega_1)$ of the acoustical time

average case(scanning method). We then find the resultant phase change and substitute it in eqns. (27) to (29) to find the required localization relationships between the phase change due to object and the sound wave frequency. This can be done if we know the expression for $\phi_t(X_i, Y_i, \omega_1)$.

CONCLUSIONS

We have done the analysis of fringe localization conditions for double exposure, multiple-beams, time average and linearized subfringe AHI using both liquid surface and scanning methods. Results from liquid surface method are the same as that from optical holography because we do not consider the time varying properties of the acoustical fields. Results from scanning method differ from the optical cases by showing the property of sound wave as long wavelength and time varying. This paper consists of theoretical works on fringe analysis for some of the experimental situations which have not been done and may serve as predictions for interference fringe patterns. AHI will have advantages over optical holographic interferometry such as the capability to photograph the internal cracks and flaws of materials nondestructively.

REFERENCES

1. M. D. Fox, et. al. , "Theory of Acoustic Holographic Interferometry", Acoustical Holography,
   vol. 5, Plenum Press, New York, (1974).

2. R. J. Collier, et. al. , "Optical Holography",
   First Ed., Academic Press, New York and London,
   p. 426-437, (1971).

3. K. Suzuki, et. al. , "Holographic Interferometry with Acoustic Waves", Acoustical Holography,
   vol. 6, Plenum Press, New York, (1975).

   H. D. Collins, "Acoustical Interferometry using
   Electronically Simulated Variable Reference and
   Multiple Path Techniques", Acoustical Holography,
   vol. 6, Plenum Press, New York, (1975).

4. R. K. Mueller, et. al. , "Sound Holograms and
   Optical Reconstruction", Applied Physics Letters,
   9:328-329, (1966).

5. K. A. Stetson, "A Rigorous Treatment of the
   Fringes of Hologram Interferometry", Optik, 29:
   386-400, (1969).

6. A. F. Metherell, "Linearized Subfringe Interferometric Holography", Acoustical Holography,
   vol. 5, Plenum Press, New York, (1974).

ACOUSTIC IMAGING TECHNIQUES FOR REAL-TIME NONDESTRUCTIVE TESTING

G.C. Knollman, J. Weaver, J. Hartog,
and J. Bellin

Lockheed Research Laboratory
Palo Alto, California 94304

INTRODUCTION

Applications of acoustical imaging techniques to non-destructive testing began to receive special attention about a decade ago. Practical systems for real-time nonintrusive ultrasonic inspection are now being developed.[1-4] Also, acoustic holographic techniques are being considered for certain nondestructive materials testing.[5-8]

In this paper we describe real-time ultrasonic imaging as applied to the nondestructive test of explosive loads. These loads are·tapered cylindrical billets in which spherical voids as small as 2.5 mm in diameter are to be detected and identified. Both single-element solid-state ultrasonic receivers and electronically scanned linear piezoelectric arrays[9] are considered, together with appropriate focusing and insonification systems. Scanning techniques and associated displays are described. Special requirements are cited such as pulse operation, focused shadow imaging, depth of focus, receiver gating, acoustic shielding, and contrast control. Two acoustic imaging systems designed for on-line inspection are presented based on helical scan.

Although the acoustic imaging methodology presented here is based on flaw detection in an object of specific geometry and composition, the concepts involved can be applied in similar fashion to the nondestructive test of any material object. Thus, explosive billets which are the present objects for rapid on-line viewing are but a specific application of our ultrasonic nondestructive testing systems to a material which is acoustically highly attenuative and of curvilinear shape.

BASIC CONSIDERATIONS

The billet to which the present forms of ultrasonic imaging system are addressed typically is a formulation of RDX with urethane base encased in a thin plastic beaker. Its configuration is a tapered cylinder about 8 cm in maximum diameter and approximately 48 cm in overall length. The well-known geometrical shape is that of a projectile. Inert formulations were utilized in all of the test programs.

Measured (nominal) acoustical properties of the explosive material are as follows: $c = 1.80 \times 10^5$ cm/sec, $\alpha = 15$ dB/cm, $\rho = 1.65$ g/cm^3. Relevant properties of the polyethylene sheath of thickness 0.127 cm are: $c = 2.20 \times 10^5$ cm/sec, $\alpha = 4$ dB/cm, $\rho = 0.94$ g/cm^3. Values for sound speed and attenuation are given for an acoustic frequency $f = 1$ MHz since this value was selected as the nominal acoustic imaging frequency. This selection was based principally on the billet material properties cited above and on the minimum flaw size (2.5 mm) to be detected.

The acoustic intensity degradation relationship in a material is given by

$$I/I_o = \exp(-2\alpha r),\qquad\qquad(1)$$

where I_o is initial intensity, I the intensity after an acoustic wave has transversed a thickness r of material, and α the attenuation coefficient in nepers per unit thickness of material. For the small acoustic paths of interest here, loss in intensity due to geometric spreading is negligible compared with the attenuation loss.

For typical piezoceramic transducers, a reasonable maximum ultrasonic output (and hence representative acoustic input for a billet) is $I_o \sim 10$ W/cm^2 for continuous operation. The threshold intensity for a piezoelectric acoustic imaging receiver is $I \sim 10^{-11}$ W/cm^2. Thus, an estimated acoustic power range is on the order of 120 dB. From Eq. (1) we establish the corresponding attenuation for 8 cm of billet material, namely $\alpha \sim 15$ dB/cm (1.7 nepers/cm). This value constitutes a rough figure for the acceptable acoustic attenuation such that ultrasonic penetration of an entire billet is feasible.

The aforementioned attenuation figure establishes an approximate upper frequency limit for acoustic operation. Measurement of attenuation versus frequency gives this limit at about 1 MHz. A lower frequency bound was selected on the basis that sound wavelength in the explosive be equivalent to minimum flaw size (for purposes of flaw detectability). This criterion gives about 0.7 MHz as the lower frequency limit. Of course, the higher the frequency, the better the resolution capabilities. Thus, the nominal acoustic imaging frequency was selected to be 1 MHz.

An ultrasonic system with a threshold sensitivity of approximately 10^{-11} W/cm^2 was developed for the present imaging problem. However, factors such as transducer directivity, acoustic losses owing to reflection at the billet boundaries, spreading losses from point defects, sound absorption in the beaker material (about 1 dB) and external immersion fluid (a few decibels), all imply that greater acoustic power input than that cited earlier may be required in practice. A level of 10 to 25 W/cm^2 has been achieved for continuous operation (i.e., 100% duty cycle). Pulse excitation, where the duty cycle might be as low as 2 to 5%, permits a piezoceramic transducer to produce ultrasonic intensities (subject to cavitation-induced limitations) up to 500 W/cm^2.

Pulsed operation was found to be mandatory for acoustic imaging of billets placed in a small reverberant tank of fluid: to eliminate standing wave interference, to allow signal enhancement by gating out spurious reflections, and to overcome acoustic losses such as those mentioned above. With pulsed ultrasonic imaging, one can obtain uniform insonification at relatively small transmitter-to-detector distances.

Three methods are possible for acoustically imaging a void in the billet of interest (or in any object). One approach involves through-transmission of sound and is designated unfocused shadow imaging, since a shadow of the void is cast directly on the receiver plane. Focused shadow imaging occurs if an acoustic lens is interposed between transmitter and void or between void and receiver (or both) with the void located at the focus of either lens (or both lenses). A third approach utilizes acoustic energy scattered from the void. Here the receiver together with an appropriate collector/focusor is in general positioned in the vicinity of

the acoustic transmitter. A transmitter with focusing lens
may also serve intermittently as an appropriate receiver.
This method is known as backscatter imaging. In all circumstances, focusors may be replaced by concave transducers.

In the first case, diffraction of the acoustic waves
about the void, and hence diffusivity of the shadow, can be
assessed by considering diffraction at a simple straight
edge. Formulas for the positions of diffraction maxima and
minima, h_{max} and h_{min}, respectively, are given by[10]

$$h_{max} = [(2m+1) \lambda b(a+b)/a]^{1/2}, \tag{2}$$

$$h_{min} = [2m\lambda b(a+b)/a]^{1/2}, \tag{3}$$

where a is the distance of the acoustic source from the
straight edge, b the distance of the receiver plane from
the straight edge, and λ the acoustic wavelength in the diffracting medium (the billet); m is zero or a positive in-
teger.

One may regard the position of the first diffraction
maximum (m = 0) as representative of shadow resolvability
R. Minimum resolution occurs with the source far removed
(a $\gg$ b), whereupon

$$R_{min} = (\lambda b)^{1/2} \tag{4}$$

If one takes R_{min} = 1.25 mm, the radius of a spherical void
of minimum size, one obtains from Eq. (4) that b $\approx$ 0.9 mm.
That is, at best minimum-size flaws will be resolvable only
if they are located within about a millimeter of the billet
surface. This unsatisfactory situation leads to the conclusion that unfocused shadow imaging is not suitable for detection of minimum-size flaws no matter their location within
the billet. We show later that focused shadow imaging, on
the other hand, is appropriate for resolution of flaws of
minimum dimensions, regardless of their position in the
billet.

For the present problem, backscatter acoustic imaging
was not considered to be as good an approach as shadow imaging. Backscatter imaging relies for success on two very important factors. First, the void must scatter a major portion of the insonification energy rather than diffract or
absorb appreciable sound. Second, a significant amount of

the scattered energy must be intercepted and introduced into an acoustic receiver for good image discernibility. Even
if a spherical void ideally would scatter all of the incident sound energy, diffracting and absorbing none, back-
scattering would occur throughout a total solid angle of
2 π steradians. Thus, to intercept a reasonable portion of
the backscattered radiation, a sizable reflector or acoustic
lens is needed. Such requirements are not economical and
generally introduce problems of image aberrations and distortions. Critical-angle reflection effects occur for acoustic rays incident obliquely on a reflector or lens. In
addition, an ultrasonic receiver casts a shadow on an acoustic reflector, and large acoustic lenses may absorb an appreciable amount of incident sound energy. Acoustic power
requirements for insonification generally are less for shadow than for backscatter imaging, a factor of considerable
importance in the present circumstance of highly attenuative material under nondestructive test.

Two techniques are commonly employed to compensate for
deleterious imaging effects owing to mismatched refractive
properties of sample and surrounding medium. One approach
involves solid, plane-parallel matching blocks which have
interior surfaces that geometrically match the exterior
surfaces of the specimen under test, and which have acoustic
refractive properties identical to those of the test piece.
A second method calls for an immersion fluid with a sound
speed identical or very nearly equal to that of the test
material. Additionally, the refractive-index matching materials should have specific acoustic impedance as close as
possible to that of the test piece (for minimization of
acoustic reflection losses) and, or course, have low sound
absorption.

The advantage of an immersion-fluid approach over the
matching-block scheme is that precision machining is not required in the former case, and acoustic coupling of external
material to the test specimen is intimate. Immersion fluids
also simplify the mechanical scanning problem. Aqueous gly-
col constitutes utilitarian immersion fluid for ultrasonic
imaging purposes.

INSONIFICATION

Our basic insonification arrangement involves a Tek-
tronix 190A oscillator, whose output is a 1-MHz, continuous-

wave signal which is shaped by an FET gate into a sine-wave
train (300 pulses/sec) whose typical duration is 35 to 50
μsec. This pulsed wave train, in turn, excites a DX-60B
transmitter, which was altered to tune 1-MHz signals.

The insonifier features a 3.8-cm by 3.8-cm plate of PZT-4
piezoelectric which is air-backed and which is contained in
a water-tight aluminum housing. A thin Mylar sheet glued
to the front of the ceramic element provides protection
against corrosion and leakage. Owing to the very low attenuation of 1-MHz acoustic energy in the immersion fluid referred to earlier, we found the necessity for a Styrofoam-
lined aluminum box between the insonifier and the insonification lens in order to reduce interference from reverberation and spurious ultrasonic energy.

Through use of an appropriate spherical acoustic-trans-
mitter lens, one realizes an appreciable increase in power
density over that for unfocused sound transmission. For the
circumstance in which the insonifier and lens combination
was employed in water with a small 1-MHz ultrasonic receiver
placed at the focal point, we found an acoustic power gain
of approximately 13 dB over that for the comparable situation with the lens absent. This gain constitutes an acoustic power increase by a factor of twenty.

If a linear array of receivers (with an appropriate lens)
is utilized for ultrasonic imaging, insonification requires
a sound source of linear extent. Thus, a linear configuration of ultrasonic transducers can be employed in conjunction with a cylindrical acoustic lens in order to concentrate ultrasonic energy along a line in the object plane.
Or an extended cylindrical transmitter is appropriate.
With all such arrangements, however, substantially greater
electrical driving power is required (in proportion to the
insonified area) than for the spherical-lens transmitter
system in order to obtain a given sound-intensity level.
For an extended array, the burden of adequate insonification
falls on the power amplifier system.

FOCUSED IMAGING

Focusing of insonification energy within an object under
nondestructive test has been introduced above. Image focusing from the receiver standpoint now involves either just
a receiver of suitable size appropriately positioned with

respect to the object, or an additional acoustic lens inter-
spersed between object and receiver. (Although one or more
focusing reflectors or an array of acoustic lenses could be
utilized, a single ultrasonic lens is deemed adequate for a
system in which resolution is limited by acoustic wavelength.)
We discuss both forms of focused shadow imaging below, con-
fining our attention to single-element receivers. Our conclusions apply as well to receiver arrays. A fluid of sound
speed equal to that in the billet is assumed to be the immersion medium.

For rapid, real-time billet inspection, our attention has
been restricted to fixed-focus (transmitter and receiver)
lenses — that is, lenses which provide sufficient depth of
focus such that they can remain stationary for focusing purposes. We chose the depth of focal field as the billet
radius. This choice does not place inordinate requirements
on the lens in order to form sharp silhouettes of minimum-
size or larger flaws anywhere within the billet. With a
total focal depth equal to the billet radius and with rotation of the billet about its longitudinal axis, all sections
lying in the billet cross-sectional plane will come within
the focal region each revolution of the billet.

Rayleigh scattering theory indicates that particulate
identities scatter appreciable acoustic energy when the
sound wavelength is of the same order as the particulate
size.[11] Thus, one may roughly regard as detectable a void
in the billet material which is on the order of a wavelength
in dimensions. Flaws of size as small as 1.8 mm in the explosive billet therefore should be readily detectable at $f = 1$
MHz. Such small voids would not necessarily be clearly resolvable. These facts were verified experimentally.

Resolution R and focal length L of an acoustic lens are
given by [12,13]

$$R = \lambda F \quad , \tag{5}$$

$$L = 4 \lambda F^2 \quad , \tag{6}$$

where λ is the acoustic wavelength in the propagation medium
and F is the relative aperture, or the ratio lens focal
length Γ to effective lens aperture D:

$$F = \Gamma/D \tag{7}$$

Both resolution and focal range relationships above were
verified in the laboratory for a full-scale billet.

Focused Source, Unfocused Receiver

Figure 1 is an illustration of ultrasonic shadow imaging involving a fixed-focus sound source and a single-element
receiver of appropriate size which is suitably positioned in
the reception field. Transmitter focal point inside the billet (shown in cross section) is as indicated. We calculate
the receiver element size d required to completely intercept
the entire cone of acoustic rays emerging from the billet.
This calculation depends on the receiver's field of view
as well as its distance s from the billet surface. Also
involved is the F-number of the transmitter lens.

The angular field of view β of a single square receiver
element is given by[14]

$$\sin \beta = \lambda/d, \tag{8}$$

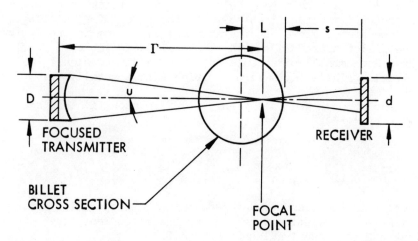

Fig. 1.   Shadow imaging with focused source and unfocused
         receiver

where λ is the wavelength of sound in the medium external to the receiver (and in the billet material). With reference to Fig. 1 we state, firstly, that $\sin \beta \geq \sin u$, where u is the indicated vertex half-angle subtended by the transmitter lens at the focal point. This requirement guarantees that the bundle of acoustic rays emerging from the billet will fall within the receiver's field of view. Assuming u to be small such that $\sin u \approx \tan u$, we write

$$\sin \beta \geq \tan u, \tag{9}$$

$$\lambda/d \geq D/2\Gamma, \tag{10}$$

where the left side of Inequality (10) arises from Eq. (8); D is the effective transmitter lens aperture and Γ is the lens focal length. Introduction of Eq. (7) on the right side of Inequality (10) yields

$$d \leq 2\lambda F, \tag{11}$$

where F is the relative aperture of the transmitter lens. Expression (11) places an upper bound on the receiver element size.

From a purely geometrical standpoint one now writes

$$d/(2s+L) \geq \tan u \tag{12}$$

by referring again to Fig. 1. Here s is the distance of the receiver element from the billet surface and L is the depth of field (taken equal to the billet radius). As before, Inequality (12) can be written in terms of F:

$$d \geq s/F + L/2F. \tag{13}$$

Substitution of Eq. (6) leads to

$$d \geq s/F + 2\lambda F. \tag{14}$$

Inequalities (11) and (14) are compatible only if s = 0 and

$$d = 2\lambda F.$$

That is, the receiver element must be placed in proximity to the billet surface and be of size given by Eq. (15) in order to intercept and detect the cone of acoustical rays emerging from the billet under ideal circumstances.

With focal depth L = 38 mm, Eq. (6) gives the requisite transmitter lens focal number F = 2.3 for the billet material; Eq. (15) then yields d = 8.3 mm. Such receiver element sizes are feasible, but are taken a little larger in practice since in reality the focal "point" is an Airy disc.

The maximum obtainable resolution of the single-element receiver is set by the thickness of the plate at its first resonant point.[15] For a PZT crystal, the minimum resolvable flaw radius in millimeters is

$$R_{min} = 1.15/f, \tag{16}$$

where f is the frequency in megahertz. With f = 1 MHz, R_{min} = 1.15 mm. Thus, a single PZT element in principle should be capable of detecting minimum-size flaws (2.5 mm diameter) in the billet. However, the actual resolution is given by Eq. (5) when applied to the transmitter lens.

Focused Source, Focused Receiver

In this section we discuss the technique of shadow imaging which involves focused insonification together with a receiver lens which is focused at the same position within the billet at which the transmitted rays are focused. An illustration of this arrangement is contained in Fig. 2. The position within the billet of the common focal point is as indicated where, as above, the focal range coincides with the radius of the billet. This dual-focus arrangement has provided the best acoustic imaging results for spherical and elongated voids situated at various positions in the interior of a billet of simulated material.

We calculated, in the previous subsection, that a focal range L = 38 mm requires F = 2.3 as determined from Eq. (6) and the wavelength of the billet material at f = 1 MHz. Corresponding object resolution is calculated from Eq. (5), R = 5 mm. Hence, given the receiver lens focal number which provides the requisite focal range in the cylindrical billet we obtain the aforementioned theoretical minimum resolvable void size.

Recalling the previous discussion on void detectability we conclude that minimum-size voids (2.5 mm) are expected to be readily detected but not sharply defined. Voids of size between minimum dimensions and the above-cited resolvable

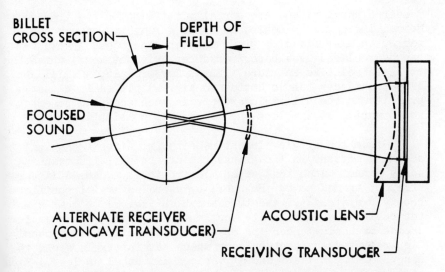

Fig. 2.   Shadow imaging with focused source and focused
          receiver

image dimensions will be imaged more distinctly. Large
voids, of course, should be clearly resolvable. Such were
the circumstances in reality.

ACOUSTIC RECEIVERS

Two basic concepts have been considered as designs of a
piezoelectric acoustic receiver for explosive billet inspection by means of ultrasonic imaging. The simpler concept
involves a single transducer element as receiver together
with mechanical motion of the receiver (and transmitter) and/
or billet to provide a two-dimensional scan of the interior
of the billet. A more complex approach is based on a fixed,
linear array of electronically scanned transducers positioned
in the focal plane of an acoustic lens. Linear or rotational motion of the billet occurs. The billet and associated
acoustic transducers are considered to be immersed in a medium of matched acoustic refractive index.

Single Element Receiver

The single-element receiver consists of a plane transducer
element and a positive acoustic lens, as depicted in Fig. 2.
Rays diverging from the focal point inside the billet are

rendered parallel and are normally incident on the transducer
Normal incidence is desirable since a flat transducer has a
narrow angular field of view. The lens and plane transducer
can be replaced by a bare transducer (without lens) which has
the same relative aperture. Thus, for example, a small concave transducer (shown dotted in Fig. 2) placed in close
prximity to the billet intercepts as much acoustic power as
would arrive at the lens and transducer combination.

Since the single-element receiver requires mechanical
scanning motions in two transverse directions, its scanning
speed cannot match that of an electronically scanned linear
array of transducers. However, by using several insonifier-
receiver systems (of the type shown in Fig. 2) to scan simultaneously several discrete sections of a billet, one can increase mechanical scanning speed significantly. The insoni-
fier-receiver systems would be spaced at intervals along the
length of the billet (and at various azimuthal angles). Each
system would scan a limited length of the billet and be adjusted to focus at the mean radius appertaining for each
particular section of billet. Sampling and multiplexing of
the outputs of the receivers would allow a common amplifier
and display system to be used. Thus, a series of mechanically scanned, single-element insonifier-and-receiver systems
acquires some of the speed of electronic sampling and yet
retains the advantages of good resolution and acoustic power
efficiency which are inherent in the doubly-focused insoni-
fier-receiver configuration.

Linear Receiving Array

An electronically scanned linear array of piezoelectric
transducers located in the focal plane of an acoustic lens i
depicted in Fig. 3. Voids in the uniformly insonified billet cast shadows which are imaged on the linear array.[9] Sequential electronic sampling of the row of small transducers
in the array generates one line of video information in the
display. Successive cross sections are similarly displayed
as the billet moves uniformly along its axis, thereby provid
ing a fast, television-type display of the interior of the
billet.

This technique for focused shadow imaging requires a un
formly insonified field, preferably collimated, as indicated
in Fig. 3. Hence, much higher acoustic power must be supplied to provide the same sound intensity that is obtainable

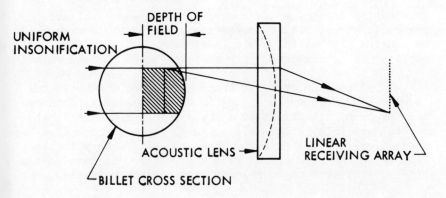

Fig. 3. Acoustic receiver consisting of electronically
        scanned linear array of transducers

with the focused insonification of Fig. 2. The insonified
field need not be rectangular in cross section but may be
focused so as to approximate a linear sheet of rays which
insonify only one line of the object field — that is, the
line being scanned by the linear receiving array. As we
noted earlier, such insonification can be obtained by use
of a rectangular, plane transducer with cylindrical lens,
or by means of a cylindrically concave transducer.

     For focused shadow imaging, the linear array requires
a large high-quality acoustic lens. We have stated in the
previous section that a relative aperture of about F/2.3
is needed to achieve the desired resolution and depth of
focus. The lens aperture should be several times larger
than the width of the linear array to avoid loss of resolution at off-axis image points. Plexiglas or polystyrene
lenses of adequate quality and aperture can be machined and
polished at reasonable cost.

     Measurements performed on several linear arrays devel-
oped by the authors have shown that the ultrasonic sensitivity of a sampled linear transducer array can approach that
of a single element. Furthermore, uniformity of response
from element to element can be held to ±5% of constant amplitude. This degree of uniformity is satisfactory for dis-

crimination of minimum-size voids in homogeneous billet
material. Thus, billet inspection by means of focused shadow imaging with either a single-element or linear-array
receiver is accomplished with state-of-the-art components.

A rectangular array of transducers can be substituted
for the linear array in Fig. 3 in order to scan a rectangular volume of the billet at a frame rate approaching that
of standard television. This approach is not recommended for
a practical inspection system because satisfactory inspection rates can be achieved with the simpler and less costly
linear array, and at much lower acoustic power levels.

Receiver Gating

Acoustic sensitivity (or noise-equivalent power input)
of a one-element receiver using a PZT ceramic transducer is
on the order of 10^{-11} W/cm^2. Sensitivity of a linear array
of transducers also approaches this level if the electronic
scanning rate is low enough (1% or less of the ultrasonic
frequency) to allow for filtering of switching transients.
To realize the aforementioned sensitivity in an inspection
system, however, one must hold the noise level resulting
from scattered sound below the receiver noise-equivalent
power level. For minimization of the scattered sound level,
elaborate acoustic shielding is required between the source
and the receiver. The shielding, in intimate contact with
a moving billet, should attenuate scattered sound by at
least 100 dB. This requirement can be relaxed if the receiver is gated on for only a short period at the arrival
time of the sound pulse. Scattered sound which arrives at
a later time may then greatly exceed the signal level without causing interference, since it is blocked from the amplifier.

The gated receiver rejects delayed signals, due to
scattered sound, which arrive via indirect paths. It also
rejects a small pulse at time zero caused by electromagnetic
pick-up from the insonifier supply. In addition, the receiver rejects pulses transmitted through the billet if the
billet cross section varies sufficiently over a longitudinal
scan to advance or delay the received signal out of the gating period. This occurrence is prevented when the transmitted pulse has a duration (at constant amplitude) equal
to the difference in propagation times through extremal billet diameters. In addition, the interval between pulses is

long enough to allow reverberations of scattered sound to
dissipate between successive pulses. This interval is approximately 2 msec in our imaging system.

Foregoing discussions apply equally to the linear receiving array if the advantages of receiver gating are to
be retained. Elements in the array are sampled sequentially
at the time of arrival of successive transmitted pulses.
Otherwise, operation of the array is identical to that of
a one-element system. However, rapid electronic scanning
(say, 10,000 elements/sec) is not compatible with the long
(2 msec) interval between pulses required for decay of reverberations. Reverberation time can be sharply reduced
if special efforts are made with regard to acoustic shielding.

MECHANICAL SCANNING AND IMAGE DISPLAY

Scanning techniques that are equally applicable to inspection of explosive billets are X-Y or "C" scan and helical scan. These techniques can involve either discrete,
focused, single-element receivers or a linear receiving
array. For each method of scan there is a corresponding
display representation. Both X-Y scanning and helical scanning are discussed below.

X-Y Scanning

Methods of X-Y scanning for a single-element receiver
require the receiver-lens combination to be aligned on the
axis of a focused insonifier with the two foci almost coincident. Insonifier and receiver are rigidly connected by a
yoke which is moved back and forth along the X-axis by means
of a reciprocating mechanism. The x-coordinate of the focal
point is a triangular function of time which sweeps across
successive lines parallel to the billet axis. Translation
of the acoustic system along both the X- and the Y-axis results in a rectangular field of scan inside the billet,
which remains stationary. Alternatively, the billet may be
transported along its axis while the acoustic system is
constrained to motion along the Y-axis. The scanning mechanism is greatly simplified if the billet is transported
along its axis, but design of the sound baffle that is required between the source and the receiver is more complex.

Electronic sampling of a linear array can replace one mechanical translation and enables a high line-scan rate. The ratio of the scan velocities along the X- and the Y-axis is such that successive scan lines are spaced at a distance no greater than the width of a detectable void in the billet. Again, motion along the Y-axis could be given either to the billet or to the acoustic system, but the predicted diameter and the focal length of the receiver lens dictate a strong preference for billet motion.

Image display for both methods of X-Y scan is identical. Voids in the billet appear as undistorted shadows projected on a plane normal to the acoustic axis. The X-Y scan cannot extend the full diameter of a billet since transmitted sound waves arrive almost at grazing incidence at the edge of the billet and are reflected. To view the entire interior of a billet, one would require a second scanning system disposed at right angles to the first. Simultaneous display of the two orthogonal X-Y scans would reveal all sectors of the billet and, in addition, would yield orthogonal coordinates for visual localization of most voids.

Optimum depth of field for focused shadow imaging in the billet extends only through the billet radius rather than through the full diameter. Thus, in X-Y scan each semicylindrical half of the billet must be scanned separately. Scan times become excessive unless a linear-array receiver is employed. Figure 4 shows schematically an approach for scanning the entire interior of a billet with a single linear receiving array. After each line is scanned, the billet is rotated 90° and simultaneously advanced along its axis by a fourth of a resolution length. The resulting display, indicated in Fig. 4, then corresponds to a shadow projection of the voids on the four sides of a rectangular prism concentric with the billet. One line is displayed in each projection per revolution of the billet. Since the electronic scanning is high, the billet can rotate continuously in a helical motion while synchronizing signals for the line sweep are generated by an optical sensor mounted on the shaft which drives the billet. Although the mechanical motion is helical, the display consists of a sequence of undistorted X-Y scans.

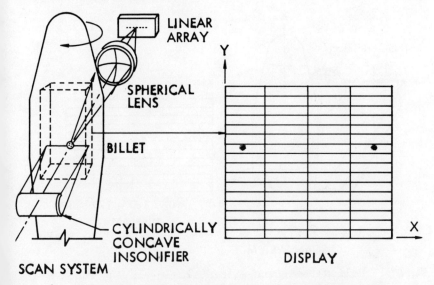

Fig. 4.   Alternate X-Y scanning with a linear receiving
          array

Helical Scanning

A helical scanning arrangement using a one-element
receiver is sketched in Fig. 5. The focused insonifier-
receiver system is mounted on a yoke which holds the acoustic axis coincident with a diameter through the billet.
Either the billet or the yoke moves parallel to the axis
of the billet (Z-axis) while the billet rotates about this
axis with constant angular speed. The resultant motion
yields a helical sweep of the acoustic axis through the
interior of the billet.

Position sensors provide deflection voltages to a
storage oscilloscope which displays the azimuthal and the
axial coordinates (ω, z) of the billet on the X- and on the
Y-axis, respectively. A typical helical display is represented in Fig. 5. The coordinate ω ranges from 0 to 360°
for a depth of field equal to one billet radius. Thus, all
voids in any one plane are displayed at least once per
revolution of the billet.

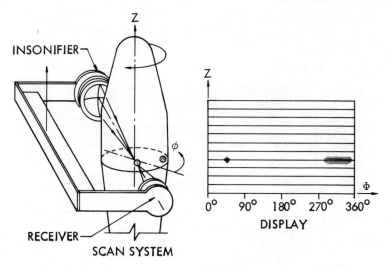

Fig. 5. Helical scanning with a single-element receiver

Voids appear as shadows projected on a cylindrical screen concentric with the billet axis, along which lies a virtual source. A void near the anatomical axis is closer to this source and subtends a larger angle than would the same void if located near the circumference of the billet. A void on axis obscures the source over all angles $0 < \omega < 360°$ and is displayed as a dark horizontal band. The z-dimension of a void is undistorted at any radial distance. Hence, the size of a spherical void is proportional to its z-dimension on the display, and its approximate radial position is indicated by its relative angular width.

Although the display for helical scan is not as easy to interpret as is that for an X-Y scan, the attainable scanning rate for a one-element receiver is much higher in helical scan by virtue of the rotational rather than the reciprocal motion. For automated inspection, where the display is of secondary importance, helical scanning in general is a convenient technique by which the entire interior of an object can be scanned with one simple acoustic system.

Rapid helical scan can be achieved by replacing the one-element receiver with a linear receiving array, as indicated in Fig. 6. The billet is insonified along a portion of its length (Δz) by a wedge-shaped beam which pro-

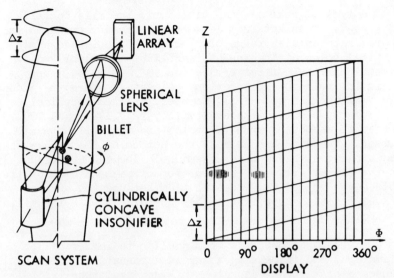

Fig. 6. Helical scanning with a linear-array receiver

vides high acoustic power density over a depth of one billet radius. This region is imaged on the linear array
by a spherical acoustic lens and is electronically scanned
200 to 300 times per revolution of the billet. As the billet rotates, it advances axially a distance Δz per revolution. For an n-element receiver array, this axial velocity
is about n times that which is obtainable with a one-element
receiver, since the electronic scanning time is negligible.

The ϕ-Z display for helical scanning with a linear receiving array is represented in Fig. 6. It is comprised of
successive angular sweeps having a width Δz which is scanned
rapidly by the linear receiving array. Voids are distorted
in the display exactly as they were in the case of Fig. 5.
Helical scan with a linear receiving array is more applicable to fast, automated billet inspection than to detailed
viewing of voids because of the high rate of scan.

For helical scanning, a section of billet was implanted
with vertical air-filled holes and spherical voids (2.5 mm
and 4.0 mm in diameter) at various depths within a billet.
Acoustic shielding consisted of urethane foam (thickness
about 2 cm). An immersion fluid of aqueous glycerol solution was used to provide an external fluid with sound speed
equal to that in the billet of simulant material. Image
distortion was thereby minimized.

A typical helical-scan ultrasonic image is contained in Fig. 7. Scan was from left to right and represents an approximate angular range of $0° < \varphi < 300°$. Shadows of vertical holes at the top of the billet are clearly discernible in their first projection ($0° < \varphi < 180°$) and again in the second projection $180°$ later, where the acoustic beam was not as sharply focused. The two spherical voids are projected at approximately $20°$, $130°$, and $200°$. Better focus occurs when the voids are closer to the receiver. The dark band across the bottom of the display is the projected shadow of a 2.5-mm hole located approximately on the billet axis. Over a portion of each revolution the hole moves out of the acoustic beam because the axis of billet rotation is not precisely coincident with the billet anatomical axis.

As the scan proceeds from smaller to larger billet diameters, the increasing acoustic attenuation of the billet causes a significant decrease in video amplitude (white level). This decrease is compensated by automatic video gain control. A cam on the vertical drive activates a potentiometer which controls the gain of the video amplifier.

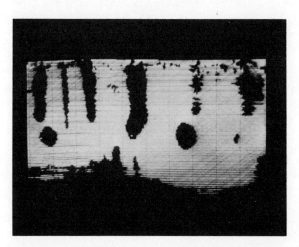

Fig. 7.   Helical scan of billet containing cylindrical and spherical voids

Scan speed for Fig. 7 was 84 rev/min or about 10 cm/min
along the billet axis. Ultimate speeds well above 15 cm/
min (maximum for the available scanner) are feasible since
fluid turbulence due to billet rotation shows no effect on
the reception of 1-MHz sound.

ULTRASONIC IMAGING SYSTEMS

Several possible acoustic imaging systems are applicable for rapid, on-line nondestructive tests of explosive
billets. A helical-scan system similar to the one described
earlier (single-element receiver) is a viable system for
billet inspection since it involves but the simplest acoustic components and is capable of achieving high scan speeds.
A helical-scan system which incorporates a linear array is
a more complex system but is capable of significantly higher
inspection rates than is the single-element arrangement.
The single-unit X-Y scanner is too slow for rapid, on-line
inspection. Practical ultrasonic imaging systems incorporating single-element and linear-array helical scan are discussed below.

Single-Element, Helical-Scan System

An ultrasonic helical scanning system for billet inspection requires substantial acoustic isolation of the
insonifier and the receiver. Acoustic shielding must be
held in intimate contact with the rapidly rotating billet.
The shield must be permanently contoured to fit the billet
or should be made sufficiently flexible so that it conforms
to a varying billet diameter. The former approach is advanced for the system sketched in Fig. 5.

The acoustic components and associated electronics for
the single-element helical-scan ultrasonic imaging system
are those variously described in this paper. However, a
spherically concave insonifying transducer (without lens)
and a smaller receiving transducer and lens are preferred.

Hybrid-Linear Array, Helical-Scan System

For high billet inspection rates, the single-element
receiver can be replaced by an n-element linear array with
associated acoustic lens. The linear scan rate could then be

increased almost by the factor n since each revolution of
the billet yields a helical display n elements wide.
Otherwise, the display is identical to that of the single-
element scanner, and shadows of voids are distorted by
radial projection as pointed out previously.

Undistorted projection of flaw shadows can be obtained
with the same linear-array components, but by means of the
arrangement suggested earlier in Fig. 4. The billet is
held, rotated, and transported by two collets which match
the billet contour and have an outside diameter approximately the same as the maximum diameter of the billet.
Thus, the billet and collets present a smooth, almost continuous cylindrical contour.

As the billet and collets rotate at constant speed,
they are pushed axially through an aperture comprised of
flexible acoustic insulation which is held in intimate
contact with the moving assembly. A rectangular aperture
in the insulation on each side of the billet provides a
shielded acoustic path from insonifier to receiver. The
insonifier is a cylindrically concave transducer focused
beyond the billet axis. The linear-array receiver is
centered vertically behind an acoustic lens which images on
the receiver, at unit magnification, a projected cross section of the billet.

For a four-faceted display, the array is scanned four times
per revolution of the billet. The four scans comprise one
line of the display. In this manner shadows of the voids
in the billet are projected on an imaginary four-sided
parallelepiped inscribed in the billet, as indicated by
the display in Fig. 4. Lens design, aperture, and acoustic
shielding requirements can be relaxed, however, if a six-
or even an eight-faceted display is adopted.

CONCLUSION

Although the real-time ultrasonic imaging systems
described herein were applied to a thick,attenuative solid
body of cylindrical shape, they have utility as well for
the nondestructive test of objects and materials of other
composition and configuration. Thin-walled tubing and pip-
ing immediately come to mind as a direct application.

Also, solid materials of any general cylindrical curvature can be readily accommodated with the basic sonovision concepts presented here. In a wider aspect, the methodology appertains to rapid on-line inspection of materials and components which are of irregular shape.

ACKNOWLEDGMENT

This investigation was supported under contract with the Naval Weapons Laboratory, Dahlgren, Virginia.

REFERENCES

1. J. E. Jacobs in Research Techniques in Nondestructive Testing (Academic Press, New York, 1970), Chap. 3.

2. M. G. Maginness and L. Kay, Radio and Elec. Eng. 41, 91 (1971).

3. B. B. Brenden, Mats. Res. and Stds. 11, 16 (1971).

4. G. Wade, ed. Acoustical Holography (Plenum Press, New York, 1972), Vol. 4, Sec. V.

5. J. L. Kreuzer and P. E. Vogel in Acoustical Holography (Plenum Press, New York, 1969), Vol. 1, Chap. 5.

6. A. F. Metherell, ed. Acoustical Holography (Plenum Press, New York, 1971), Vol. 3, Sec. III.

7. B. P. Hildebrand and H. D. Collins, Materials Res. and Stds., 12, 23 (1972).

8. G. C. Knollman and J. L. Weaver, J. Appl. Phys. 43, 3906 (1972).

9. G. C. Knollman, J. L. Weaver, and J. J. Hartog in Acoustical Holography (Plenum Press, New York, 1974), Vol. 5, p. 647.

10. G. S. Monk, Light, Principles and Experiments (McGraw-Hill, New York, 1937), p. 169.

11. R. B. Lindsay, Mechanical Radiation (McGraw-Hill, New York, 1960), p. 104.

12. A. E. Conrady, Applied Optics and Optical Design (Dover, New York, 1957), p. 126.

13. K. J. Habell and A. Cox, Engineering Optics (Pitman and Sons, London, 1958), p. 102.

14. T. F. Hueter and R. H. Bolt, Sonics (John Wiley and Sons, New York, 1955), p. 65.

15. J. E. Jacobs, Materials Evaluation 25, 41 (1967).

PERFORMANCE OF AN ULTRASOUND CAMERA TUBE UTILIZING PYRO-

ELECTRIC CONVERSION LAYERS

John E. Jacobs

Northwestern University

Evanston, Illinois 60201

The Sokoloff type ultrasound camera tube uses a piezoelectric target as the basic ultrasound-to-charge pattern
converting layer. The scanning beam interaction with the
conversion surface is such that only the energy in the
piezoelectric element during the time interval the scanning
beam is touching that element is effective in forming the
visual signal. This characteristic operation has precluded
for all practical purposes the use of pulsed ultrasound
radiation in systems utilizing the Sokoloff Tube.[1] The work
reported here involves the use of an ultrasound sensitive
pyroelectric target which provides image storage in the form
of a spatial temperature differential related to the absorbed
ultrasound energy. The signal output produced by the
scanning beam is directly related to the stored temperature
differential existing on the extended layer pyroelectric
material which is produced by absorption of the ultrasound
in the target material. It is only recently that highly
sensitive pyroelectric materials in a readily managed form
have become available.[2] The ability to store the incident
ultrasound energy for subsequent scanning permits utilization of the advantageous duty cycle relationship with regard
to the interaction of ultrasound with biological organisms.[3]
It should be emphasized that in this type of ultrasound
camera tube the signal at each picture point is, under
proper conditions, integrated continuously which serves to
increase the efficiency of detection. The practical limitation on obtaining this idealized operation is set by the
thermal loss associated with the supporting structure for
the pyroelectric conversion layer.

PYROELECTRIC ELEMENTS AS CAMERA TUBE TARGETS

The major effort in the development of pyroelectric detectors has been toward their use as detectors of infrared radiation. Although reference to the pyroelectric nature of tourmaline dates back to the eighteenth century, it has been only in the past decade that pyroelectricity has received attention as a detector of infrared, x-ray, and ultraviolet radiation. The majority of work reported has utilized triglycine sulphate, a crystalline substance.[4] The work reported here has resulted from the recent recognition that pyroelectricity exists in polymers in such an order of magnitude as to permit construction of a device whose sensitivity is theoretically in the order of 42 volts per watt absorbed.[5]

The pyroelectric detector is a thermovoltaic device consisting of a pyroelectric material whose spontaneous polarization and dielectric constant change with the temperature. It requires no bias once the detector has been poled with the DC field, and is not limited by Johnson noise of the pyroelectric material, but only by the noise of the reading beam. An added advantage of the pyroelectric material when used as the target for a television-type camera tube is that its resistivity is high enough to permit charge storage. Since the potential observed in a pyroelectric material occurs during temperature changes, it is important that the electrical resistivity of the material be great enough to permit retention of this potential until it is read by the scanning beam. The pyroelectric element in a sense parallels the piezoelectric element, the only difference being that the observed potential is a function of the temperature of the material in contrast to the piezoelectric element in which the observed potential is a function of the mechanical stress applied to the element. In fact, all pyroelectric materials exhibit piezoelectricity, although all piezoelectric materials do not exhibit pyroelectricity. It should be emphasized that the performance of the tube reported here depends on the storage of the detected ultrasound in the form of thermal energy, in contrast to the conventional storage of electrical charge.

BEHAVIOR OF PYROELECTRIC TARGET CAMERA TUBES

When a single domain of a pyroelectric material is

subject to a change in temperature, there is a corresponding change in the spontaneous polarization. In this domain, comprised of a thin slice of the material with the polarization axis perpendicular to the faces, the change is equivalent to an increment in surface charge. This increment is directly proportional to the pyroelectric coefficient of the material, its area, and its change in temperature. When an element is used as the target in a television camera tube, the signal current derived from the element is given by Equation 1. Equation 1 has important connotations with regard to the operation of the ultrasound image converter, since it shows that the signal responds only to the changes in the radiation level that result in changes in the elemental temperature. This has important consequences, first that the background radiation corresponding to the ambient temperature has no effect so that the normally encountered contrast of thermal images is greatly enhanced, and second that once the target has come into thermal equilibrium with a stationary object in the scene, the temperature will remain constant and no signal will be derived. This means then that incident radiation must be modulated, which is accomplished in the ultrasound image conversion tube discussed here by pulsing the ultrasound radiation.

$$i_s(t) = P \frac{\delta A}{\delta t} \left[T_m(t) - T_m(t - \tau_f) \right] \qquad \underline{\text{Equation 1}}$$

$i_s(t)$ = video signal derived from target

P = pyroelectric coefficient of material

δA = area of element of the target

δt = element scan time

T_m = element temperature

τ_f = integration time

In operation of the tube it is necessary to provide a second source of stabilization for the target to assure charging of the element during the time the pyroelectric potential is present. In the tubes reported here, this was accomplished by overscanning the target and utilizing the

secondary emission fall back much in the manner of the
iconoscope. Since the potential generated by the pyroelectric element is directly proportional to the temperature
differential, the final charge accumulated by the element
will be related to the maximum temperature attained by the
element during absorption of the incident ultrasound. This
potential is effectively stabilized by the secondary emission
characteristics, and is subsequently read by the scanning
beam. In some of the infrared camera tubes utilizing pyroelectric elements, the stabilization mechanism is by means
of an ion beam produced in the tube by controlled pressure
devices.

ULTRASOUND DETECTING TARGET CONSTRAINTS

As was noted above, the signal output of the television
camera tube is proportional to the temperature differential
of the pyroelectric film. This temperature differential is
produced by the ultrasound energy being absorbed in the film
itself. The maximum potential produced in the pyroelectric
material is dependent upon the temperature differential
which in turn is dependent not only upon the amount of thermal
energy absorbed from the ultrasound beam but also on the
loss of this thermal energy into the supporting structure.
Ideally, the pyroelectric detecting element should be coupled
to the ultrasound energy by a material transparent to the
ultrasound energy and at the same time a perfect thermal
insulator. For the purposes of the experiments reported
here, ordinary Pyrex glass is used for the faceplate. The
films used for these experiments were of such a thickness
that they absorbed approximately 10% of the incident energy
at a frequency of 3 Mhz. No attempt was made to match the
thickness of the pyroelectric layer to an integral number
of wavelengths, thus obtaining better absorption, however,
some experiments were run whereby a thin layer of colloidal
graphite was applied between the pyroelectric layer and the
faceplate to obtain more absorption of the ultrasound.
These tubes exhibited increased sensitivity in the order of
some 10 dB. Work is continuing to optimize the conversion
of the ultrasound energy into a temperature differential
for subsequent detection by the pyroelectric receiver.

For the tubes to be effective as receivers for pulsed
ultrasound it is necessary that the thermal storage capabil-

ity of the composite structure be optimized for the desired
duty cycle. Indications to date are, as anticipated, that
the storage time attained in this manner is a critical
function of the thermal characteristics of the composite
faceplate. A number of possibilities exist for modification
of the faceplate to achieve the desired end. Some of the
methods reported in connection with ultrasound optical
systems for use with piezoelectric image converters that
involve a fluid filled region within the tube appear to have
considerable merit with regard to the faceplate structure.[8]
It should be recognized that the pyroelectric material used
in these investigations is a vacuum stable film which is
quite pliable and hence could serve as the electron beam
scanned surface of an integral lens system.

CONSIDERATIONS REGARDING SENSITIVITY

As was noted above, the material used for these studies
(PVF_2) has a pyroelectric coefficient such that approximately
40 volts/watt absorbed is available. In the tubes reported
here, the measured sensitivity was such that a peak pulse
level of approximately 10^{-3} watts was observed experimentally. Since it is known that with the return beam scan and
secondary stabilization of the targets used for these studies
the minimum discernible potential on the target is in the
order of a few tens of millivolts, it may be seen that the
operating tube is at the present time less than 1% efficient.

The use of the pyroelectric converting layer removes
one of the major limitations to high frequency operation of
the Sokoloff-type image tube. In the Sokoloff-type tube,
maximum sensitivity and resolution capability occurs when
the piezoelectric converting layer is operated at the first
resonant point. As the frequency is increased, the thickness
of the layer to resonate at the higher frequencies decreases.
Inasmuch as the piezoelectric voltage for a given incident
pressure (and hence the ultrasound energy input) is proportional to the thickness of the piezoelectric converter, the
net effect is that the sensitivity of the tube decreases
with the increased frequency. This, coupled with the normal
absorption characteristics of the material (approximately
equal to the frequency squared) means that the system has a
net loss as frequency increases such that performance
above 20 Mhz is not too practical.

Quite in contrast, the pyroelectric materials exhibit
an increasing loss with frequency, and since it is the conversion of ultrasound energy to thermal energy that forms
the basis for the operation of the tube, the pyroelectric
ultrasound imaging systems increase in sensitivity approximately as the square of the frequency. The layer does not
exhibit resonance effects, hence a given layer of pyroelectric material (assuming the proper faceplate coupling
configuration) will maintain sensitivity across a wide
frequency band. This advantage of thermal receivers has
been recognized for some time in that attempts have been
made to produce ultrasound imaging devices utilizing thermal
receivers such as thermocouples. The relatively low sensitivity of the thermoelectric effect as well as the low
resistance of the thermal junction has precluded construction of a working system based upon the thermocouple
detector.[7]

PERFORMANCE OF THE SYSTEM

All the tubes constructed utilized PVF_2 films of sixty
microns thick. These films were polarized at a potential
of approximately 10^6 volts/cm. They were then checked as
targets of a thermal sensitive camera tube utilizing a
heated tungsten spiral. The sensitivity measured under these
conditions was such that the tungsten spiral could be seen
when its temperature was $15°C$ above ambient. Following this
test, the films were attached to a variety of transmitting
faceplates of varying thicknesses. Figures 1-3 illustrate
typical responses obtained from the targets fabricated in
this manner.

SUMMARY

The pyroelectric effect in easily handled polymers
permits the construction of an electron beam scanned ultrasound image converter tube which is responsive to pulsed
ultrasound energy. The characteristics of the pyroelectric
element functioning as the target of an ultrasound sensitive
electron beam scanned tube are such as to permit introductio
of image storage in the system. A number of factors influen
the ability to achieve perfect storage, and at this time
appear to be amenable to technological trade-offs. The

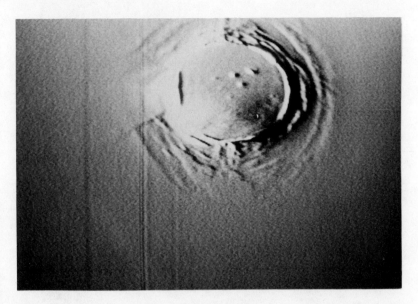

FIGURE 1   Stored image obtained with pyroelectric layer
polarized such that scanned surface assumes positive polarity
when irradiated with ultrasound energy. During pulsed ultrasound irradiation, image element attracts secondary electrons
to equalize ultrasound induced potential. Following cessa-
tion of ultrasound radiation, this ultrasound related negative charge repels the scanning beam and produces the positive or white signal shown. Frequency of ultrasound 3.6 Mhz.
Pulse duration 2×10^{-3} seconds. Pulse repetition rate
5 pulses/second. Faceplate thickness 3.2 mm. Graphite
conversion layer 10 microns. Pyroelectric conversion layer
thickness 50 microns. Minimum hole size resolved 3 mm.
Image formed by shadowing.

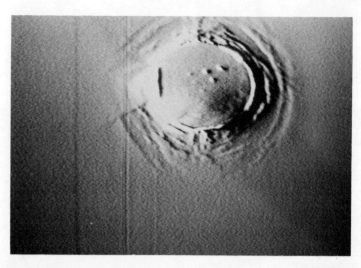

FIGURE 2   Same configuration as Figure 1 except frequency
of ultrasound is 10 Mhz.

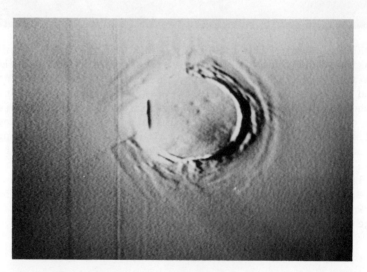

FIGURE 3   Same conditions as Figure 1 except pyroelectric
layer polarized such that scanned surface assumes negative
potential when irradiated by ultrasound.

principal advantage of the tube in addition to the storage feature is that the sensitivity increases with increasing frequency, inasmuch as the conversion of ultrasound to thermal energy (and hence signal output) of the tube varies as the frequency squared.

It is thought that the simplicity of the converter tube in its present form coupled with the potential technological improvements in the faceplate configuration to increase sensitivity will result in widespread application of this tube configuration to the visualization of extended area ultrasound fields.

This work supported in part by PHS Grant GM08522-13.

BIBLIOGRAPHY

1. Jacobs, J. E., "Ultrasonic Imaging Devices", chapter in Advances in Image Pickup and Display Devices, Vol. V, (B. Kazan, ed.) Academic Press, New York (In press).

2. Buchman, P., "Pyroelectric and Switching Properties of Polyvinylidene Fluoride Film", Ferroelectrics, 5, pp. 39-43, (1973).

3. Ulrich, U. D., "Ultrasound Doseage for Non-therapeutic Use in Human Beings - Extrapolations from a Literature Survey", IEEE Trans. of Biomedical Eng, BME-21 #1, pp. 48-51, (January, 1974).

4. Beerman, H. P., "Improvement in the Pyroelectric Infrared Radiation Detector", Ferroelectrics, 2, pp.123-128 (1971).

5. McFee, J. H., Bergman, J. G., and Crane, G. R., "Pyroelectric and Non-linear Optical Properties of Poled Polyvinylidene Fluoride Films", IEEE Trans. Sonics and Ultrasonics, SU-19, pp. 305-313, (1972).

6. Smyth, C. N., "The Ultrasound Camera - Recent Considerations", Ultrasonics 4 #1, pp. 15-21, (January, 1966).

7. Fry, W. J. and Dunn, F., "Ultrasonic Absorption in Microscopy and Spectroscopy", Proc. of Symposium on Physics and Non-destructive Testing, Southwest Research Institute - Publishers, pp. 33-59, (1972).

ANALYSIS OF THE RESPONSE OF PIEZOELECTRIC

RECEIVERS IN ACOUSTIC IMAGING SYSTEMS

Mahfuz Ahmed

Zenith Radio Corporation

Chicago, Illinois 60639

Resonant piezoelectric elements are being used as receivers in a number of acoustic imaging systems. These include the Sokolov[1] and pulse-echo systems[2], and others which use a linear or matrix transducer array[3,4]. The piezoelectric elements are highly desirable since they provide a higher sensitivity tnan any other type of acoustic signal detector when the charge induced on the surface by tne acoustic pressure is read directly with an electronic circuit. The sensitivity of these receivers is limited only by thermal noise which is the best signal detection condition (in terms of signal-to-noise) possible. By comparison, the acoustic signal in the laser-scanned systems[5,6] (the detector is the gold-coated surface of a thick nonresonant Plexiglas plate) is detected through the phase modulation of a scanning laser beam and its subsequent demodulation by a knife-edge or grating. This is a relatively inefficient process, which is shot noise limited, and the sensitivity of the laser-scanned system is, therefore, worse than those employing a piezoelectric element by several orders of magnitude[7]. Although a piezoelectric element (faceplate) is also used in the Sokolov system, the charge on the faceplate is read by sensing the secondary emission obtained from a scanning electron beam. This is also an inefficient technique and the sensitivity of the Sokolov system is on the same order of magnitude as that of the laser-scanned device.

The direct conversion of acoustic signals to electric signals by a piezoelectric element, however, has its drawbacks. Since the acoustic velocity in PZT is roughly twice that in (say) Plexiglas, the diffraction limited resolution obtained with a PZT faceplate is worse than that obtained with a Plexiglas faceplate by about a factor two. Furthermore, variations in the phase response of PZT occur more rapidly and, this can introduce artifacts in the image. These artifacts will be referred to as aberrations. As shown later on, the aberrations can be reduced considerably by employing a smaller numerical aperture.

The best theoretical resolution that can be obtained with the piezoelectric receiver, and the aberrations introduced by it in the image, are investigated here by deriving the phase and amplitude response of the receiver to an angular spectrum of plane waves. Computer plots of the responses are shown and their effect on resolution and aberration is calculated. It should be noted that these curves are obtained by assuming the piezoelectric plate to be of finite thickness but to be infinite in the other two dimensions. This analysis has been found to give results in good agreement with experimental data obtained with piezoelectric plates operating at 5 MHz.[8]

Due to additional reflections at the sides, tiny (1 mm x 1 mm) elements in a transducer array may act as resonant cavities rather than as a resonant plate, with modes of acoustic wave propagation somewhat different from that considered here. The numerical values represented by the response curves presented here may not, therefore, correspond to the actual response of the tiny elements. Nevertheless, the conclusions derived from these curves have very general validity and would be appliable to transducer arrays.

In a typical imaging system, the piezoelectric receiver which may be an array of tiny elements or a

faceplate, has water on one side and air on the other.
The acoustic energy, containing information about the
object being viewed, strikes the receiver-water interface. It can be resolved into its Fourier components
consisting of a two-dimensional spectrum of plane waves.
Each plane wave incident on the receiver causes a sinusoidal displacement ripple to propagate across the boundaries with water and with air. The generation of the surface ripple by the plane wave is shown in Fig. 1. The
sum of all ripples generated by the entire spectrum of
plane waves forms the final image. In a piezoelectric
material the sinusoidal surface ripples induce sinusoidal
waves of surface charge which propagate along the boundary. The sum of the "charge ripples" now constitutes the
final image. The resolution and the aberrations present
in the image depends on the phase and amplitude distortion of the "charge ripples" relative to the incident plane
waves. It is useful to consider this in somewhat more
detail.

The resolution of an imaging system may be defined
as the minimum separation Δx at which two point sources
can be resolved. It is governed by the width of the

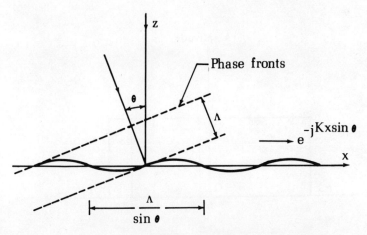

Fig. 1. Generation of a sinusoidal displacement ripple
by an incident acoustic plane wave.

object's angular plane wave spectrum that can be faithfully detected by the imaging system. Considering a two-
dimensional model, the angular plane-wave content of a
point source about the angle θ = 0 has a flat amplitude
spectrum. As shown in Fig. 1, a plane wave incident at
an angle θ causes a "charge ripple" pattern of the form
exp(-j K x sin θ), where K is the propagation constant of
the plane wave and the time dependence has been left out
for simplicity. An ideal receiver would be one in which
the peak amplitude of the "charge ripple" is the same for
any angle of incidence of a plane wave component of the
field. Furthermore, the phase relationship among the
ripples at any point on the surface would be the same as
that of the original plane wave components that created
them. Considering a flat amplitude response in sin θ
applicable to a line source, as shown in Fig. 2, the total
image field f(x), is given by

$$f(x) \propto \int_{-1}^{+1} \exp(-j K x \sin \theta') \, d(\sin \theta) \propto \frac{\sin K x}{K x} \qquad (1)$$

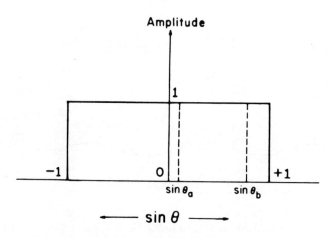

Fig. 2. Angular plane wave spectrum of a line source

which is shown in Fig. 3. The minimum separation distance Δx, mentioned earlier, is defined to equal half the separation between zeros by the so-called Rayleigh criterion[9]. In this case $\Delta x = \Lambda/2$, where Λ is the wavelength in water.

It should be noted that in the case of a point source, the image field corresponding to (1) is proportional to $J_1(x)/x$, where $J_1(x)$ is the first order Bessel function. The resolu- in this case is $\Delta x = 1.22 \ (\Lambda/2)$.

In any practical system the response is limited to angles θ such that $\theta_a \leq \theta \leq \theta_b$ (for instance). If the response is uniform within that range the image field is given by

$$f(x) \propto \int_{\sin \theta_a}^{\sin \theta_b} \exp(-j \ K \ x \ \sin \theta) \ d \ (\sin \theta)$$

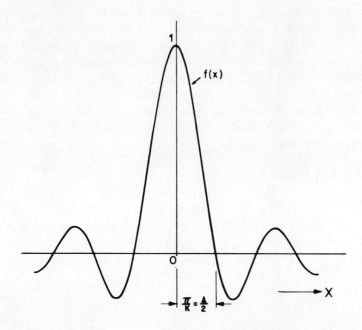

Fig. 3. Amplitude distribution in the image of a line source.

$$\propto \frac{\sin[(K\,x/2)(\sin\theta_b - \sin\theta_a)]}{(K\,x/2)(\sin\theta_b - \sin\theta_a)} \tag{2}$$

The response is thus similar in character to that of Fig. 3, but with increased width

$$\Delta x = \frac{\frac{1}{2}\Lambda}{\frac{1}{2}(\sin\theta_b - \sin\theta_a)} \tag{3}$$

In the more general case of nonuniform response, no simple conclusions can be drawn and the image field must be calculated by an integral of the form[10]

$$f(x) \propto \int \widetilde{H}(\sin\theta)\,\exp(-j\,K\,x\,\sin\theta)\,d(\sin\theta) \tag{4}$$

This integral can be evaluated quite readily by numerical techniques using a computer for any arbitrary complex function $H(\sin\theta)$. The intensity distribution is then given by $f(x)\,f^*(x)$, where the * denotes the complex conjugate. This function $\widetilde{H}(\sin\theta)$, which represents the relative phases and amplitudes of the "charge ripples" resulting from plane waves incident on the receiver, is a function of the properties, thickness and size of the receiver. It can be conveniently expressed as

$$\widetilde{H}(\sin\theta) = H(\sin\theta)\,\exp[j\,\phi(\sin\theta)] \tag{5}$$

in order to explicitly show the amplitude and phase part of $\widetilde{H}(\sin\theta)$. The image distortion brought about by the phase term $\phi(\sin\theta)$ is what has been called (above) the aberration. However, not all forms of $\phi(\sin\theta)$ cause unavoidable distortion. Thus if $\phi(\sin\theta)$ is written as

$$\phi(\sin\theta) = a\,\sin\theta + b\,(1-\cos\theta) + \phi_r(\sin\theta) \tag{6}$$

it may be shown readily that the first two terms represent the phase shift due to a lateral and longitudinal (defocus)

displacement of the image, respectively. The residual
term ϕ_r (sin θ) is responsible for the unavoidable distortion or aberration. According to the Maréchal criterion[11]
the aberration may be considered negligible when the
"mean-square deformation" $(\Delta \phi_r)^2$ does not exceed
0.45 rad. Here $(\Delta \phi_r)^2$ is defined as $\overline{\phi_r^2} - (\overline{\phi_r})^2$, with the
bars denoting average values.

The amplitude and phase responses, $H(\theta)$ and $\phi(\theta)$,
of the PZT-4 vacuum interface of a PZT-4 plate are
shown in Figs. 4 to 10. Since the density of air is very
small compared to that of PZT-4, these curves also
represent the response of the PZT-4-air interface quite
accurately. The parameter 'm' indicates the thickness

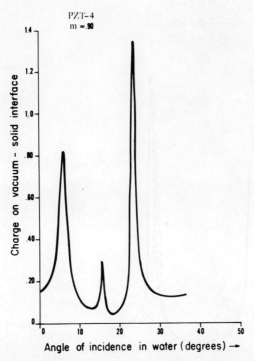

Fig. 4.   Amplitude of surface charge on PZT-4 - vacuum
          interface. 'm' is the thickness expressed in
          half-wavelengths of the fast wave.

of the plate in terms of the number of half-wavelengths
of the "fast" wave in the piezoelectric material. The
term "fast" wave will be explained shortly. These
curves are derived from an analysis of the complex
modes of wave propagation within a piezoelectric material, resulting from the interaction between acoustic
and electric fields and from the special crystallographic
symmetry of the material. PZT-4, for instance, has the
same elasto-electric matrix as the class 6 mm[12], which
is characterized by the 5 elastic coefficients c_{11}, c_{12},
c_{13}, c_{33} and c_{44}, the 3 piezoelectric coefficients e_{15},
e_{31} and e_{33}, and the dielectric constants ϵ_1 and ϵ_3. The
details of the analysis are too lengthy to be presented
here and will be published elsewhere.

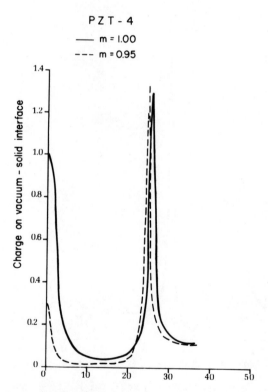

Fig. 5. Amplitude of surface charge on PZT-4 -
        vacuum interface.

Stated briefly, it was found that two types of acoustic waves can propagate in the piezoelectric medium. The velocity of each wave is dependent on the angle of incidence of the plane wave in water which generated them. One of the waves has a velocity of roughly 4×10^3 m/sec and will be referred to as the _fast_ wave, while the other has a velocity of roughly 1. 8 $\times 10^3$ m/sec and will be referred to as the _slow_ wave. In each case the displacement vector makes an angle other than 0° or 90° with the direction of propagation. Each wave is, therefore, a combination of a longitudinal and a shear mode, the exact amount of each being again a function of the angle

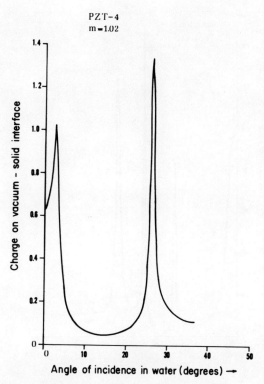

Fig. 6. Amplitude of surface charge on PZT-4 vacuum interface.

of incidence. The fast wave is, however, mostly longitudinal while the slow wave is mostly shear. Figure 11
shows the deviation of the fast and slow waves from being a purely longitudinal and a shear mode. It is seen
that this deviation represented by the angles τ_1 and τ_2 is
very small -- on the order of a few degrees. Except for
the fact that the fast and slow waves are not purely
longitudinal or shear, and that the velocity of each wave
is a function of the angle of incidence, the derivation of
the response is very similar to that of the isotropic
faceplate[13].

It is seen from Figs. 4 to 7 that the amplitude response has a very sharp peak at the angle of incidence

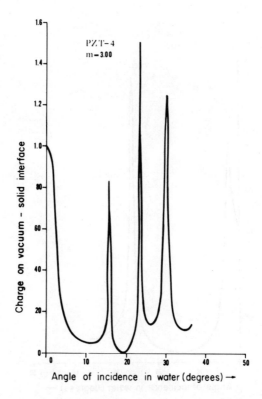

Fig. 7. Amplitude of surface charge on PZT-4-
vacuum interface.

corresponding to a resonance of the plate, and drops off very rapidly on either side. The number of peaks depends on the thickness of the plate, the thicker plates being resonant at several angles of incidence. At the same time, as shown in Figs. 8 and 9, the phase not only varies quite rapidly from 0° to some angle (the first critical angle being at 20° when m ~ 1. 0, m being the thickness expressed in half-wavelengths of the fast wave, as defined earlier) but undergoes quantum-like jumps at specific angles of incidence which depend on the plate thickness.

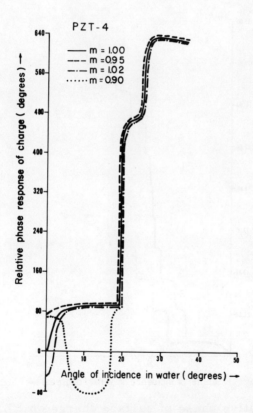

Fig. 8. Relative phase of surface charge on PZT-4-vacuum interface. 'm' is the thickness expressed in half-wavelengths of the fast wave.

Figure 10 shows the phase correction that can be obtained
by refocussing ⌈this is equivalent to removing a term of
the type $b(1-\cos\theta)$ mentioned in equation (6)⌉ and the residual phase ϕ_r which contributes to image aberration.

The nature of the aberration can be understood better
by considering Fig. 12 which shows the intensity distribution, obtained from a numerical calculation of (4), over
the angular range $-20° \le \theta \le 20°$. Here, curve 1 corresponds to a flat amplitude response with no phase shift

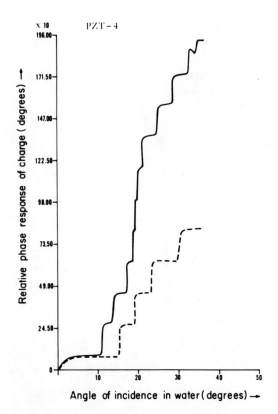

Fig. 9.   Relative phase of surface charge on PZT-4
          vacuum interface.   The solid line corresponds
          to m = 7 while the dotted line corresponds to
          m = 3.

and is given by $[\sin(K \times \sin 20°)/K \times \sin 20°]^2$. Using
the Rayleigh criterion we find that this gives the best
resolution. Curve 2 corresponds to a plate whose thickness is represented by m = 0.95. The amplitude response
between 0°-20° (Fig. 5) at this thickness is not as peaked
as in the other cases, and the phase response (Fig. 8) is
also relatively flat. Using a Rayleigh-like criterion, i.e.
a resolution element being taken to equal the width of the
main lobe corresponding to the 40% intensity points, the
resolution in this case may be considered to be $\sim 3 \Lambda$.
However, when two such curves are placed side by side,
spaced by the resolution element, as in the Rayleigh

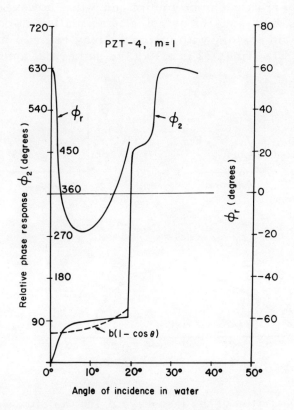

Fig. 10. Relative phase ϕ_2 and residual phase ϕ_r of
surface charge when m = 1.

criterion, instead of obtaining just two peaks the large
secondary and tertiary lobes would give rise to additional
ripples which would introduce artifacts in the image.
This has already been referred to as aberration. In each
of the curves 2, 3 and 4 the residual phase ϕ_r was used
rather than the phase response shown in Fig. 8. Curve 5
corresponds to m = 1.0 and the phase ϕ_2, shown in
Fig. 10, rather than ϕ_r. Comparing 5 and 3 (which corresponds to m = 1.0 and the residual phase ϕ_r) it is clear
that removal of a term such as b(1-cos θ) from the
phase ϕ_2 actually corresponds to a focussing operation
since resolution is improved.

The aberration can be minimized with, however,
some loss in resolution by going to a smaller numerical
aperture. In analogy with optics we may refer to the
term (sin θ_b - sin θ_a)/2 in (3) as the numerical aperture

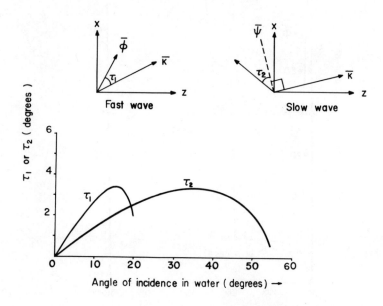

Fig. 11. Deviation of the fast wave $\bar{\phi}$ and the slow wave
$\bar{\psi}$ from being a purely longitudinal or a purely
shear wave.

N. A.   This is shown in Fig. 13, where the curves correspond to an angular aperture $-7° \leq \theta \leq 7°$. Again, curve 1
is given by $[\sin(K \text{ x } \sin 7°)/K \text{ x } \sin 7°]^2$, and 2 and 3
correspond to $m = 1.0$, $\phi = \phi_r$, and $m = 1.0$, $\phi = \phi_2$,
respectively. Side lobes are missing and the best resolution (curve 2) is $\sim 7.5 \Lambda$. Again we see that removal of
the term $b(1-\cos \theta)$ from the phase response corresponds
to focussing.

We have seen that a flat amplitude response with no
phase shift gives the best resolution without any aberration. It is possible to improve the response curves by
electronic phase and amplitude equalization (sometimes
referred to as inverse filtering), with, however, some
loss in sensitivity. This has not been simulated here and
could be the subject of an interesting study.

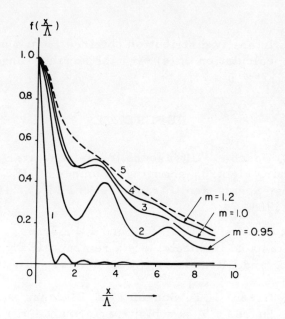

Fig. 12.   Intensity distribution in the image of a line
            source when its plane wave components are fil-
            tered by the response of the receiver. The
            curves are obtained through a numerical calcu-
            lation of (4) over the angular range $-20° \leq \theta \leq 20°$.

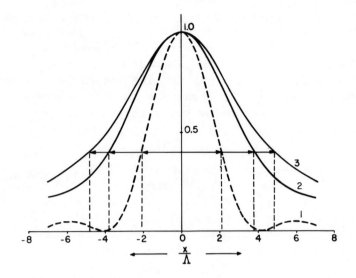

Fig. 13.   Intensity distribution obtained from a numerical
calculation of (4) over the angular range
$-7° \leq \theta \leq 7°$.

REFERENCES

1. J. E. Jacobs, "Ultrasound image converter systems
   utilizing electron-scanning tecnniques," IEEE,
   Trans. Sonics and Ultrasonics, SU-15, p. 146,
   July 1968.

2. K. R. Erickson, F. J. Fry and J. P. Jones,
   "Ultrasound in medicine - a review," IEEE, Trans.
   Sonics and Ultrasonics, SU-21, p. 144, July 1974.

3. P. S. Green, L. F. Schaefer, E. D. Jones and
   J. R. Suarez, "A new high-performance ultrasonic
   camera," in Acoustical Holography Vol. 5,
   P. S. Green ed., Plenum Press, New York, (1974),
   p. 493.

4. M. G. Maginness, J. D. Plummer and J. D. Meindl,
   "An acoustic image sensor using a transmit-receive

array," in Acoustical Holography Vol. 5, P. S. Green ed., Plenum Press, New York (1974), p. 619.

5. R. L. Whitman, M. Ahmed and A. Korpel, "A progress report on the laser scanned acoustic camera," in Acoustical Holography Vol. 4, G. Wade ed., Plenum Press, New York (1972), p. 11.

6. A. Korpel, L. W. Kessler and P. R. Palermo, "Acoustic microscope operating at 100 MHz," Nature, 232, p. 110 (July 1971).

7. A. Korpel and L. W. Kessler, "Comparison of methods in acoustic microscopy," in Acoustical Holography Vol. 3, A. F. Metherell ed., Plenum Press, New York (1971), p.23.

8. R. C. Addison, "A progress report on the Sokolov tube utilizing a metal fiber faceplate," in Acoustical Holography Vol. 5, P. S. Green ed., Plenum Press, New York (1974), p. 659.

9. M. Born and E. Wolf, Principles of Optics, 3rd edition, Pergamon Press, New York (1965), p. 333.

10. J. W. Goodman, Introduction to Fourier Optics, McGraw-Hill, New York (1968), pp. 48-54.

11. M. Born and E. Wolf, ibid. Chapter 9.

12. D. A. Berlincourt, D. R. Curran and H. Jaffe, "Piezoelectric and piezomagnetic materials and their function in transducers," in Physical Acoustics Vol. 1A, W. P. Mason ed., Academic Press, New York (1964), p. 174.

13. M. Ahmed, R. L. Whitman and A. Korpel, "Response of an isotropic acoustic imaging faceplate", IEEE, Trans. Sonics and Ultrasonics, SU-20, p. 323, October 1973.

STANDARD PHANTOM OBJECT FOR MEASUREMENTS OF GRAY SCALE

AND DYNAMIC RANGE OF ULTRASONIC EQUIPMENT *

Anant K. Nigam

C.B.S. Laboratories

Stamford, Conn. 06905

ABSTRACT

The present study was undertaken to design a Standard Phantom Object for measurements of gray scale and dynamic range of Ultrasonic Imaging equipments in accordance with the test procedures contained in the National Science foundation Experimental R&D Incentives Program, Experiement No. 5, NSF 73-34 (Revised May 17, 1974). Three Phantom Object geometries are studied and the required specifications are discussed in reference to high-resolution pulse-echo imaging techniques. The test objects are experimentally evaluated based on their performance with variations in the numerical aperture and pulse-spectrum of an imaging system as well as the surface finish and temperature of the Phantom Object. Performance at other frequencies and imaging modalities is discussed. Data is also included on the velocity and reflectance (including temperature effects) of various engineering materials of which the test Phantom Objects were fabricated during evaluation.

INTRODUCTION

The National Science Foundation Office of Experimental R&D Incentives' Experiment No. 5 outlines performance specifications for advanced ultrasonic imaging equipment for applications in Medical diagnosis[1]. The specifications

---

* This work was done under contract NSF-C-796, for the National Science Foundation.

contain the performance criteria for resolution, field of
view, image artifacts, dynamic range, gray scale, etc.,
provided by the equipment. The present paper outlines the
design of a Standard Phantom Object (SPO) that may be
utilized in testing the criteria in respect to dynamic
range and gray scale for a pulse-echo type imaging equipment. The Phantom when scanned across its surface is required to provide a succession of echoes whose amplitudes
vary in a consistent manner over a range in excess of 60dB.
If the equipment under test can image the entire Phantom,
its dynamic range capability is established. Furthermore,
by comparing the range in brightness in the displayed
image with the known range of echo strengths produced by
the Phantom, the gray scale capability of the equipment is
also determined. The latter feature can also be employed
in optimizing the equipment performance for preferential
gray scale expansion as may be desirable for certain diagnostic applications.

Equipment testing requirements as outlined in Ref. 1
determine the additional constraints on the design of the
Standard Phantom Object. An apparent constraint is on the
physical size of the phantom which should be smaller than
the field of view of the equipment. Performance constraints
require consistency and repeatibility with long-term storage and with the allowed temperature fluctuations during
operation (36 - 37.5°C).

The overall goal of the present task is to determine
a phantom object design that is simple to fabricate
and utilize in the clinical environment. It is expected
that in designing a phantom, consideration will have to be
given to the equipment characteristics. This includes the
center frequency f_O and spectral content $S(f)$ of the ultrasound pulse generated by the equipment, the capture aperture
of the transducer and the distance at imaging (the latter
two are represented by the numerical aperture $f_\#$ at imaging).
A broad range of values for these equipment parameters is
expected[1], accordingly the present testing philosophy has
been to select and optimize a phantom design for one set
of equipment parameters and to quantify the variations in
performance with variations in the parameter values. The
intent is to provide sufficient data to enable construction
of a phantom object for any particular pulse-echo equipment
with reasonable assurance of phantom performance. It is in
keeping with this intent that only commercially available

TABLE 1

Equipment Parameter Values. Expected range and values
selected for experimental testing

		Expected Range	Selected Value
Center Frequency, f_o	MHz	1.0–15.0	5.0, 10.0
Resolution, depth	mm	0.2–0.5	0.2–0.3
Transverse directions	mm	Rayliegh	0.4–1.0
Sidelobe Suppression	dB	40–60	60–70
Field of View each direction	cm	4–15	6–7
Aperture $f_{\#}$		1.5–4.0	1.7, 3.5
Maximum S/N	dB	60–80	80

materials which exhibit repeatibility in performance over
time have been considered here for phantom construction.

The chosen values of the equipment parameters employed
in testing are summarized in Table 1. A pulse-echo system
with these values was fabricated in the laboratory. The
equipment employs a novel apodization technique which is
simple to implement and provides high resolution with substantial sidelobe suppression[2] as summarized in Appendix A.

Appendix B lists the ultrasonic properties of certain
commercially available plastic materials which exhibit
acceptable repeatible behavior, ease of machining, weak
temperature dependence, and good shelf life. These
materials appear suitable for phantom construction.

In the following section we present three different
phantom designs which appear useful. These are experimentally evaluated to determine the optimal design.

The change in performance of the phantom with variations in the values of the equipment parameters is studied

experimentally and analytically. In the analytical analysis
we briefly attempt to identify the different mechanisms
which contribute to the overall effect. Each mechanism has
a different functional relationship with the equipment parameters and this is assessed as described in Appendix C. In
the experimental analysis we observe if the overall effect
is linear over the range of equipment parameters and then
attempt to quantify[3] the functional relationship of each
mechanism with the equipment parameters.

A modification of the phantom design for the commonly
encountered special case of sector scanning is mentioned in
the last section.

USEFUL PHANTOM GEOMETRIES

1. Sequence of discrete Reflectors

The echo strength at normal incidence from a flat reflector is given by the reflectivity coefficient

$$R= (Z_1 - Z_2)/ (Z_1+Z_2) \tag{1}$$

where Z_1 and Z_2 are the impedances of the reflector and the
surrounding medium.

With the proper choice of materials it is theorctically
possible to achieve an entire range of values of R between
0 and 1; i.e., an infinite dynamic range of echoes may be
generated. In practice, however, the equality $Z_1=Z_2$ (for
R=0) is not achieved because the impedances are complex
functions of frequency and temperature and sizeable variations in both are permitted during equipment tests.

The received echo strengths from several reflectors is
shown in Fig. 2. The absolute echo strength is displayed in
relation to its standard variation as the transducer is
scanned across about 20mm of the reflector surface. The
observed small variance is attributed to the fact that the
imaging aperture is large (F/3.5-F/1.7, as is to be expected in testing advanced imaging system) and that the reflector surfaces are smooth (milled).

Note also that only a 20dB dynamic range is achieved by
the materials employed. Three special materials[4] were also
tested and the echo strengths received from these materials

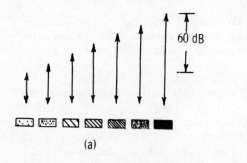

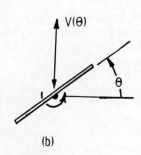

(a)

(b)

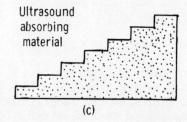

Ultrasound
absorbing
material

(c)

FIG.1 Useful Phantom
Geometries.

(a) Sequence of discrete
reflectors
(b) Rotating Reflector

(c) Stepped Wedge

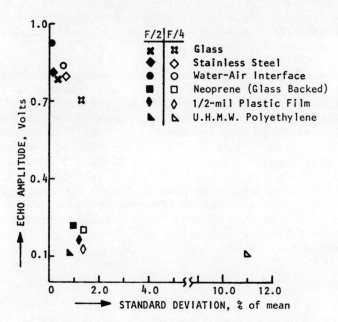

FIG.2 Performance of the Discrete Reflector Geometry

were generally 20-30 dB below the ideal reflector (water-
air interface, see Fig. 2), however, this data is considerably more sensitive to frequency and temperature as mentioned
above.

It is concluded that a phantom object consisting of
discrete reflectors (Fig. 1a) provides only about 20-25 dB
dynamic range at the brightest end of the gray scale. Also
that for this range the performance of the phantom object is
largely independent of frequency, spectrum and numerical
aperture. The echo strengths are also insensitive for small
misalignments (see section below).

2. Rotating Reflector

A larger dynamic range (about 70dB) is achieved by a
rotating reflector (see Fig. 1b). A typical example of a
glass reflector is shown in Fig 3. The data are largely ind-
pendent of all equipment parameters except the numerical
aperture and this variation may be readily quantized. The
uniformity in response with inclination angle is also good.
This is again attributed to the surface finish of the reflector and the affects of angular averaging due to a large aperture system. The data suggests that a numerical aperture
less than about F/4-F/5 may exhibit non-uniformities which
are greater than desired.

Although this phantom geometry provides the necessary
60dB dynamic range, there is a serious drawback to its use
in the field (clinic). For proper utilization, the scanning
mechanism of the transducer would have to be disconnected
while the deflection in one axis of the video display would
require synchronization with the rotation of the reflector
(or vice versa). This is not practical and even impossible
with certain equipment in a clinical setting (for systems
with electronic scanning this also does not result in a full
testing of the equipment). However, because of the simplicity of the phantom object, it is strongly recommended for
equipment testing during the developmental stages in the
laboratory.

3. Stepped and Tapered Wedges

Fig. 1c shows the stepped wedge phantom geometry. The
wedge is made of an ultrasound absorbing material so that the
echoes from the back surface of the wedge at each step loca-

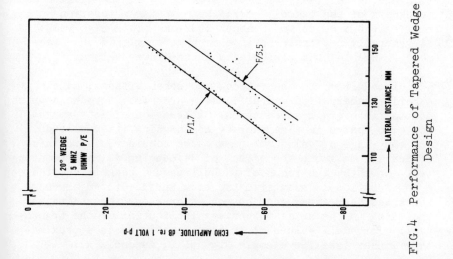

FIG. 4 Performance of Tapered Wedge
Design

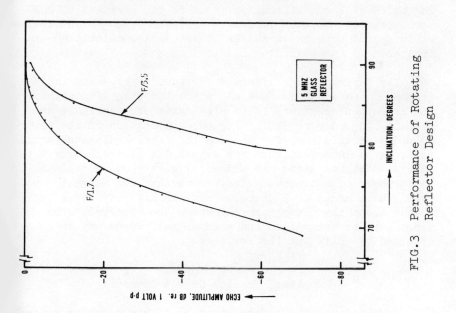

FIG. 3 Performance of Rotating
Reflector Design

tion are successively of lower strength. If the maximum
thickness of the wedge is 3cm, the material absorption
coefficient α needs to be about 10dB/cm. This high value
is generally not achieved in common materials in the frequency range 1-15MHz (especially at the lower values).

In addition to absorption, two other effects, namely
spherical aberration and defocusing are also introduced which
add to the overall attenuation. Aberration effects spread
the echo energy in time thereby diminishing its amplitude.
For the range of values of parameters considered here this
effect is small but the effect of defocusing is more pronounced. In this, the focus of the ultrasound beam is shifted to
a new location Δz away from the back surface ($\Delta z \to 0$ as wedge
thickness decreases) and the echo decreases proportionately
according to the axial intensity distribution of the transducer. For small values of Δz this dependence is approximately linear (see for example Figs 9&10, Appendix A).

Within the guidelines of the specifications[1], higher
resolution (i.e., higher frequency) equipments are expected
to possess proportionately smaller fields of view. Hence
a tapered wedge design is preferred over the step wedge
design. The tapered wedge additionally introduces ray bending which causes a lateral shift Δx in the incident ultrasound beam at the back surface. The echo strength is expected to depend both on Δx and Δz and these are calculated as
indicated in Appendix C.

The tapered wedge offers other advantages-- the entire
length of the wedge is "useful" since the step discontinuities are not present and it is also easier to fabricate and
polish. In order to facilitate the calibration of the tapered wedge at different frequencies and other equipment parameters, it is important that the echoes diminish linearly with
lateral distance along the wedge. Fig. 4 shows a typical
performance of a tapered wedge. Note that the largest echo
strength is about 20-25dB below maximum and only small variations about the linear approximation are observed. The
wedge surface are milled and greater uniformity may be
obtained by polishing the surfaces.

4. Optimal Phantom Object Geometry

As mentioned above, the response of the rotating reflec
tor is a function primarily of the numerical aperture of the

system and largely independent of all other equipment and experimental variables in the allowed range[1]. For other reasons mentioned above, however, its applicability in the field is restricted. Nonetheless it appears ideal for laboratory testing.

For field use, a combination of the discrete reflector and tapered wedge phantom geometry appears as the most likely candidate. The discrete reflector (consisting of 3-4 reflectors) provides echoes in the brightest 20 dB range of the gray scale. The tapered wedge provides the remaining 40-45 dB of the required range of echoes. The combined design is fabricated in one piece, about 5 cm long to facilitate field testing.

As mentioned above, the discrete reflector-portion of the phantom is largely independent of all equipment parameters. On the other hand, the analysis in Appendix C suggests that the performance of the tapered wedge is related to several equipment parameters which we shall investigate below.

The typical data in Fig 4 may be approximated by the slope α_t of the straight line approximation. In the case of Fig 4, $\alpha_t = 4.8$ dB/cm where the length dimension refers to the distance along the back surface of the wedge. As mentioned in the preceeding section

$$\alpha_t = \alpha_a + \alpha_b + \alpha_c \qquad (2)$$

where the subscripts a, b and c refers to the absorption, ray bending and aberration affects respectively. Each α_i is, in general, a function of the equipment and material parameters, namely f_o, $S(f)$, $f_\#$, refractive index n and absorption coefficient α of the wedge material, and temperature.

The purpose of the experimental testing is to determine that the cumulative effect on α_t in the range of values of these parameters is linear. It is also the intent to quantify the individual functional relationships with the goal that sufficient data is available so that when a new equipment, having different values of f_o, $S(f)$ and $f_\#$, is encountered it may be possible to select the wedge material (n), and wedge angle to fabricate the phantom object with assurance that the object will provide the necessary 60dB dynamic range and the distribution of echo strengths across its

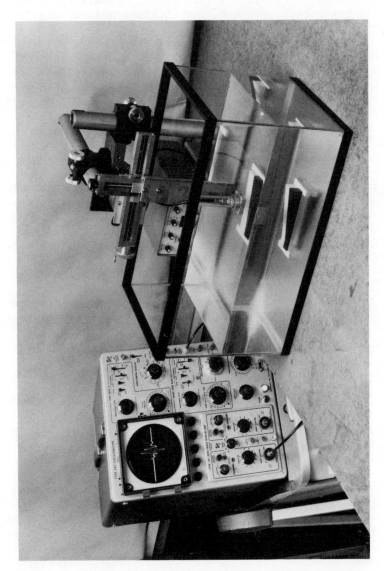

FIG. 5 Experimental Setup

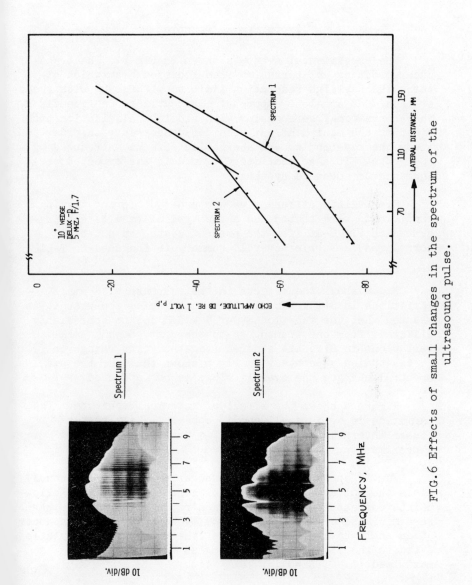

FIG.6 Effects of small changes in the spectrum of the ultrasound pulse.

surface is linear (to within about ± 1.5dB).

EXPERIMENTAL RESULTS

The experimental setup is shown in Fig 5. The transducer produces an aberration-free focussed beam which at-
tains the Rayleigh resolution limit to within 30% with greater than 70dB sidelobe suppression as outlined in Appendix A.
The aberration-free focusing with sidelobe elimination provided by the equipment was considered necessary in order to
study the ray-bending and aberration effects introduced by
the wedge. It also realistically models the expected performance of advanced systems.

Over thirty different wedges under different testing
conditions were tested and a few typical results are presented in Fig 6-10 (for the present, data has not been
normalized). A brief overall summary is included in Table
II.

Fig 6 shows the effects in the response with small
variations in $S(f)$. To simplify the problem somewhat it is
assumed that the function $S(f-f_o)$ would not be appreciably
different from one transducer to the next so only small
variations in $S(f)$ are studied (together with change in f_o
to be studied seperately). Fig 6 shows that for the small
spectral changes the overall slope is not affected.

Figure 4 shows the effect of numerical aperture. The
variation in α_t is small and it appears likely that a
linear interpolation for other $f_\#$s can serve as a good
approximation.

Fig. 7 shows data for two wedges under two test conditions the solid points show usual conditions and the hollow
points correspond to the case when ray bending effects have
been minimized. This is achieved by scanning the transducer
at an angle to the rear surfaces (the value of the angle is
noted in the figure) at these angles the echo strength is
maximized (i.e., $\Delta z=0$).

Fig. 8 shows a preliminary result at two frequencies.
In this case it is seen that the surface finish on the wedge
is not adequate for the higher frequency (subsequently
eliminated by polishing). Table II summarizes data on
different wedges. It is noted that in all cases except for

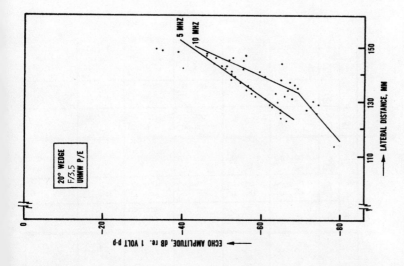

FIG.8 Performance at different
frequencies.

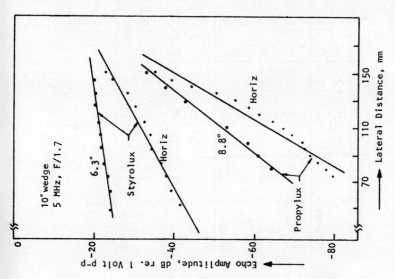

FIG.7 Approximate estimate of contributions from ray bending effects.

TABLE II

SUMMARY OF RESULTS

MATERIAL	F#	MAX ECHO DB BELOW 1VP.P			RANGE DB			VARIATION DB MAX			SLOPE DB/CM			TEST DIFFERANCE DB/CM		
WEDGE ANGLE		5°	10°	20°	5°	10°	20°	5°	10°	20°	5°	10°	20°	5°	10°	20°
DELUX	F/1.7		33.7	27.8		43.0	41.5		0.5	0.3		3.7	7.8			
	F/3.5		35.7	28.0		39.6	51.0		0.8	0.8		5.2	12			
ETHYLUX	F/1.7		30.5			47.0			2.5			6.4				
	F/3.5		37.7			36.5			2.5			5.4				
NORYL	F/1.7	23.7	26.7		25.0	41.1		1.0	1.5		2.8	3.9		0.6	1.3	
	F/3.5	21.7	30.0		12.1	24.5		0.5	1.5		1.4	2.5		0.0	0.3	
PROPYLUX	F/1.7	36.0	35.0		30.9	44.7		0.8	1.0		3.0	5.9		1.0	1.4	
	F/3.5	28.8	37.9		25.0	40.3		0.3	1.3		2.9	4.9		0.1		
STYROLUX	F/1.7		22.7			18.9			1.0			2.0			1.9	
	F/3.5															
UHW P/E	F/1.7			30.5			29.9			1.0			4.6			
	F/3.5			33.5			33.4			2.3			5.0			

Delux-D wedges, a straight line approximation to the data
can be made. Noryl and Propylux wedges show the greatest
uniformity and appear to be the two most optimum materials.
Noryl seems better suited for higher-frequency/large field-
of-view equipment whereas Propylux for lower frequency or
small field of view equipment

DISCUSSIONS

For brevity, only some typical data have been presented
here and the more interested reader is referred to the complete report being submitted to the National Science Foundation Office of Experimental R&D (which additionally contains recommendations on measurement techniques for
resolution, artifacts, transmitted power and other performance parameters). It is hoped that the limited data presented here illustrates our overall conclusions as summarized below.

For the preferred wedge materials (Noryl and Propylux)
the overall α_t is uniform across the wedge for a dynamic
range of 40-45 dB over the expected range of values (see
Table I) of the parameters (even better uniformity may be
achieved with molded wedges).

Additionally, the changes in α_t with f#, frequency and
S(f) are small and it appears that linear interpolations of
data can be established over the range of the expected
values of these parameters. The absorption effect α_a is
primarily a linear function of frequency (only). α_b is
most strongly influenced by wedge orientation, positioning
inaccuracies and temperature. The dependence, however, is
small for the selected materials (Noryl and Propylux). The
approximate relative contributions from this parameter,
based on data such as that in Fig. 8, is displayed in the
last column in Table II. α_c is also a function of the wedge
angle, and an approximate linear dependence with the latter
is indicated for the expected range of values.

The individual relationships may be quantified by controlled
tests which are currently in progress[3]. The testing set-up
is ideally suited for these since it is possible to vary
only one equipment parameter at a time. Tests at different
frequencies will reveal the size and functional dependence
of α_a which is related to α . Comparison of the tapered
wedge data with that obtained from a stepped wedge (under

the condition $\Delta z = 0$) at different $f_\#$s reveal the functional dependence of α_b. Comparison with tests at two temperatures will establish the variations with the refractive index of the the material.

Of special mention is the behavior of the wedge with an imaging equipment employing sector scanning. It is seen that if small metallic pins (of diameter slightly larger than the resolution element) are attached to the back of the wedge periodically across its length, the echoes from the pins are linearly graded (although of slightly higher magnitude) with α_t being the same as that from the back surface of the particular wedge under linear scanning.

The complete report with measured dependences of α_b and α_c with the equipment parameters is expected to be shortly available from the National Science Foundation.

ACKNOWLEDGEMENTS

The author gratefully acknowledges the painstaking and careful data collection by Mr. Albert Zemola, the several stimulating discusstions with Dr. William E. Glenn and the overall project co-ordination by Mr. Kenneth R. Solomon. This study was supported under contract NSF-C-796 from the National Science Foundation.

APPENDIX A

Apodization of Focussed Transducers

A large aperture transducer excited uniformly across its radiating surface produces strong sidelobes.[5] These may be described as "edge effects" insofar as a transducer of infinite surface does not produce sidelobes. For a finite aperture transducer the effect is approximated by minimizing the velocity discontinuity at the edges (apodization). A known technique for this is selective metalization of the transducer whereby the peripheral region is excited at a lower level than the central region thereby achieving a degree of apodization. This process is expensive to implement and because of·practical limitations on number of transducer elements, rarely achieves full apodization.

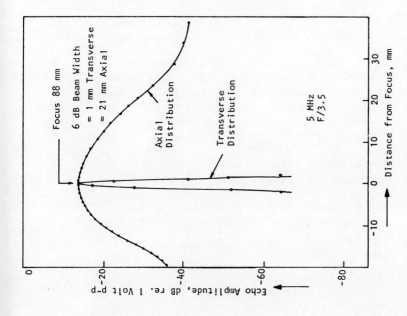

Fig. 10. Axial and transverse beam patterns of the F/3.5, 5 MHz transducer.

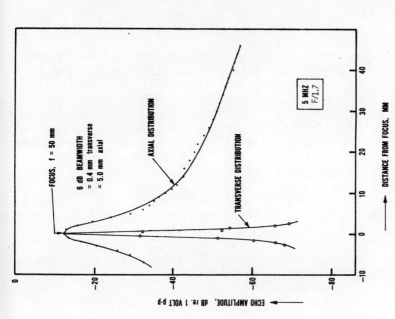

Fig. 9. Axial and transverse beam patterns of the F/1.7, 5 MHz transducer.

We describe here a new apodization technique which appears more practical and useful. This technique is most suited for apodizing large aperture focussed transducers.

The refractive index n of common plastic materials with respect to water is less than unity. A focussing lens made of these materials possesses a concave geometry. The plastics also show significant ultrasound absorption. Thus these lenses, being thicker at the edges, also give rise to various levels of apodization. We present below the parameters which determine the choice of a particular lens material to achieve a known level of apodization.

For a plano-concave lens, aberration-free focussing is achieved[6] by machining the concave surface according to the ellipse

$$(Z-a)^2/a^2 + r^2/b^2 = 1 \tag{3}$$

where
$$a = f/(1+n) \tag{4}$$
$$b = f\sqrt{[(1-n)/(1+n)]} \tag{5}$$

where z and r are the axial and radial co-ordinates, f the focal length and n the refractive index of the material of the lens. The thickness of the lens at any radius r is given by

$$z = a\sqrt{[1 - (1- r^2/b^2)]} \tag{6}$$

If α denote the attenuation coefficient of the lens material and the transducer is, as usual, excited uniformly across its surface, the velocity distribution at the surface of the lens is $\exp(-\alpha a\sqrt{1-r^2/b^2})$

The transmitted beam pattern for this velocity distribution can be geometically calculated and the various parameters α, a, b can be optimized for sidelobe suppression. Below, we present an optimization based on the approximation

$$\exp[-\alpha a\sqrt{(1-r^2/b^2)}] \approx \exp[-\alpha ar^2/2b^2] \tag{7}$$

where, by neglecting higher order terms we have approximated the velocity distribution as gaussian. The transmitted beam pattern for gaussian distributions of the form $\exp(-r^2/R^2)$ is known[7] and it is seen that the sidelobes are completely suppressed if one requires that

$$R = sA \tag{8}$$
$$1.5 <s< 0.5$$

TABLE III

DILATATION WAVE VELOCITIES AT 2MHZ

MATERIAL	VELOCITY. 22°C SHEET	ROD	DIFF REF.1 AT 0.5MHZ	DV/DT, ROD	n, 22°C 2MHZ
	m /SEC	M/SEC	%	M/S-°C	SHEET
CELCON	2374		+0.0	-7.0	0.6279
DELUX D	2402	2451	+2.7	-4.7	0.6202
H.I.P/S	2166		-0.2	-3.0	0.6876
MARLEX	2365	2367	-2.6	-5.7	0.6298
NORYL	2193	2246	-3.0	-2.6	0.6797
NYLON 101	2609	2364	-0.1	-3.1	0.5719
PROPYLUX	2612	2656	+1.1	-4.7	0.5689
PLEXIGLAS	2740	2721	+2.2	-1.4	0.5437
STYROLUX	2326	2317	+0.4	-7.4	0.6417
SYNTACTIC FOAM	2479	2519	+2.4	-0.4	0.6013
UHMW/PE	2239	2302	+1.2	-5.8	0.6654

where A is the half-aperture. For the lower values of s the sidelobe suppression is better[7] but the lens gain is reduced whereas for s > 1.5 sidelobe suppression is not completely adequate. Under the present hypothesis the lens design require from eqs 3-7 and above.

$$\alpha \doteq 16.36 \, f_{\#}(1-n^2)/s^2 f \qquad (9)$$

For the specific applications to diagnostic ultrasound imaging relationship between α and n required in eq. (9) is in the range of values of common plastic materials. With $f_{\#} = 2$, n = 0.6 (see Appendix B) it is found that for f = 30 cm, 2.1dB/cm > α > 0.5 dB/cm for 0.6 > s > 1.2 whereas for f = 5cm 12.3dB/cm > α > 3.1 dB/cm over the range of s. These values of α are attained by the plastic materials over the range 1 - 15MHz in frequency which are employed in diagnostic ultrasound imaging.

Two lenses were constructed according to eq. (9) for providing f#s of 1.7 and 3.5 respectively in conjunction with a 5MHz transducer. The resulting focal spot patterns for these transducers are shown in Figs 9 and 10.

APPENDIX B

Ultrasound Wave Velocities in Plastics

The propagation velocity of dilatation waves in several plastic materials was measured at 2MHz and two reference temperatures by employing the sing-around technique. The basic sing-around equipment for these measurement was bought from a commercial source.[8]

Measurement inaccuracies are estimated at less than $\pm$ 5 m/sec (about $\pm$ 0.3%) in velocity. For each test, the dilatation velocity is measured both in the specimen and the surrounding water bath and the refractive index of the material is also calculated.

Test samples of different plastic materials were made from commercially available[9] rod and sheet-stock. The measured velocities are shown in Table III. For comparison, earlier data by folds[10] is also shown. The latter data was taken at 0.5MHz on similar plastics ordered from the same supplier[9].

Differences between rod and sheet data as well as those with the earlier data (extrapolated to the same temperature) are small. This tends to suggest that the velocity is a weak function of frequency and other variables such as the different molding processes employed by the manufacturer between rod and sheet stock as well as other time-related phenomena.

APPENDIX C

Aberrations Caused by Refraction at an Inclined Plane

Consider a focussed transducer, focal length fo, immersed in a medium, medium 1, with another medium, medium 2, placed at an inclination ϕ with the axis of the transducer and at a distance d from it as shown in Fig.11. Let r and z be the co-ordinate axis as shown and θ represent the conver-

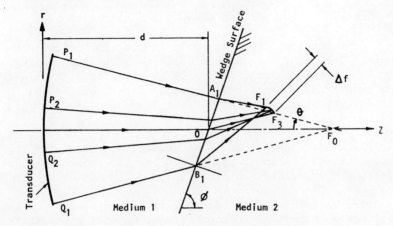

Fig. 11. Ray bending and aberration effects at an inclined interface.

gence angle of a particular ray at the hypothetical point F which represents the focus of the transducer in the absence of medium 2. From geometrical optics, the co-ordinates (F_r, F_z) of the focus in medium 2 are found as

$$F_r = (f_o-d)C_1\,[(m_b+m_a)\tan\Phi-(m_b-m_a)\tan\Theta-2m_am_b] \qquad (10)$$

$$F_z = d +(f_o-d)C_1[2\tan\Phi-(m_b+m_a)-(m_b-m_a)\tan\Theta/\tan\Phi] \qquad (11)$$

where

$$C_1 =.\tan\Theta\tan\Phi/(m_b-m_a)[\tan^2\Phi-\tan^2\Theta] \qquad (12)$$

$$m_b = [\tan\Phi\cos(\Phi-\Theta)-C_2]/[C_2\tan\Phi+\cos(\Phi-\Theta)] \qquad (13)$$

$$m_a = [\tan\Phi\cos(\Phi+\Theta)-C_3]/[C_3\tan\Phi+\cos(\Phi+\Theta)] \qquad (14)$$

$$C_2 = \sqrt{[n^2-\cos^2(\Phi-\Theta)]} \qquad (15)$$

$$C_3 = \sqrt{[n^2-\cos^2(\Phi+\Theta)]} \qquad (16)$$

For paraxial rays, the small angle limit is easily deduced, if it is recognized that

$$\lim_{\Theta\to 0} C_1=C_2\cos\Phi[C_2\tan\Phi+\cos\Phi]^2/2\tan^2\Phi[C_2^2+\cos^2\Phi] \qquad (17)$$

From this Δx and Δz (see text) can be calculated which may be utilized in quantifying the ray bending effect.[3]

REFERENCES

1.   "Perspectives for Ultrasonic Imaging in Medical Dia-
gnosis," Report of the National Science Foundation Survey
Team on Ultrasonic Imaging, Nov 1973, Appendix A. See also
Experimental R&D Incentives Program, "Experiment No. 5,
Medical Instrument Experiment," NSF 73-34 (Revised May 17,
1974)

2.   Sidelobe suppression is considered necessary for the
testing of the phantom designs. It is also necessary in
modelling the performance of advanced equipment

3.   The more interested reader is referred to the complete
report being submitted to the National Science Foundation
Office of Experimental R&D Incentives, which includes actual
quantification values. This report additionally contains
recommendations for measurement of other equipment parameters such as resolution, transmitted power, artifacts etc.

4.   Samples 6063-71-19, 1606-73-192 and 1606-73-194, kind-
ly supplied by Mr. S. Caprette, B.F. Goodrich Company,
Brecksville, Ohio

5. See for example, P.M. Morse and K. Uno Ingard, Theoretical Acoustics McGraw-Hill, N.Y. (1968)pp366-394

6.   The original discovery is attributed to R. Descartes,
"Discoursde la méthode et essais de cette méthode,"
Dioptrique, discours VIII, lere édition, leyde 1637.

7.   L. Filipczynski and J. Etienne, "Spherical Focussing
Transducers with Gaussian Surface Velocity Distribution,"
Proc. Vibr. Prob. 13, 117-136 (1972) Polish Academy of
Sciences

8.   Ultrasonic Engineering Co., ltd, Tokyo, Japan

9.   Westlake Plastics Company, Lenni, PA.

10.   D.J. Folds, "Experimental Determination of Ultrasonic
Wave Velocities in Plastics, Elastomers and Syntachc Foam as
a Function of Temperature," J. Acoust. Soc. Am. 52, 426 (1972)
See also "Design, Construction and Evaluation of Liquid and
solid Ultrasonic Lenses" NCSL 117-72, Naval Coastal Systems
Laboratory, Panama City, Florida, pp A1-A7 (1972).

POWER MEASUREMENT TECHNIQUES APPLIED

TO IMAGING SYSTEMS

P. C. Pedersen and D. A. Christensen

Department of Bioengineering

University of Utah, Salt Lake City, Utah 84112

ABSTRACT

Techniques for accurate measurements of peak power, average power, and spatial intensity distribution are described, especially with respect to large surface transducers such as commonly used in acoustical imaging devices. The radiation force technique, using a reflecting target and a microbalance, was used as a primary standard to measure peak and average power. A secondary standard in form of an electronic power meter was developed for displaying average power output. The sound field was mapped at several distances from the transducer by means of a scanning miniature hydrophone, and the intensity field for square and circular transducers was studied theoretically using computer generated field maps.

INTRODUCTION

The widespread support and interest which have been given to the development of ultrasonic diagnostic devices is for a large part due to the presumed freedom of radiation hazards. Although most irradiation studies indicate that the intensity levels where toxic effects occur are far above the intensity levels used for diagnostic purposes [1], [2], [3], this is still an area where there is need for further research, and as the range of medical applications in ultrasound expands, the need for accurate measurement techniques are becoming increasingly more important.

The objective in most dosimetry studies is to develop

711

a set of ultrasound field parameters which most directly
are related to the biological effects of the ultrasound, but
until a complete understanding of the interaction of ultrasound with tissue and cells has been achieved, we may not
know whether we have devised the perfect set of exposure
parameters. Several suggestions as to which set of parameters would be necessary in order to fully quantify the
ultrasound field can be found in the literature [4], [5],
[6].

Based on these sources, we decided upon the following
field parameters to characterize the ultrasound field:
1. Average power, total and in mW/cm^2.
2. Peak power, total and in mW/cm^2.
3. Spatial distribution of the ultrasound field.
4. Frequency.

The objective of this project has therefore been to
reliably determine these parameters. In this paper we will
only deal with the techniques for determining the parameters
given in items 1 and 3. The temporal peak power then can
be found when duty cycle and average power are known. For
irregularly shaped excitation pulses the peak power can be
found by first determining the transducer conversion coefficient, k_c, from $P_{sound} = k_c \cdot V_{exc}^2$, measure the maximum
value of V_{exc} during the pulse and then calculate the peak
power, using the expression for P_{sound}.

SECTION I - RADIATION FORCE MEASUREMENTS

Discussion

The radiation force technique appears to be the most
appropiate technique to use as a primary standard for measuring total average power. The method is sensitive, relatively easy to use and the measured force is related directly to total average power by a factor which is based on
known physical constants, and thus the method has become the
most widely used and accepted method for determining total
average power.

Despite these factors, the radiation force technique
should not be used without due theoretical and practical
considerations, which will be discussed in the following.

The radiation force equation, although simple, is a

result of a quite complicated derivation [7], based on the assumption of a continuous plane wave in a nonviscous liquid with constant compressibility. Looking at the continuous wave assumption, we concluded, that as a steady state is reached very fast, the fact that we use pulsed ultrasound with pulse widths 50 μsec to 400 μsec and not CW ultrasound, does not introduce any error in the radiation force equations, but when very short pulses are used (1 - 2 μsec), this might be of concern. With the frequencies (1 - 10 MHz), used in medical ultrasound, water can be considered a non-viscous fluid, and for the power level used for diagnostic purposes water has a constant compressibility. The assumption of greater concern is the plane wave assumption. Only in the far field, where no interference occurs, is the plane wave assumptions completely fulfilled, but since the near field extends to approximately 8 meters from the 4 in. x 4 in. transducer at 3 MHz used in the Holosonics unit, our measurements were restricted to the near field. To the authors' knowledge, there is no literature describing the radiation force in the near field, but some unpublished experimental results [8] seem to indicate that the radiation force equations might hold also for a non-plane wave. However, our faith in the measurements stems mainly from the fact that the field close to the surface of a square transducer is a quite good approximation to a plane wave (see section III). Further discussion of the assumptions of the radiation force equations is found in [9].

The practical considerations include the various possible interfering factors such as target design, multiple reflections, surface tension, temperature effects, etc., which all are potential error sources. These will all be discussed under the descriptions of the experiment.

In the literature are described several methods of measuring total average power of small diagnostic ultrasound transducers by means of radiation force, for example by measuring horizontal displacement of a suspended target [10], vertical displacement of a float [6], or by measuring radiation force on a reflecting target [11] or on an absorbing target [12]. These techniques appear all to be accurate and sensitive, but none could be applied directly to the large surface transducers, such as commonly used in acoustical imaging devices.

The relationship between sound intensity and radiation

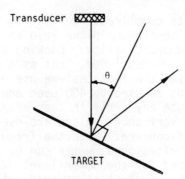

Reflecting target: $F = 2 \frac{I}{c} \cos^2 \theta$

For $\theta = 22.5^0$: $F/I = 117$ mg/W

Figure 1.   Relationship between sound intensity
and radiation force

force is shown on Fig. 1. For a reflecting target, the

force is: $F = 2\frac{I}{c} \cos^2 \theta$, where I is the sound intensity,

c is the velocity of sound in water and θ is the incident
angle, as shown. Thus, F/I = 117 mg/w at 20°C and θ =
22.5°.

Experiment

Figure 2 shows a schematic diagram of the radiation
force measurement system. The transducer is mounted at
the top of the tank, radiating downwards. A trough-shaped,
reflecting aluminum target large enough to intercept the
whole beam is suspended by 4 thin wires from a light-weight
target hanger, which is attached to the pan of the Mettler
semi-microbalance, model H51. The balance is a special
version which allows weighing below the balance through a
hole underneath the pan. The balance is placed on a balance
table, which helps minimize the effect of air currents in
the room.

Figure 3 shows the water tank with target and baffle
in greater detail. The water tank is 15 in. x 15 in. x
11 in., holding about 37 liters or 10 gallons of distilled

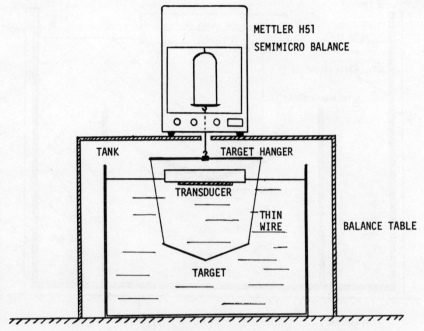

Figure 2.   Schematic diagram of the radiation
force measurement system

water. The target is made of 2 thin aluminum plates with
a small airgap in between, and being airbacked, the target
has excellent reflecting characteristics for any incident
angle. The distance between the target and the transducer
is approximately 12 cm.

All the inner surfaces of the tank were lined with
butyl rubber, a good ultrasound absorber. To completely
prevent sound from hitting the target more than once, we
designed a butyl rubber-covered baffle, shaped as a cylinder
with a square cross-section. After the sound is reflected
from the target, it hits the water surface, from where it
is reflected to the absorbing walls. Further reflection
from these walls is trapped between the baffle and the walls
of the tank. To further scatter the sound, glass wool is
placed around and inside the baffle. The incident angle
of 22.5° was chosen because it allowed a design which would
completely preclude multiple reflections of sound from the

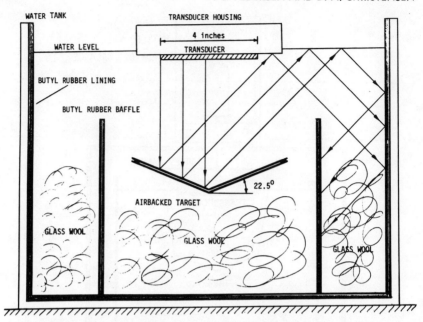

Figure 3.   Schematic drawing of water tank
with target and baffle

target for the available water tank.   To prevent corrosion
of the target and the aluminum frame for the baffle, the
surfaces exposed to water were sprayed with a thin layer of
lacquer.

Figure 4 shows the actual water tank, lined with butyl
rubber.   In the middle is seen the baffle with glass wool
inside and around.

Figure 5 shows a close-up of the experimental set-up,
as it looks during measurements.   The transducer housing
and target hanger can be seen between water tank and the
balance table.

Precautions

Possible sources of error and various practical considerations in connection with radiation force measurements
will be discussed in the following.

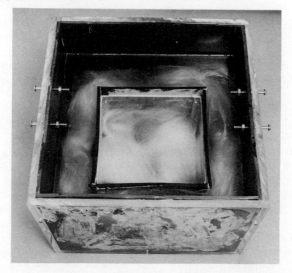

Figure 4.   Water tank with baffle

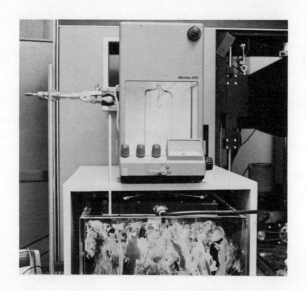

Figure 5.   Experimental set-up

Multiple reflections. When using a reflecting target a significant source of error is multiple reflections from the target, which exists when sound can be reflected back to the target more than once. However, our design of water tank and baffle has essentially eliminated this error source.

Target Geometry. We have discovered that when the target intercepts only part of the beam, and the result then is corrected to correspond to the total beam power, significant errors are likely to appear. We hypothesized that the error source is partly due to inhomogenities in the field, and thus the intercepted part of the beam does not become a representative part of the total beam. Another error source may arise from the fact that it is difficult to determine the effective beam area at the target. In our study we performed a series of measurements, in which we only intercepted approximately half the beam. The target here was a cone-shaped, air-backed target. After correcting the results to correspond to the total beam power, our results were invariably 22-24% too high compared with our target intercepting the whole beam.

Target Reflecting Characteristics. There are several reasons why we considered a reflecting target superior to an absorbing target.
1. It is difficult, if not impossible, to construct a perfectly absorbing target.
2. The radiation force from an absorbing target is only half of that from a reflection target.
3. At higher intensities, heating of an absorbing target would take place, causing changes in buoyancy and convection streaming in the water.
4. An absorbing target is often heavier and bulkier than a reflecting target, and the larger mass makes it more temperature sensitive.

As a reflecting target, the air-backed target is superior to a simple aluminum target for the following reasons:
1. With an air-backed target one achieves an almost total reflection (99.991%) independent of the incident angle.
2. In studies with a simple aluminum target we found that sound was transmitted through the target even above the critical angle for longitudinal waves, ca. 14°, and the critical angle for shear waves, ca. 30°. The problem

was alleviated when we changed to an air-backed target.

Surface Tension. We found that surface tension would produce significant errors of random nature due to stickiness between the water surface and the suspension wires. Adding surfactant, 1 gram of dodecyl sodium sulfate per gallon of water, removed the problem.

Hydrodynamic Streaming. The effect of hydrodynamic streaming is negligible. We examined the effect of this phenomenon by placing a very thin polyethylene film just above the target and performed a series of measurements. No deviation from previous results could be detected.

Temperature Effects. We found temperature effects to be negligible. Over a couple of hours of dissipating full power in the tank, there was 0.1°C or less temperature rise in the tank, and calculations as well as experiments showed no effect.

Results

Figure 6 shows some of our results based on radiation force measurements on the Holosonics Acoustical Imager. The curves show total average power as a function of the peak-to-peak voltage of the excitation pulse for several pulse widths. The curves fit parabolas quite well, and the deviations, which can be noted, are due to an irregular pulse shape for some settings of the pulse amplitude and pulse width controls, which makes the determination of the effective pulse amplitude difficult.

The surface area of the transducer is 92 cm^2, which for full pulse width gives a maximum average power density of 74 mW/cm^2.

SECTION II - ELECTRONIC POWER METER

Although the radiation force measurement as described in Section I is invaluable as a primary standard and gives a reliable technique for determining total average power, this is by no means a quick and convenient method. In dosimetry studies one might wish to perform experiments for different combinations of pulse width and pulse amplitude settings, corresponding to different combinations of peak

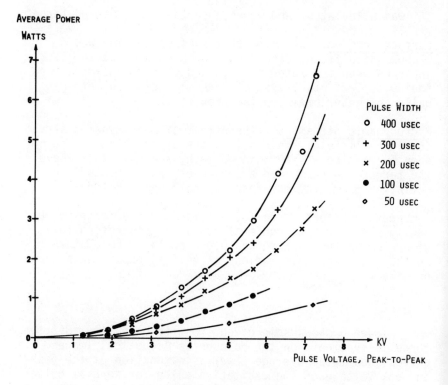

Figure 6.   Acoustical power based on radiation
force measurements

power and average power, and it would be very impractical
indeed to perform the radiation force experiment for each
new setting of the power controls. In a clinical setting
it would be just as impractical to do radiation force
measurements as a routine check of the total maximum output
power.

    This gave us the main rationale for the development of
a secondary standard in form of an electronic power meter,
which in digital form displays the real time average power
output for any setting of the pulse amplitude and pulse
width controls. Figure 7 shows a block diagram of the power
meter. The input signal comes from 1:100 capacitive voltage
divider, directly from the output of the transducer power
amplifier. Because a non-linear circuit like a detector

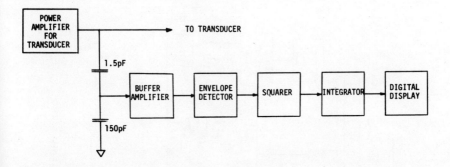

Figure 7. Block diagram of electronic power meter

would build up a dc voltage on the 150 pF capacitor in the
voltage divider if it were connected directly, a very high
slew rate buffer amplifier is inserted before the envelope
detector.

Since the sound intensity is proportional to the square
of the excitation voltage, the pulse envelope is fed to a
squaring circuit. The integrator produces a dc voltage
proportional to the average sound power. Except for the
display, the power meter is mounted directly on the transducer power amplifier in order to keep a short signal path
for the high frequency input signal.

Figure 8 shows the schematic diagram of the power meter.
Maximum input signal is 35V peak, producing a maximum output
signal from the buffer amplifier of 10V. The required slew

rate is $\frac{d}{dt}$ (10 cos ωt) $\simeq$ 200V/μsec for a frequency of 3 MHz,

a requirement, which is adequately fulfilled by the LH0024
operational amplifier. The 1N4454 in the detector is a
high speed switching diode. To minimize the effect of the
junction capacitance, and thereby making the detector respond
faster, a "keep-alive" current is supplied via 100KΩ to -15V.

Only part of the external circuitry for the CA3091D
squaring circuit is shown. The 4pF and 100pF capacitors are
included to improve the frequency response. The 5KΩ zero
adjustment is available at the front panel of the display.

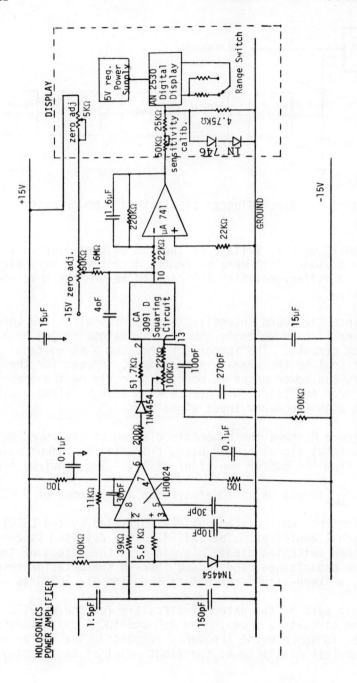

Figure 8.   Schematic diagram of power meter

The pulse repetition frequency of the acoustical imager is 60 Hz, and with a time constant of 0.35 sec the integrator produces a completely smooth dc signal, proportional to the sound power.

The two 3.3V zener diodes, 1N746, in the display, is for overload protection.

Figure 9 illustrates the performance of the power meter. The graph is constructed from a series of measurements, where we recorded radiation force and power meter readings simultaneously. The horizontal axis shows average power, based on the measured radiation force, and the vertical axis shows the corresponding power meter reading. The performance of the power meter was tested for many different pulse widths, of which only 400 μsec, 300 μsec and 100 μsed are shown here. The correlation is very good for all pulse widths, with a maximum error of ca. 2%, which occurs at the highest pulse amplitude setting.

The following four points characterize the performance and use of the electronic power meter:
A. The power meter displays total average power output for any setting of the pulse amplitude and pulse width controls.
B. The power meter measures accurately the power output even for irregularly shaped excitation pulses, in which cases the power would be difficult to calculate based on oscilloscope measurements of the excitation pulse.
C. The power meter measures power directly to the transducer, and thereby it eliminates possible errors or changes in the control system and the pre- and power-amplifier system.
D. The power meter is a useful instrument for any researcher doing dosimetry with pulsed or CW ultrasound, as well as in a clinical setting.

SECTION III - STRUCTURE OF THE ULTRASOUND FIELD

As mentioned earlier, in order to characterize the ultrasonic field completely, one must know the field structure of the ultrasonic beam, and the information is applicable in dosimetry studies as well as in clinical use. In dosimetry studies one must be aware of the possibility of localized intensity peaks in the sound field. This is

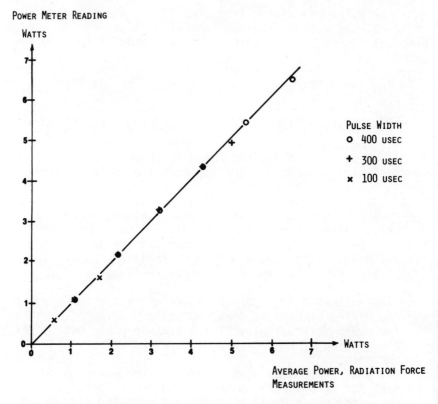

Figure 9.   Power meter reading versus radiation force

especially critical when one uses a small specimen or bacteria culture in the dosimetry studies. In clinical use of an ultrasonic device it should be the spatial peak intensity of the field which is used in evaluating the safety of the device.

In this part of the project we examined the field experimentally by mapping it underneath the 4 in. x 4 in. transducer by means of a miniature hydrophone as well as studied the field theoretically. The theoretical study of the field was intended to help determine which transducer size and shape will give the most homogeneous field, as well as to evaluate the ratio of peak intensity to average intensity at various distances from the transducer.

For the experimental mapping a Helix hydrophone, type 101 subminiature hydrophone S/N 3, was used. The hydrophone is virtually omnidirectional at 3 MHz.

Figure 10 shows a schematic of the experimental set-up for mapping the ultrasonic field under the 4 in. x 4 in. transducer. Two X-Y recorders were used, one to record the curves on paper, one to move the hydrophone. The output from the hydrophone was displayed on one trace on a dual trace oscilloscope, and the vertical position of the other trace was controlled by the 10KΩ potentiometer. While the hydrophone is slowly moved through the ultrasound field underneath the transducer, and the pulse amplitude on the oscilloscope varies proportional to the sound pressure, a person will turn the potentiometer so that the trace is always kept right on the top of the hydrophone pulse. Besides letting the voltage from the potentiometer arm deflect the trace on the oscilloscope, the voltage is also used as the Y-input to the X-Y recorder on which the curves are recorded.

Figure 11 is a photo of the complete experimental set-up. In the middle of the picture is seen the X-Y recorder which moves the hydrophone. The mechanical attachment is

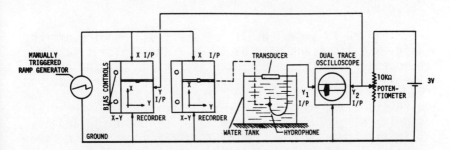

Figure 10. Experimental set-up for mapping the ultrasonic field under the transducer

Figure 11.   Experimental set-up for mapping sound field

clearly seen. The vertical rod near the transducer holds
the hydrophone and is calibrated in distances from the transducer to the tip of the hydrophone. In the right part of
the picture is the X-Y recorder, on which the curves are
recorded.

Figure 12 shows results of the field mapping of the
4 in. x 4 in. transducer at 3 MHz. The field is fairly
homogeneous, with a high intensity "ridge" at the edges of
the field. Note that the ripples are of smaller size and
width in the field map 9 cm. from the transducer compared
with the field map 18 cm. from the transducer. The field
mapping also reveals that the level of the field appears
slightly lower on one side of the transducer than on the
other. It should be emphasized, that these field maps display sound pressure and not intensity, which means, that

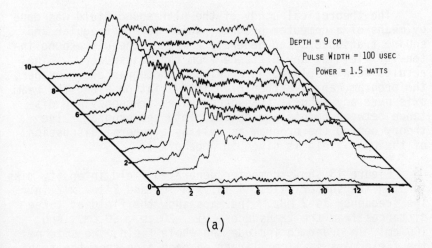

DEPTH = 9 CM
PULSE WIDTH = 100 USEC
POWER = 1.5 WATTS

(a)

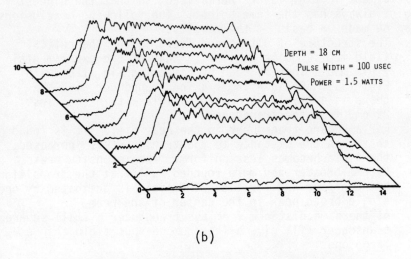

DEPTH = 18 CM
PULSE WIDTH = 100 USEC
POWER = 1.5 WATTS

(b)

Figure 12.   Field mapping with hydrophone

the variations in the field will appear larger in a field
intensity map like the computer generated field intensity
maps, discussed later.

The theoretical study of the ultrasound field was done
by means of a computer program, which for rectangular and
square transducers can generate a map of the ultrasound intensity field at any distance from the transducer.   The
results were displayed on a CRT and photographed. Further,
the program can give the intensity variation along the beam
axis over any range. Transducer size, frequency and distance between points are entered into the program. The
theory behind the program as well as a general discussion
of the near field is given in Appendix I.

Figure 13 shows computer generated field intensity maps
for two transducer sizes, 4 in. x 4 in. and 2 in. x 2 in.
The frequency is 3 MHz. The maps show the field at three
distances from the transducer, viz. 20 cm.; 50 cm., and
100 cm. In order to include the whole field, the size of
the field maps is 1 cm. larger on each side than the transducer as shown in Fig. 14. It should be noted that all the
field maps in Fig. 13 are normalized, so that the highest
point always has the same height. The following conclusions
can be made on basis of the results:
1.   The structure of ultrasound field in the theoretical
     field map is quite similar to that of the experimental
     field map. This is seen by comparing Fig. 12b) with
     Fig. 13a). For a 4 in. x 4 in. transducer at 3 MHz
     there is a fairly homogeneous field within the first
     25 cm. from the transducer.
2.   For a square transducer in the near field, it is found
     that when the distance to the transducer is increased,
     the field becomes less uniform. The intensity peaks
     become broader and more rounded, until at the transition
     between near and far field the field is dominated by one
     single broad peak in the center of the beam.
3.   At the same distance from the transducer a small square
     transducer will give a less homogeneous field than a
     large one.

Figure 15 shows the peak intensity of the ultrasound
field relative to the average intensity for a 4 in. x 4 in.
transducer at 3 MHz. Only 50 cm. away from the transducer
the peak intensity, which appears in each "corner" of the
the field (see Fig. 13), is 100% higher than the average

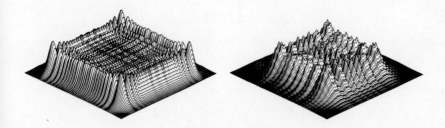

(a) 4 in. x 4 in. z = 20 cm. (d) 2 in. x 2 in. z = 20 cm.

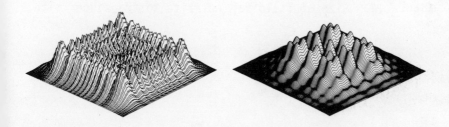

(b) 4 in. x 4 in. z = 50 cm. (e) 2 in. x 2 in. z = 50 cm.

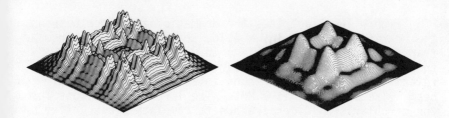

(c) 4 in. x 4 in. z = 100 cm. (f) 2 in. x 2 in. z = 100 cm

Figure 13. Field intensity maps for a 4 in. x 4 in.
transducer and a 2 in. x 2 in. transducer at
20 cm., 50 cm. and 100 cm. from the transducer.

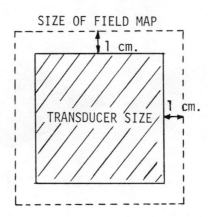

Figure 14. Size of field map relative to transducer size

intensity. These peaks are very localized, however, and the
main part of the field has much lower intensity peaks.
Nonetheless, the peak intensity of the field should be an
important factor in evaluating the safety of an ultrasonic
diagnostic device.

Figure 16 illustrates the dramatic difference between
the near field of a 4 in. x 4 in. transducer and a circular
transducer with a diameter = 4 in. The graphs show the
theoretical sound intensity variation along the beam axis
relative to I_0, the mean sound intensity over the total
beam. The frequency is 3 MHz. The large fluctuations for
the circular transducer are due to complete constructive
and destructive interference along the beam axis. In comparison, the fluctuations for the square transducer, especially close to the transducer, are quite small. Note that
for the circular transducer the axial field fluctuates a-
round a mean value of $2 I_0$, indicating a very non-uniform
field with a much higher intensity near the beam axis than

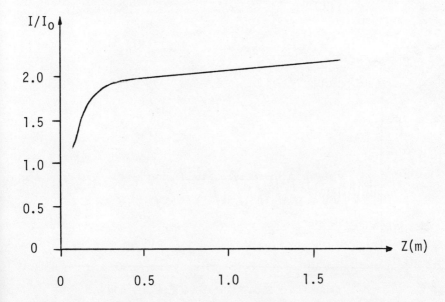

Figure 15.   Peak intensity relative to average intensity
             for 4 in. x 4 in. transducer

over the rest of the beam, while for the square transducer
the fluctuations occur around a mean value of I_0. These
results demonstrate that a square transducer produces a far
more uniform field than a circular one of similar size and
therefore should be preferred for imaging systems.

The uniformity of the square transducer has importance
in another context: The radiation force equation is based
on the assumption of a plane wave field, an assumption
which normally is fulfilled in the far field. Figure 16
illustrates how much closer we are to that assumption with
a square transducer than with a circular transducer of
similar size.

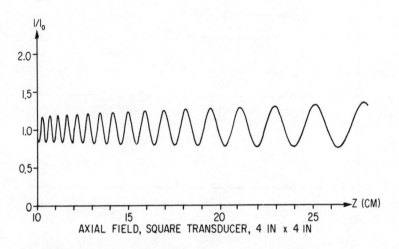

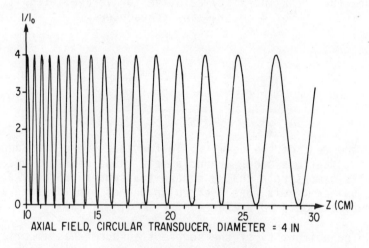

Figure 16.   Comparison of large size square and
circular transducers

CONCLUSIONS

The conclusions, which can be drawn from this study of power measurement techniques, can be summarized in the following points.

1. For radiation force measurements an air-backed reflecting target is preferable. The total beam should be intercepted even for large surface transducers. Multiple reflections is a significant error source, for which the necessary precautions should be taken.

2. The eletronic power meter has proven to be a reliable, accurate secondary standard for continuously monitoring total average power.

3. In order to obtain a uniform field in the near field zone, one should use a square transducer rather than a circular one. Reducing the distance to the transducer surface as well as selecting a large square transducer rather than a small one improves the uniformity of the field.

ACKNOWLEDGEMENTS

The authors express appreciation to Jack Guthrie, Electrical Engineering, University of Utah, for excellent craftsmanship in making the mechanical devices for this project. We would like to thank Dwight Potter, Medical Physics, University of Utah Medical Center, for great help in setting up the experiments and Dr. Robert Anderson, Radiology, University of Utah Medical Center and Dr. Curtis Johnson, Department of Bioengineering, University of Utah, for many fruitful and stimulating discussions.

This work was supported in part by a grant from HEW. No. 1 RO1 RL 01007 01, and the rest by the Department of Bioengineering, University of Utah.

Appendix I

The Near Field of Ultrasound Transducers

Exact description of the field generated by transducers and other acoustical radiators is a complicated topic. Hence, theoretical calculations have been based on the assumption of uniformly, sinusoidally vibrating pistons.

Using this idealized field model, the ultrasonic beam can be divided into two distinct zones, the near field or the Fresnel zone and the far field or the Fraunhofer zone. (For rectangular transducers it has been suggested [13] to divide the field into three zones, the very near field, the near field and the far field).

For a circular transducer the near field is characterized by complete or almost complete constructive and destructive interference, which produces pressure variation of large amplitude as described in section III. In the near field of a square transducer the interference is not complete due to lack of symmetry of the transducer boundaries with respect to the beam axis, resulting in smaller pressure fluctuations and a more uniform field.

In the far field of an ultrasound beam the interference phenomena do   no longer take place, and the field is characterized by a plane wave.

The theory regarding interference assumes a continuously excited sound source, but for pulse widths in the range 50 μsec to 400 μsec as is encountered in this study, the theory is applicable with a very good approximation.

For circular transducers the transition between near and far field occurs at an axial distance, $d_c = r^2/\lambda$, [14], where r is the radius of the transducer. For a square transducer the near field extends out to a distance from the transducer, $d_{sq} = b^2/2.88\lambda$, [13], where b is the length of the side of the transducer.

With the transducer sizes commonly used in imaging devices one will normally operate in the near field. For example, at a frequency of 3 MHz in water the far field begins at 50 cm. for a circular transducer with a diameter of 30 mm or a square transducer with side length 27 mm.

Circular transducers

Descriptions of the interference phenomena, the near field and the far field of circular transducers are commonly found in the literature, for example, [15]. The axial intensity variation in the near field of a circular transducer [13] is $I(z) = 4 \cdot I_0 \cdot \sin^2 \frac{k}{2} \left[(z^2+r^2)^{\frac{1}{2}} - z \right]$

where $I_0 = \frac{1}{2} \rho c \mu_0^2$

ρ = density of propagating medium

c = wave velocity

μ_0 = amplitude of particle velocity

$k = \frac{2\pi}{\lambda} = \frac{\omega}{c}$

Square and rectangular transducers

The sound field due to rectangular or square transducers is discussed extensively in 2 papers [13], [16], and the calculations in the following pages are based on these two sources.

Using Freedmans's technique [16] and given a transducer with the dimensions a and b, the center of which has the coordinates (0,0,0) (fig. 17), then the sound pressure, P_M, at M = (x,y,z) is given by:

$$P_M = \frac{jc\rho\mu_0}{\lambda R\alpha^2} \times \exp[j(\omega t - kz)] \times$$

$$\int_{\alpha(y - \frac{b}{2})}^{\alpha(y + \frac{b}{2})} \exp(-j\frac{\pi}{2}v^2)\,dv \times \int_{\alpha(x - \frac{a}{2})}^{\alpha(x + \frac{a}{2})} \exp(-j\frac{\pi}{2}v^2)\,dv$$

where

$$R = (x^2 + y^2 + z^2)^{\frac{1}{2}} \cong z[1 + \frac{x^2 + y^2}{2z^2}]$$

$$\alpha^2 = 2/(\lambda z)$$

P_M can be written as:

$$P_M = \frac{j\rho c \mu_0}{2} \times \cos\gamma_0 \times \exp[j(\omega t - kz)] \times$$

$$(A_y - j B_y)(A_x - jB_x)$$

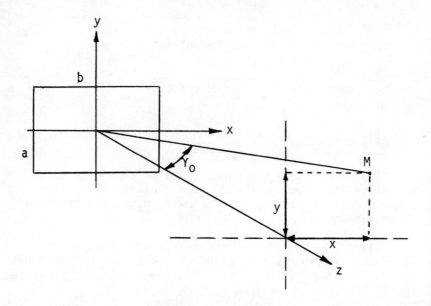

Figure 17.   Geometrical representation of an
acoustical field point

$$A_y - j\, B_y = \int_{\alpha(y - \frac{b}{2})}^{\alpha(y + \frac{b}{2})} \exp\left(-j\, \frac{\pi}{2}\, v^2\right) dv$$

$$= \int_0^{\alpha(y + \frac{b}{2})} \cos\left(\frac{\pi}{2}\, v^2\right)\, dv - \int_0^{\alpha(y - \frac{b}{2})} \cos\left(\frac{\pi}{2}\, v^2\right)\, dv$$

$$-j\left[\int_0^{\alpha(y + \frac{b}{2})} \sin\left(\frac{\pi}{2}\, v^2\right)\, dv - \int_0^{\alpha(y - \frac{b}{2})} \sin\left(\frac{\pi}{2}\, v^2\right)\, dv\right]$$

The integrals are now in the form of the Fresnel integrals:

$$C(x) = \int_0^x \cos(\frac{\pi}{2}t^2)dt$$

$$S(x) = \int_0^x \sin(\frac{\pi}{2}t^2)dt$$

This gives the following expressions for A_y and B_y:

$$A_y = C[\alpha(y + \frac{b}{2})] - C[\alpha(y - \frac{b}{2})]$$

$$B_y = S[\alpha(y + \frac{b}{2})] - S[\alpha(y - \frac{b}{2})]$$

Similarly we find for A_x and B_x:

$$A_x = C[\alpha(x + \frac{a}{2})] - C[\alpha(x - \frac{a}{2})]$$

$$B_x = S[\alpha(x + \frac{a}{2})] - S[\alpha(x - \frac{a}{2})]$$

In this study we will only be concerned with amplitude:

$$|P_M| = \rho c \ \mu_0 \ \frac{\cos \gamma_0}{2} \sqrt{(A_x^2 + B_x^2)(A_y^2 + B_y^2)}$$

The acoustical intensity at M is $I_M = |P_M|^2/(2 \rho c)$ and with $I_0 = \frac{1}{2} \rho c \ \mu_0^2$ we get $I_M = I_0 \frac{\cos^2\gamma_0}{4} (A_x^2 + B_x^2)(A_y^2 + B_y^2)$

An estimate of the error of this expression is given in [16] and as a rule of thumb the expression is reasonably accurate for $Z > b$, where b is the largest side of the transducer.

The series expansions and the rational approximations, which were used in the computer program, were taken from [17].

For $|x| < 0.4$ we used the following expressions to prevent underflow:

$$C(x) = x - \frac{\pi^2}{40} x^5 \qquad S(x) = \frac{\pi}{6} x^3 - \frac{\pi^3}{336} \cdot x^7$$

For $0.4 \leq |x| \leq 2.2$ the series expansions were used:

$$C(x) = \sum_{n=0}^{\infty} \frac{(-1)^n \left(\frac{\pi}{2}\right)^{2n}}{(2n)! \ (4n+1)} \ x^{4n+1}$$

$$S(x) = \sum_{n=0}^{\infty} \frac{(-1)^n \left(\frac{\pi}{2}\right)^{2n+1}}{(2n+1)! \ (4n+3)} x^{4n+3}$$

For values of $|x|$ larger than 2.5, the error in $C(x)$ and $S(x)$ becomes excessive, and the following rational expansion were used:

$$C(x) = 0.5 + f(x) \times \sin\left(\frac{\pi}{2}x^2\right) - g(x) \times \cos\left(\frac{\pi}{2}x^2\right)$$

$$S(x) = 0.5 - f(x) \times \cos\left(\frac{\pi}{2}x^2\right) - g(x) \times \sin\left(\frac{\pi}{2}x^2\right)$$

where $f(x) = \dfrac{1 + 0.926 \ x}{2 + 1.792 \ x + 3.104 \ x^2}$

and $g(x) = \dfrac{1}{2 + 4.142 \ x + 3.492 \ x^2 + 6.670 \ x^3}$

which gives an error < 0.2%. The last formulas for $C(x)$ and $S(x)$ accepts only positive arguments, but one can use the relations:

$$C(-x) = -C(x); \qquad S(-x) = -S(x).$$

References

[1]   Woodward, B. et al., "How safe is diagnostic sonar?" Br. J. Radiol., Vol. 43, 1970, p. 719-725.

[2]   Dunn, F., Summary paper for session on "Effects on tissue and organs," Interaction of Ultrasound and Biological Tissues, p. 103-107.

[3]   Fry, F.J. and Dunn, F., "Interaction of ultrasound and tissue," Interaction of Ultrasound and Biological Tissue, p. 109-114.

[4]   Myers, G.H. et al., "Preliminary proposal on ultra-

sonic diagnostic dosimetry measurements," <u>Interaction of Ultrasound and Biological Tissue,</u> p. 207-210.

[5]   Smith, S.W. and Steward, H.F., "A discussion of ultrasonic intensity measurements," <u>Interaction of Ultrasound and Biological Tissue</u>, p. 211-214.

[6]   Kossoff, G., "Calibration of Ultrasonic Therapeutic Equipment," Acustica, Vol. 12, 1962, p. 84-90.

[7]   Borgnis, F.E., "Acoustical radiation pressure of plane compressional waves," Rev. Mod. Phys., Vol. 25, 1953, p. 653-664.

[8]   Rooney, J.A., personal communication.

[9]   Sokollu, A., "On the measurement of ultrasound intensity in tissue: Reservations of radiation force and the case for calorimetry," <u>Interaction of Ultrasound and Biological Tissue,</u> p. 165-170.

[10] Wells, P.N.T. et al., "Milliwatt ultrasonic radiometry," Ultrasonics, July-September, 1964.

[11] Hill, C.R. et al., "Calibration of ultrasonic beams for biomedical applications," Phys. Med. Biol., Vol. 15, No. 2, 1970.

[12] Rooney, J.A., "Determination of acoustic power outputs in the microwatt-milliwatt range," Ultrasound in Med. & Biol., Vol. 1, 1973, p. 13-16.

[13] Marini, F. and Rivenez,F., "Acoustical fields from rectangular ultrasonic transducers for nondestructive testing and medical diagnosis," Ultrasonics, November, 1974.

[14] Hill, C.R., "Ultrasound dosimetry," <u>Interaction of Ultrasound and Biological Tissues</u>, p. 153-158.

[15] Heuter, Th. F. and Bolt, R.H., <u>Sonics</u>, p. 59-72.

[16] Freedman, A., "Sound field of a rectangular piston," JASA, Vol. 32, February 1960, p. 197-209.

[17] <u>Handbook of mathematical functions</u>, NBS, Applied Mathematics Series 55, ninth printing.

LIST OF PARTICIPANTS

R. Addison
American Optical
39 Bailey Road
South Berlin, MA 01549

Mahfuz Ahmed
Zenith Radio Corp.
6001 W. Dickens
Chicago, IL 60639

G. Alers
Rockwell International
1049 Camino Dos Rios
Thousand Oaks, CA 91360

P. Alais
University of Paris
5 pl Jussieu
Paris V France

Henry Allen
Stanford Electronics Lab
McC 58
Stanford, CA 94305

Karena Arnold
UC Livermore LL
P. O. Box 808
Livermore, CA

Arturo Arriola
Naval Undersea Center
San Diego, CA 92132

J. Attel
Stanford University
Stanford, CA 94305

Donald Baker
University of Washington
Center for Bioengineering
Seattle, WA 98195

Robert Bard
Haimonides Medical Center
4802 10th Avenue
Brooklyn, NY 11219

Kenneth Bates
Stanford University
Stanford, CA 94305

D. Bernardi
Picker Corporation
333 State Street
North Haven, CT 06473

P. K. Bhagat
University of Kentucky
Lexington, KY 40506

Jason Birnholz, M. D.
Department of Radiology
Stanford University Medical Center
Stanford, CA 94305

D. Bonnet
Battelle Inst., Am Roemerhof
6Frankfurt 90
Germany

Newell Booth
Naval Undersea Center
Code 6513
San Diego, CA 92132

Herman P. Briar
Aerojet Solid Propulsion
10215 Malaga Way
Rancho Cordova, CA 95670

Ernest N. Carlsen, M. D.
Loma Linda University
Loma Linda, CA 92354

Henry H. Chaskelis
Naval Research Lab
4555 Overlook Ave. SW
Washington, D. C. 20375

Doug Christensen
University of Utah
Bioengineering Department
Salt Lake City, UT 84112

Agustin Coello-Vero
UC Santa Barbara
Santa Barbara, CA 93106

James H. Cole
TRW
1 Space Park
Redondo, CA 90278

Carter Collins
Inst. Med. Sci.
2232 Webster Street
San Francisco, CA 94115

H. D. Collins
Holosonics, Inc.
2950 George Washington Way
Richland, WA 99352

L. Douglas Clark
Varian
611 Hansen Way
Palo Alto, CA 94303

Steve Cowen
Naval Undersea Center
San Diego, CA 92132

P. Das
RPI
Troy, NY 12181

D. A. DeLise
University of Arizona
Tucson, AZ 85721

Gerald Denny
UC San Diego
3265 Jemez Drive
San Diego, CA 92117

Reiner Diehl
KAE
Postfach
Bremen, West Germany

F. P. Diemer
Office of Naval Research
800 Quincy Street
Arlington, VA 22217

John Dreyer
Dreyer Laboratories
9854 Zig Zag Road
Cincinnati, OH 45242

Francis Duck
Mayo Clinic
200 1st Street SW
Rochester, MN 55901

H. Eagle
CalTrans
5900 Folsom Blvd.
Sacramento, CA 95819

John A. Edward
General Electric Company
P. O. Box 1122 SP4-39A
Syracuse, NY 13201

Ken Erikson
McDonnell Douglas
700 Royal Oaks
Monrovia, CA 91016

K. F. Etzold
RCA Laboratories
Princeton, NJ 08540

N. H. Farhat
University of Pennsylvania
200 South 33rd Street
Philadelphia, PA 19174

Stephen Farnow
Stanford University
114 Tennyson
Palo Alto, CA 94301

John B. Farr
Western Geophysical Company
P. O. Box 2469
Houston, TX 77001

David Feinstein
University of Pennsylvania
200 South 33rd St.
Philadelphia, PA 19174

W. R. Fenner
Aerospace Corporation
P. O. Box 92957
Los Angeles, CA 90009

John J. Flynn, M. D.
UC Irvine
Department of Radiology
Orange County Medical Center
Orange, CA

D. G. Foeller
BWB
AM RHEIN
Koblenz, Germany 54

Martin D. Fox
University of Connecticut
Box U-157
Storrs, CT 06268

John D. Fraser
Stanford University
Stanford, CA 94305

R. Gawlik
RCA Laboratories
Princeton, NJ

Carl Gerle
P. O. Box 99991
San Diego, CA 92109

Albert Goldstein
KU Medical Center
Kansas City, KS 66103

Lou Ann Granger
Bendix Electrodynamics
15825 Roxford Street
Sylmar, CA 91342

Charles Green
Naval Undersea Center, Code 2533
San Diego, CA 92132

Philip S. Green
Stanford Research Institute
333 Ravenswood
Menlo Park, CA 94025

Robert Green
ACTRON
700 East Royal Oaks
Monrovia, CA 91016

James F. Greenleaf
MAYO Foundation
Rochester, MN 55901

Dan Griffin
Naval Undersea Center
San Diego, CA 92132

K. Gruenewald
Dornier System
P. O. Box 648
799 Fr' hafen, Germany

J. Havlice
Stanford University
Stanford, CA 94305

M. L. Henning
Plessey Co. LTD
Templecombe, So Merset
England

Francis J. Higgins
Naval Coastal System Lab
Panama City, FL 32401

B. P. Hildebrand
Battelle NW, Battelle Blvd.
Richland, WA 99352

Ronald Hileman
Unirad
P.O. Box 37002
Denver, CO 80239

Joseph Hirsch
Naval Undersea Center
San Diego, CA 92132

Pauline Hirsch
State Health Dept.
1350 Front Street
San Diego, CA 92101

Hoffmann
F. Hoffman-LaRoche VI 1Phy
Basle, Switzerland

D. R. Holbrooke
Children's Hospital
P. O. Box 3805
San Francisco, CA 94118

Paul Hoeller
Fraunhofer Gesellschaft University
66Saarbruecken, Germany

A. E. Holt
Babcock and Wilcox
RT 1, Box 1260
Lynchburg, VA 24505

John F. Holzemer
SRI
333 Ravenswood
Menlo Park, CA 94025

Charles Hottinger
Stanford University
92 A Escondido Village
Stanford, CA 94305

Lee L. Huntsman
Center for Bioengineering
University of Washington
328 A. R. L. (FL-20)
Seattle, WA 98195

Bernard Hurley
RCA Labs
Princeton, NJ 08540

J. E. Jacobs
Northwestern
Evanston, IL 66201

Steven A. Johnson
MAYO Foundation
Rochester, MN 55901

Ed Karrer
Hewlett-Packard
1501 Page Mill Road
Palo Alto, CA 94304

P. N Keating
Bendix Research Labs
Civic Center Drive
Southfield, MI 48075

H. Keyani
Santa Barbara, CA 93105

Gordon S. Kino
Stanford University
867 Cedro Way
Stanford, CA 94305

James W. Knutti
Stanford University
58 McCullough
Stanford, CA 94305

J. K. Krohn
HSVA
Bramfelderstr. 164
2 Hamburg, West Germany

J. M. LaCrotta
Pratt & Whitney
Aircraft Road
Middletown, CT 06457

Kenneth Lakin
University of Southern California
Seaver Science Center
Los Angeles, CA 90007

John Landry
UC Santa Barbara
Santa Barbara, CA 93106

Lansiart
C. E. A.
GIF-SUR-YVETTE
Paris 91190 France

Lewis Larmore
Office of Naval Research
800 North Quincy
Arlington, VA 22217

R. A. Lemons
Stanford University
Stanford, CA 94040

Wing Pun Leung
W. W. Hansen Lab, Stanford Univ.
Stanford, CA 94305

Melvin Linzer
National Bureau of Standards
Washington, D. C. 20234

Chester Loggins
Naval Coastal Systems Lab
Panama City, FL 32401

M. Lutkemeyer
KAE, Postf. 448545
28 Bremen 44, West Germany

Albert Macovski
Stanford Univ., Elect. Eng. Dept.
Stanford, CA 94305

M. G. Maginness
Stanford University
Stanford, CA 94305

K. Marich
SRI, 333 Ravenswood
Menlo Park, CA 94025

Sam Maslak
Hewlett-Packard
1501 Page Mill Road
Palo Alto, CA 94304

Ron McKeighen
Searle Analytic
2000 Nuclear Drive
Des Plaines, IL 60018

B. J. McKinley
UCLLL
P. O. Box 808
Livermore, CA 94550

James Meindl
Stanford University
Stanford, CA 94305

Charles Merhib
Army Materials and Mechanical
Research Center
Watertown, MA 02172

R. S. Mezrich
RCA Labs
Princeton, NJ 08540

Moretti
Cen. Saclay
Gif-sur-Yvette-91190
Paris, France

Professor Moriamez
University of Valencienne
le Mont Holy
59326 Valenciennes
France 59326

Rolf K. Mueller
University of Minnesota
9707 Manning Avenue
Stillwater, MN 55082

H. A. Muller
Muller-BBM GmbH
Herzogspitalstr. 10
Munchen-2
Germany D-8000

C. Nianois
University of Pennsylvania
4039 Chestnut
Philadelphia, PA 19104

Anant K. Nigam
CBS Labs
High Ridge Road
Stamford, CT 06905

K. Nitadori
Oki Electric Ind. Co., LTD.
550-5 Higashiasakawa
Hachioji, Tokyo, Japan 193

Stephen Norton
Stanford University
2288 Williams Street
Palo Alto, CA 94306

Murry Ohtsuka
Canon Holosonics
Tokyo, Japan

Torkil Olsen
Smith Kline Inst.
3077 South Court
Palo Alto, CA 94306

C. D. Payne
Home Office, Woodcock Hill
Sandridge, St. Albans
Herts, England

Peder C. Pedersen
University of Utah
Bioengineering Dept.
Salt Lake City, UT 84112

Soo-Chang Pei
UC Santa Barbara
Santa Barbara, CA 93106

Ronald C. Petersen
Holosonics
4340 Redwood Way, Suite 128
San Rafael, CA 94903

Per Pettersen
SIMRAD, Box 111
Horten, Norway 3190

David J. Phillips
Duke University
Dept. Biomedical Engr.
Durham, NC 27701

E. J. Pisa
McDonnell Douglas
Monrovia, CA 91016

Patrick O. Prendergast
Naval Ship R&D Center
Annapolis, MD 21402

N. B. Proctor
Tuboscope
P. O. Box 808
Houston, TX 77001

H. A. F. Rocha
General Electric
Rm 680 – Bldg. 37
Schenectady, NY 12345

Lewis H. Rosenblum
Picker Corp.
333 State Street
North Haven, CT 06473

Robert Rosenfeld
Eastman Kodak
Kodak Park, Bldg. 59
Rochester, NY 14650

G. A. Rost
1630 State College
Anaheim, CA 92806

C. Routh
Naval Undersea Center, Code 3524
San Diego, CA 92132

Ben Saltzer
Naval Undersea Center
San Diego, CA 92132

William Samayoa
MAYO Foundation
Rochester, MN 55901

Larry Schlessal
UC Santa Barbara
6571 Trig
Goleta, CA

Schmitz
TzfP
Saarbrucken
66 Saarbrucken, Germany

Gunter W. Schubart
Schubart Company
33 Harenauer Street
Wiesbaden, West Germany 6202

Bernard Shacter
Nigms/NIH
Bethesda, MD 20014

William Sheldon
Northrop
5316 Horsham
Westminster, CA

N. Sheridon
XEROX Corp.
3180 Porter Drive
Palo Alto, CA 94304

C. S. De Silets
Stanford University
Microwave Lab
Stanford, CA 94305

John Simonds
Eastman Kodak
Kodak Park, B-81
Rochester, NY 14650

E. R. Silverstein
Cubic Corporation
9233 Balboa Avenue
San Diego, CA 92123

Roy A. Smith
TRW
One Space Park
Redondo Beach, CA 90278

Eugene M. Spurlock
Stanford Research Institute
Menlo Park, CA 94025

Blaine Stauffer
ACTRON
700 Royal Oaks Drive
Monrovia, CA 91016

D. Steputis
ACTRON
700 Royal Oaks Drive
Monrovia, CA 91016

Karyl-Lynn Stone
Visual Acoustics
Ocean Highland
Magnolia, MA 01930

W. Ross Stone
5703 Soledad Mountain Rd.
La Jolla, CA 92037

J. R. Suarez
Stanford Research Institute
Menlo Park, CA 94025

Gary W. Sullivan
UC Irvine
4954 Claremont Drive
San Diego, CA 92117

Jerry Sutton
Naval Undersea Center
San Diego, CA 92132

G. E. Stewart
Aerospace Corporation
2824 Lakeside
Orange, CA 92667

R. Bruce Thompson
Rockwell International
P. O. Box 1085
Thousand Oaks, CA 91360

J. Tiemann
General Electric
P. O. Box 8
Schenectady, NY 12301

Neal Tobochnik
UC Los Angeles
Los Angeles, CA 90024

Torguet
Thomson CSF
Corbeville
Orsay 91 France

Paul W. Trainer
UC San Diego
3265 Jemez
San Diego, CA 92117

G. Tricoles
General Dynamics
P. O. Box 81127
San Diego, CA 92115

Nie But Tse
UC Santa Barbara
8547 Imperial Hwy. No. 73B
Downey, CA 90242

Robert Tyce
Scripps Institute
P. O. Box 1529
La Jolla, CA 92037

O. T. von Ramm
Duke University
Durham, NC 27706

David Vilkomerson
RCA Labs
Princeton, NJ 08540

Glen Wade
UC Santa Barbara
Santa Barbara, CA 93106

Keith Wang
University of Houston
Houston, TX 77004

E. E. Watson
Applied Research Lab
Box 30
State College, PA 16801

Thomas M. Waugh
Stanford University
Stanford, CA 94305

A. Waxman
Searle
2270 Martin Avenue
Santa Clara, CA 95050

R. D. Weglein
Hughes Research
2011 Malibu Canyon Road
Malibu, CA 90265

H. J. Whitehouse
Naval Undersea Center
San Diego, CA 92132

Dave Wilson
Hewlett-Packard
1501 Page Mill Road
Palo Alto, CA 94304

Donald C. Woods
Del Mar Engineering
6901 W. Imperial Hwy
Los Angeles, CA 90045

John Wreede
McDonnell-Douglas
700 Royal Oaks
Monrovia, CA 91016

Joel Young
Naval Undersea Center
San Diego, CA 92132

Leslie M. Zatz, M. D.
959 Mears Court
Stanford, CA 94305

G. Zilinskas
Bendix
11600 Sherman Way
North Hollywood, CA 91605

Marvin C. Ziskin, M. D.
Department of Radiology
Temple University Medical School
Philadelphia, PA 19140

Louis T. Zitelli
Varian Associates
611 Hansen Way
Palo Alto, CA 94303

INDEX